Yearbook on Space Policy

More information about this series at http://www.springer.com/series/8166

Yearbook on Space Policy

Edited by the
European Space Policy Institute

Director: Jean-Jacques Tortora

Edward Burger • Giulia Bordacchini

Yearbook on Space Policy 2017

Security in Outer Space: Rising Stakes for Civilian Space Programmes

Springer

Edward Burger
European Space Policy Institute
Vienna, Austria

Giulia Bordacchini
European Space Policy Institute
Vienna, Austria

ISSN 1866-8305 ISSN 2197-9405 (electronic)
Yearbook on Space Policy
ISBN 978-3-030-05416-8 ISBN 978-3-030-05417-5 (eBook)
https://doi.org/10.1007/978-3-030-05417-5

This Springer imprint is published by the registered company Springer Nature Switzerland AG
The registered company address is: Gewerbestrasse 11, 6330 Cham, Switzerland

Foreword

Increasingly, space-faring nations have to deal with a wide range of security challenges threatening the sustainable operation of space activities. In this sense, security in outer space is of growing importance not only for operators but also for public and private entities using space-based data and services for their operations. This increasingly challenging security situation has been acknowledged by the global community at large, recognising that space security can only be achieved as the outcome of a global effort. The European Union upholds "the protection and resilience of critical European space infrastructure" as a flagship objective of the Space Strategy for Europe and stresses the importance of cooperation at European and also international levels, in particular with the USA. And across the Atlantic, space has always held a prominent defence and national security dimension, making space security of critical importance.

In considering the above, ESPI's research team focused greatly on space security matters in 2017 and 2018. In this thread, we have recently engaged in research on security in outer space from European as well as transatlantic perspectives (provided in this Yearbook's Part II). We likewise focused our annual Autumn Conference in September 2018 on "Security in Outer Space: Rising Stakes for Civilian Space Programmes". Bringing together satellite operators, SMEs, European and American institutions and think tanks, the latest Autumn Conference served as a platform for fresh insights on security in outer space and the potential of transatlantic relations to address its challenges. Held over 2 days, the conference covered the following three focus areas: (1) the State of Play in Space Security Strategies: Space Assets in an Evolving Environment; (2) Forging Ahead with European Space Security Efforts; and (3) Rethinking Transatlantic Cooperation: towards STM. And in order to both highlight and share these findings, we have ultimately chosen to reflect this activity in the annual theme of the 2017 Yearbook.

Beyond this scope, plenty of developments outside the topic of space security have of course also happened during the Yearbook's reporting period. This content is laid out in Part I, which provides a comprehensive overview of the economic, political, technological and institutional trends that affected space activities in the past year. It is prepared in-house by ESPI, and while its perspective is European, it also provides a comparative review of space developments around the world. In 2017, we note in particular the private space sector, which continued to evolve at a fast pace with the consolidation of a number of private initiatives in the wake of the so-called new space approach, while major progress also occurred in space exploration, in particular on the Moon and towards Mars as well as in the discovery of exoplanets beyond our solar system. The first part moreover includes a collection of ESPI thought papers, covering selected topics of growing importance such as suborbital spaceflight, the multiplication of spaceports, super-heavy-lift launch vehicles, collaboration with China and the delimitation of outer space. Overall, the aim of this section is to offer the reader an overview of the development of several key features of European and global space activities during the reporting period.

The second part of the Yearbook begins with the Proceedings of ESPI's 12th Autumn Conference, held in late September 2018, which discussed the growing importance of security in outer space and the stakes for civilian space programmes in the public and private sectors. Part II additionally explores the rising stakes for Europe in security in outer space as well as perspectives on transatlantic relations in security in outer space.

The third part of the Yearbook serves as an archive of 2017 space activities prepared with ESPI's in-house database, including a complete launch log, a chronology of major policy and related events, data on ESA Member States as well as a comprehensive bibliography of space law and policy publications produced during the reporting period.

In closing, I would like to thank the contributors of the materials featured in Part II of the Yearbook as well as the ESPI staff that have been instrumental in its overall production.

ESPI, Vienna, Austria Jean-Jacques Tortora

Acronyms

A

A3R	Arkyd 3 Reflight spacecraft
AAD	Advanced Air Defence
ABS	Asia Broadcast Satellite
ADPC	Asian Disaster Preparedness Center
AEGIS	Autonomous Exploration for Gathering Increased Science
AG	Aktiengesellschaft
AGRHYMET	Agriculture, Hydrology and Meteorology Regional Center
AIA	Atmospheric Imaging Assembly
AIM	Asteroid Impact Mission
Airbus D&S	Airbus Defence and Space
AIS	Automatic Identification Satellites
AIST	Advanced Industrial Science and Technology
ALR	Austrian Aeronautics and Space Agency
AMESD	African Monitoring of the Environment for Sustainable Development
AMS	Alpha Magnetic Spectrometer
APAC	China and other Asia Pacific
ARISE	Agricultural Resources Inventory and Survey Experiment
ASAP	Austrian Space Applications Programme
ASAT	Anti-Satellite
ASEAN	Association of Southeast Asian Nations
ASI	Agenzia Spaziale Italiana (Italian Space Agency)
ASL	Airbus Safran Launchers
ASPERA-3	Mars Express Analyzer for Space Plasmas and Energetic Atoms
ATK	Alliant Techsystems Inc.
ATV	Automated Transfer Vehicle
AWE	AWE Management Limited
AWS	Automatic Weather Stations

B

BDS	BeiDou Navigation Satellite Systems
BELSPO	Belgian Federal Science Policy Office
BHRS	Belgian High Representation for Space Policy
BIS	Business, Innovation and Skills
BMD	Ballistic Missile Defence
BMVIT	Austrian Federal Ministry for Transport, Innovation and Technology

C

CAD	Computer-Aided Design
CALET	CALorimetric Electron Telescope
CAPE	Crop Acreage and Production Estimation
CASC	China Aerospace Science and Technology Corporation
CAST	China Aerospace Science and Technology Corp.
CATHALAC	Water Center for the Humid Tropics for Latin America and the Caribbean
CBERS	China–Brazil Earth Resources Satellite
CCP	Central Committee of the Communist Party of China
CD	Conference on Disarmament
CDOP 3	Third Continuous Development and Operations Phase
CDRA	Carbon Dioxide Removal Assembly
CDTI	Centre for the Development of Industrial Technology
CEC	Consortium for Educational Communication
CELAC	Community of Latin American and Caribbean States
CENI	Commission Électorale Nationale Indépendante
CEOS	Committee on Earth Observation Satellites
CERSGIS	Centre for Remote Sensing and Geographic Information Services
CET	Centre for Education Technology
CFAS	Federal Commission for Space Affairs
CFOSat	Chinese-French Oceanography Satellite
CGWIC	China Great Wall Industry Corporation
CHEOPS	CHaracterising ExOPlanet Satellite
CHF	Swiss franc
CIET	Central Institute of Educational Technology
CILSS	Comité permanent Inter-Etats de Lutte contre la Sécheresse dans le Sahel (Ghana)
CIS	Communications, Intelligence & Security
CLARREO	Climate Absolute Radiance and Refractivity Observatory
CMA	Governing Body of the Paris Agreement
CME	Coronal Mass Ejection

CMSA	China Manned Space Agency
CNES	Centre National d'Études Spatiales (French Space Agency)
CONAE	Argentinian Space Agency
CONCORDi	European Commission's biennial Conferences on Corporate R&D and Innovation
COP	Conference of the Parties
COPUOS	United Nations Committee on the Peaceful Uses of Outer Space
COSPAR	Committee on Space Research
COSTIND	Commission for Science, Technology and Industry for National Defense
CRESDA	Centre for Resources Satellite Data and Application
CRISM	Compact Reconnaissance Imaging Spectrometer for Mars
CRS	Commercial Resupply Services
CSA	Canadian Space Agency
CSDP	Common Security and Defence Policy
CSE	Centre de Suivi Ecologique (Senegal)
CSES	China Seismo-Electromagnetic Satellite
CSLCA	Commercial Space Launch Competitiveness Act
CSS	Chinese Space Station
CTX	Context Camera
CubeSats	Cube Satellites

D

DAMPE	Dark Matter Particle Explorer
DARS	Digital Audio Radio Service
DBS	Direct Broadcast Services
DECU	Development and Educational Communication Unit
DJEI	Department of Jobs, Enterprise and Innovation
DLR	Deutsches Zentrum für Luft- und Raumfahrt (German Aerospace Center)
DoD	Department of Defence
DRDO	Defence Research and Development Organisation
DSC	Decision Support Center
DSCOVR	Deep Space Climate ObserVatoRy
DSI	Deep Space Industries
DTH	Direct To Home

E

EBIT	Earnings before Interest and Taxes
EBITDA	Earnings before Interest, Taxes, Depreciation and Amortization
ECA	Evolution Cryotechnique type A
ECSS	European Cooperation for Space Standardization
EDM	ExoMars Entry, Descent and Landing Demonstrator Module
EDT	Electrodynamic Tether
EELV	U.S. Evolved Expendable Launch Vehicle Program
EIB	European Investment Bank
EIC	East India Company
EIF	European Investment Fund
EIT	Extreme ultraviolet Imaging Telescope
ELIRG	Extremely Luminous Infrared Galaxies
ELV	European Launch Vehicle
EM	Exploration Mission
EMEA	Europe, the Middle East and Africa
EMMRCs	Educational Multimedia Research Centres
EO	Earth Observation
EON-MW	Earth Observing Nanosatellite-Microwave
EPS-SG	European Polar System Second Generation
ERG	Exploration of Energization and Radiation in Geospace
ESA DG	ESA Director General
ESA	European Space Agency
ESEC	ESA Space Security and Education Centre
ESM	European Service Module
ESMD	Exploration Systems Mission Directorate
ESPI	European Space Policy Institute
ESSO	Earth System Science Organization
ETC	Emergency Telecommunications Cluster
EU	European Union
EUISS	EU Institute for Security Studies
EUMETSAT	European Organisation for the Exploitation of Meteorological Satellites
EUTELSAT	European Telecommunications Satellite Organization
EVE	EUV Variability Experiment

F

FAA	Federal Aviation Administration
FASAL	Forecasting Agricultural output using Space, Agrometeorology and Land based observations

FCT	Foundation for Science and Technology
FFG	Austrian Research Promotion Agency
FFL	Fondation Follereau Luxembourg
FLPP	Future Launchers Preparatory Programme
FOCAC	Forum on China–Africa Cooperation
FSS	Fixed-Satellite Service
FY	Fiscal Year

G

GAFAs	Google, Amazon, Facebook, Apple
GCSP	Geneva Centre for Security Policy
GDP	Gross Domestic Product
GEO	Geostationary Earth Orbit
GEO	Group on Earth Observations
GEOSS	Global Earth Observation System of Systems
GERD	Gross Domestic Expenditure on R&D
GFDRR	Global Facility for Disaster Reduction and Recovery
GGIM	Global Geospatial Information Management
GmbH	Gesellschaft mit beschränkter Haftung
GMT	Greenwich Mean Time
GNI	Gross National Income
GNSS	Global Navigation Satellite Systems
GOES-R	Geostationary Operational Environmental Satellite R
GOLF	"Global Oscillations and Low Frequency" instrument
GOVSATCOM	Governmental Satellite Communications
GPS	Global Positioning System
GRaND	Gamma Ray and Neutron Detector
GSA	European GNSS Agency
GSLV	Geosynchronous Satellite Launch Vehicle
GSRT	General Secretariat for Research and Technology
GSSAP	Geosynchronous Space Situational Awareness Program
GTO	Geosynchronous Transfer Orbits

H

HAT	Human African trypanosomiasis/Sleeping sickness
HDTV-EF2	High Definition TV Camera—Exposed Facility 2
HFA	Hyogo Framework for Action
HMI	Helioseismic and Magnetic Imager
Hot DOG	Hot, Dust-Obscured Galaxy

HR	High-Resolution
HRE	Human and Robotic Exploration
HSO	Hungarian Space Office
HSTI	Human Space Technology Initiative
HTV	H-2 Transfer Vehicle

I

I&B	Information and Broadcasting
IAA	International Academy of Astronautics
IAC	International Astronautical Congress
IADC	Inter-Agency Space Debris Coordination Committee
IAEG-SDGs	UN Statistical Commission's Interagency Expert Group
IARI	Indian Agriculture Research Institute
IASC	Inter-Agency Standing Committee
ICBM	Intercontinental Ballistic Missile
ICG	International Committee on Global Navigation Satellite Systems
ICIMOD	International Centre for Integrated Mountain Development
ICoC	Draft International Code of Conduct for Outer Space Activities
ICRC	International Committee of the Red Cross
ICS	Information and Communication Systems
ICT	Information and Communications Technology
IEA	International Energy Agency
IEV	Intermediate Experimental Vehicle
IFIs	International Financial Institutions
IGMA	International GNSS Monitoring and Assessment
IGS	International GNSS Service
IISL	International Institute of Space Law
IKAR	Interdepartmental Committee for Space Affairs
ILS	International Launch Services
IMF	International Monetary Fund
IMU	Inertial Measurement Unit
INCOIS	Indian National Centre for Ocean Information Services
INTA	National Institute of Aerospace Technology
IODC	Indian Ocean Data Coverage
IOs	Regional Organizations and International Organizations
IoT	Internet of Things
IPP	International Partnership Programme
IR	Intermediate Result
IRIS	Interface Region Imaging Spectrograph
IRNSS	Indian Regional Navigation Satellite System
ISC	International Satellite Company Limited

ISED	Innovation, Science and Economic Development
ISIS	Islamic State
ISO	International Organization for Standardization
ISRO	Indian Space Research Organisation
ISS	International Space Station
ITAR	International Traffic in Arms Regulations
ITU	International Telecommunication Union
IUCAA	Inter-University Centre for Astronomy and Astrophysics
IUVS	Imaging UltraViolet Spectrograph
IXPE	Imaging X-ray Polarimetry Explorer

J

JAXA	Japan Aerospace Exploration Agency
JIRAM	Jovian Infrared Auroral Mapper
J-PAL	Abdul Latif Jameel Poverty Action Lab
JPSS	Joint Polar Satellite System
J-SSOD	JEM Small Satellite Orbital Deployer
JUICE	Jupiter Icy moon Explorer

K

K2	Kepler 2
KARI	Korea Aerospace Research Institute (Korean Space Agency)
KITE	Kounotori Integrated Tether Experiment

L

L2	Earth-Moon Lagrange
LAXPC	Large Area X-ray Proportional Counter
LEE	Latching End Effectors
LEO	Low Earth Orbit
LRO	Lunar Reconnaissance Orbiter
LTS	Long-Term Sustainability
LULC	Land Use and Land Cover

M

MARSIS	Mars Advanced Radar for Subsurface and Ionospheric Sounding
MAVEN	Mars Atmosphere and Volatile EvolutioN
MDA Corp.	MacDonald, Dettwiler and Associates Ltd.
MDGs	Millennium Development Goals
MDI	Michelson Doppler Imager
Melco	Mitsubishi Electric Co.
MEO	Medium Earth Orbit
MERLIN	Methane Remote Sensing LIDAR Mission
MESA	Monitoring for Environment and Security
MESSENGER	MErcury Surface, Space ENvironment, GEochemistry and Ranging
Metop	Meteorological Operational Satellite
Metop-SG	Metop Second Generation
MEXT	Ministry of Education, Culture, Sports, Science and Technology
MFG	Meteosat First Generation
MHI	Mitsubishi Heavy Industries
MIT	Massachusetts Institute of Technology
MIUR	Ministry of Education, University and Research
MMO	Mercury Magnetospheric Orbiter
MOD	Ministry of National Defense
MOKV	Multi-Object Kill Vehicle
MOM	Mars Orbiter Mission
MOSDAC	Meteorological and Oceanographic Satellite Data Archival Centre
MoU	Memorandum of Understanding
MPO	Mercury Planetary Orbiter
MRO	Mars Reconnaissance Orbiter
MSF	Médecins Sans Frontières
MSG	Meteosat Second Generation
MSL	Mars Science Laboratory
MSM	Methane Sensor for Mars
MSS	Mobile Satellite Service
MTG	Meteosat Third Generation
MTM	Mercury Transfer Module

N

NASA	National Aeronautics and Space Administration
NATO	North Atlantic Treaty Organization
NCERT	National Council of Educational Research and Training
NCSTE	China's National Centre for Science and Technology Evaluation

NDAA	National Defense Authorization Act
NDCs	Nationally Determined Contributions
NEC	Nippon Electric Company
NEO	Near-Earth Orbit
NextSTEP	Next Space Technologies for Exploration Partnerships
NGA	National Geospatial-Intelligence Agency
NGCV	Next-Generation Crew Vehicle
NGO	Non-governmental Organization
NNRMS	National Natural Resources Management System
NOAA	National Oceanic and Atmospheric Administration
NOW	Netherlands Organisation for Scientific Research
NRO	National Reconnaissance Office
NRSC	National Remote Sensing Centre
NSA	National Security Agency
NSC	National Space Council
NSC	Norwegian Space Centre
NSO	Netherlands Space Office

O

OBIA	Object-Based Image Analysis
OCO	Orbiting Carbon Observatory
ODA	Official Development Assistance
OECD	Organisation for Economic Co-operation and Development
OHB	Orbitale Hochtechnologie Bremen
OOF	Other Official Flows
OPEC	Organization of the Petroleum Exporting Countries
ORU	Orbital Replacement Units
OST	Outer Space Treaty

P

PACE	Plankton, Aerosol, Cloud, ocean Ecosystem
PAD	Prithvi Air Defense
PAROS	Prevention of an Arms Race in Outer Space
PAS	Polish Academy of Sciences
PES	Payment for Ecosystem Services
PHA	Potentially Hazardous Asteroids
PLA	People's Liberation Army
PLATO	Planetary Transits and Oscillations of stars

PND	Portable Navigation Devices
PNTAB	Position, Navigation and Timing Advisory Board
POLSA	Polish Space Agency
PPP	Public–Private Partnership
PPWT	Draft Treaty on the Prevention of the Placement of Weapons in Outer Space and of the Threat or Use of Force against Outer Space Objects
PRM	Period Reduction Manoeuvre
PROBA	PRoject for OnBoard Autonomy
PRS	Public Regulated Service
PSA	Programme on Space Applications
PSLV	Polar Satellite Launch Vehicle
PS-TEPC	Position-Sensitive Tissue Equivalent Proportional Chamber
PUMA	Preparation for the Use of MSG in Africa programme

Q

QZSS	Quasi-Zenith Satellite System

R

RCM	RADARSAT Constellation Mission
RCMRD	Regional Centre for Mapping of Resources for Development
REDD+	Reducing Emissions from Deforestation and Forest Degradation
RHESSI	Reuven Ramaty High Energy Solar Spectroscopic Imager
RKV	Redesigned Kill Vehicle
RLV	Reusable Launch Vehicle
ROSA	Romanian Space Agency
Roscosmos	Roscosmos State Corporation
RRS	Regional Radiocommunication Seminars
RSCC	Russian Satellite Communications Company

S

SAARC	South Asian Association for Regional Development
SAB	Security Accreditation Board
SAB	Space Advisory Board

SAC	Space Applications Center
SACRI	Space and Advanced Communications Research Institute
SAF	Satellite Application Facilities
SAHEL	Sub-Saharan initiative for Telemedicine
SAOCOM	Satellites for Observation and Communications
SDA	Space Data Association
SDGs	Sustainable Development Goals
SDO	Solar Dynamics Observatory
SDP	Space for Development Profile
SEI	Stockholm Environment Institute
SEP	Solar electric propulsion
SEPP	Space Environment Protection and Preservation
SEPs	Solar Energetic Particles
SERI	State Secretariat for Education, Research and Innovation
SES	Single European Sky
SES	Société Européenne des Satellites
SHARAD	Shallow Subsurface Radar
SHLLV	Super-Heavy-Lift Launch Vehicle
SIA	Satellite Industry Association
SIETs	State Institutes of Educational Technology
SIG	Spatial Informatics Group
SIS	Space Infrastructure Security
SITE	Satellite Instructional Television Experiment
SLS	Space Launch System
SMILE	Solar Wind Magnetospheric Ionospheric Link Explorer
SMPAG	Space Mission Planning Advisory Group
SNC	Sierra Nevada Corporation
SNSB	Swedish National Space Board
SOHO	SOlar and Heliospheric Observatory
SpaceX	Space Exploration Technologies
SRC	Space Research Centre
SRON	Netherlands Institute for Space Research
SS/L	Space Systems/Loral
SSO	Sun-synchronous orbit
SST	Space Surveillance and Tracking
SSV	Space Service Volume
STEM	Science, Technology, Engineering and Mathematics
STEREO	Solar TErrestrial RElations Observatory
STSC	Scientific and Technical Subcommittee

T

TCBM	Transparency and Confidence-Building Measures
TDP	Technology Demonstration Programme
TEU	Treaty on European Union
TGO	ExoMars Trace Gas Orbiter
THAAD	Terminal High Altitude Area Defense system
THEMIS	Thermal Emission Imaging System
TRAI	Telecom Regulatory Authority of India
TRAPPIST	Transit Planets and Planetesimals Small Telescope

U

U.S. MDA	Missile Defense Agency
U.S.	United States of America
UAE	United Arab Emirates
UAV	Unmanned Aerial Vehicle
UGC	University Grants Commission
UK	United Kingdom
ULA	United Launch Alliance
UN	United Nations
UNCOPUOS	United Nations Committee on the Peaceful Uses of Outer Space
UNCTAD	United Nations Conference on Trade and Development
UNDP	United Nations Development Programme
UNDSS	United Nations Department of Safety and Security
UNFCCC	United Nations Framework Convention on Climate Change
UNGA	United Nations General Assembly
UNGIWG	United Nations Geographic Information Working Group
UNIDIR	United Nations Institute for Disarmament Research
UNISPACE	United Nations Conference on the Exploration and Peaceful Uses of Outer Space
UNOOSA	United Nations Office for Outer Space Affairs
UNSDI	United Nations Spatial Data Infrastructure
UN-SPIDER	United Nations Platform for Space-based Information for Disaster Management and Emergency Response
USAID	U.S. Agency for International Development
USAT	Ultra-Small Aperture Terminals

V

VAST	Vietnam Academy of Science and Technology
VHR	Very High Resolution
VIR	Visible and Infrared Mapping Spectrometer
VKO	Aerospace Defence Forces
VSAT	Very Small Aperture Terminals

W

WFP	World Food Programme
WG	Working Group
WGP	World Gross Product
WISE	Wide-field Infrared Survey Explorer
WRC	World Radiocommunication Conference
WRS	World Radiocommunication Seminar

Contents

About the Authors

Giulia Bordacchini is a Resident Fellow at the European Space Policy Institute (ESPI) in Vienna, Austria. Giulia joined ESPI shortly after completing her Master of International Relations at the University of Rome *La Sapienza* as well as the Master Course in Space Institutions and Policies with the Italian Society for International Organizations (SIOI) and the Italian Space Agency (ASI). Giulia holds a Bachelor of Political Science and International Relations and was also a Trainee at ALTEC, dealing with suborbital spaceflights and spaceport capabilities and export control, in Turin. Currently, Giulia's other works at ESPI deal with the policy and business issues related to in-orbit services in the new space economy.

Edward Burger is a Research Fellow at ESPI. Edward joined the ESPI team in February 2018 after interning at the space markets consulting firm Euroconsult, in Paris. In 2018, he worked on the present publication as well as the *Space Policies, Issues and Trends in 2017–2018* report. In December 2017, he completed a Master in the Law of Space Activities and Telecommunications at the *Université Paris Sud* in France. He has also participated in the ECSL Summer Course on Space Law and Policy (2014, held at the premises of the ITU in Geneva, Switzerland) and the ISU Space Studies Program (2016, hosted by the Technion in Haifa, Israel). Previously, Edward worked as a Project Officer at the Geneva Centre for the Democratic Control of Armed Forces.

Part I
The Year in Space 2017

Chapter 1
Global Space Policies and Programmes

1.1 Global Political and Economic Trends

1.1.1 Global Economic Outlook

The "World Economic Situation and Prospects" report is the United Nation's leading publication in the annual discussion of current economic trends and prospects. Its 2016 reporting painted a bleak picture, and the global economy looked to be held back by both slow growth and reduced international trade. However, 2017 has seen a turnaround, and it was estimated that global economic growth reached 3%, representing a large step up from the 2.4% of 2016 and also the greatest global growth rate since 2011. Furthermore, about two-thirds of countries have seen increased growth from the sluggishness which characterised 2016, and globally levels are expected to hold at 3% into 2019.[1]

WGP growth in developed economies increased from 1.6% in 2016 to 2.2% in 2017, with decreases to 2.0% in 2018 and 1.9% in 2019 estimated. As of 2017, synchronised growth is seen across all of the major developed economies. In Europe, both the Union and the Euro Area saw growth improvements over 2016–2017, with the new members (EU-13) in particular showing significant growth from 2.9 to 4.2%. This growth among Eastern European and Baltic EU members is attributed to improvements in productivity and capital accumulation. The EU-15 rate also rose to 2.0% in 2017 from 1.8% in 2016, although for all aforementioned regions, decreases are forecasted in 2018. Across the Atlantic, the same trend is seen, where the USA also saw a rise from 1.5% in 2016 to 2.2% in 2017, but with a drop to 2.1% expected

[1] "World Economic Situation and Prospects 2018." 11 Dec. 2017. United Nations, 15 Mar. 2018 <https://www.un.org/development/desa/dpad/wp-content/uploads/sites/45/publication/WESP2018_Full_Web-1.pdf>.

E. Burger, G. Bordacchini, *Yearbook on Space Policy 2017*, Yearbook on Space Policy, https://doi.org/10.1007/978-3-030-05417-5_1

in 2018. Likewise, Japan has seen growth in output from 1.0% in 2016 to 1.7% in 2017, and again a decrease in 2018 to 1.2%.[2]

Overall, transition economies follow a pattern of overall general growth each year in the same period, from a rate of 0.4% in 2016 to an estimated 2.4% in 2019. However, looking at individual regions, different fluctuations are observed: South-Eastern Europe, for example, demonstrates an opposite pattern from the developed economies, with a drop from 2.9% in 2016 to 2.5% in 2017, followed by an estimated growth toward 3.2% in 2018 and on to 3.3% in 2019. The Commonwealth of Independent States and Georgia saw significant growth from a rate of −2.4% in 2015 to 0.3% in 2016 and again significant growth to a rate of 2.2% in 2017. Developing economies overall, however, have also seen a general rise from 3.8% in 2016 to 4.3% in 2017, with further growth estimated at a rate of 4.6% 2018 and 4.7% in 2019, and as a group they are considered to remain the primary drivers of growth worldwide in 2017. South America's rate of −2.7% in 2016 turned up to 0.4% in 2017 and is also forecasted to reach 1.8% in 2018. The African continent is also demonstrating an overall fast upward trend, with its 2016 rate of 1.7% increasingly to 3.0% in 2017 and again to 3.5% in 2018.[3]

The recent strengthened activity seen in global averages and the developed countries has not been observed across all parts of the world. The growth, and in some regions' cases rapid growth, in world gross product is largely attributed to (in order of the scale of contribution to the change in world gross product growth) the USA, Japan, the Euro Area and Canada. In addition, a third of 2017's global growth is also attributed to cyclical activity in Argentina, Brazil, Nigeria and Russia. Despite these general improvements, though, effects from the low investment and productivity growth rates resulting from the global financial crisis are still observable in wage growth, debt levels and policy uncertainty, which themselves are holding back aggregate demand. And despite the general growth rate improvements of 2017, the prospects are not secure regarding the Sustainable Development Goals (SDGs), which aim to eradicate extreme poverty and create decent work for all. In fact, despite the overall positive global trends, forecasts reflect decreased growth for some regions, including for some of the poorest countries. And due to the decrease in GDP per capita in West, Central and Southern Africa, as well as Latin America and the Caribbean in 2016, SDG progress has even been reduced in some countries.[4]

[2]Ibid.

[3]Ibid.

[4]Ibid.

1.1.2 *Political Developments*

1.1.2.1 Geopolitics

North Korea's Weapons Testing in 2017

North Korea's missile and nuclear testing activities continued into 2017 with a series of missile tests including three intercontinental ballistic missile (ICBM) tests and its sixth nuclear test.[5] The ICBM flight tests were the first for the state led by Kim Jung-un, with the first test announced by North Korea on 4 July 2017, as the USA celebrated its national Independence Day; North Korea state television highlighted that the country is now "a full-fledged nuclear power that possesses [a] powerful inter-continental ballistic rocket capable of hitting any part of the world". And as a follow-up in the media, Kim Jung-un also reportedly stated that the USA would not be pleased by this "package of gifts".[6] The third test flight in November flew 960 km and reached an altitude of 4500 km, thousands of kilometres higher than the International Space Station, and is considered capable of reaching any point in the continental USA.[7] The development marks a steady growth in North Korea's technical capability, while the rhetoric between the state, its neighbours and the USA has remained aggressive. It is however not clear how the global community can respond to de-escalate the situation. Further complicating the matter, deployment of the American-built Terminal High Altitude Area Defense (THAAD) system began in South Korea in early 2017 but sparked significant diplomatic tension with China.[8] The latter does not consider the system to be a strong deterrent and has concerns about whether THAAD's radar is capable of reaching far into its own territory.[9] And for its part, international relations experts are growing sceptical of China's ability to deter North Korea.[10,11] And this in itself further raises the question of if a military response is the only means capable of deterring the state's behaviour, and US President Donald Trump has repeatedly made clear that this would indeed be an

---

[5]"North Korea: A Timeline of Missile and Warhead Tests in 2017." Time.com, 26 Feb. 2018 (retrieved) <http://time.com/5040375/north-korea-nuclear-missile-tests-2017/>.

[6]Campbell, Charlie. "How North Korea's ICBM Test Could Bring Negotiators Back to the Table." 5 July 2017. Time.com, 26 Feb. 2018 <http://time.com/4844829/north-korea-missile-icbm-negotiation/>.

[7]*Op. cit.* – "North Korea: A Timeline of Missile and Warhead Tests in 2017."

[8]Westcott, Ben and Suk, Lauren. "China, South Korea end year-long diplomatic feud over missile system." 31 Oct. 2017. CNN, 26 Feb. 2018 <https://edition.cnn.com/2017/10/31/asia/china-south-korea-thaad/index.html>.

[9]"South Korea to deploy more THAAD units after North Korea ICBM launch." 29 July 2017. Reuters, 26 Feb. 2018 <https://www.reuters.com/article/us-norhtkorea-missiles-thaad/south-korea-to-deploy-more-thaad-units-after-north-korea-icbm-launch-idUSKBN1AE02L>.

[10]Albert, Eleanor. "The China–North Korea Relationship." 28 Mar. 2018. Council on Foreign Relations, 26 Feb. 2018 <https://www.cfr.org/backgrounder/china-north-korea-relationship>.

[11]Fisher, Max. "Bad News, World: China Can't Solve the North Korea Problem." 6 Sep. 2017. The New York Times, 26 Feb. 2018 <https://www.nytimes.com/2017/09/06/world/asia/china-north-korea-nuclear-problem.html?mcubz=0&_r=0>.

option.[12] Lastly, the situation is unique in that it is one of the few current issues around which China, Russia and the USA are generally aligned, all having a common goal of de-escalation.[13,14]

US President Trump's First Year in Office

US President Donald Trump's presidency, for which he was inaugurated on 20 January 2017, has been a major media focus in 2017, with his actions, nominations, appointments, the close role of his family in the White House and public statements having caused significant controversy.[15] Major examples include the withdrawal from the Paris Climate Agreement and his stance in general toward climate change, recognition of Jerusalem as the Israeli capital and an increase in the use of drones (especially with regard to civilian casualties).[16,17] From the perspective of diverse interest groups, his actions have been often met with either great satisfaction or outrage, although all agree that the stakes are extremely significant. The American space community was, for its part, excited by President Trump's reestablishment of the National Space Council, which was previously disbanded in 1993 and has been given an ambitious mandate of maintaining American space leadership.[18] However, some actions have even resulted in direct resistance from other branches of government; for example, all three iterations of his ban on travel to the USA from certain majority-Muslim nations were fought against by federal

---

[12] *Op. cit.* – "The China–North Korea Relationship."

[13] "Russia, China Urge Peaceful Resolution to North Korea Tensions." 25 Dec. 2017. VOA News, 26 Feb. 2018 <https://www.voanews.com/a/russia-china-north-korea-tensions/4178038.html>.

[14] "Explaining U.S. Policy Toward North Korea." (Interview transcript, with NPR host David Greene) 13 Sep. 2017. NPR, 26 Feb. 2018 <https://www.npr.org/2017/09/13/550607412/explaining-u-s-policy-toward-north-korea>.

[15] Lewis, Tanya. "A Year of Trump: Science Is a Major Casualty in the New Politics of Disruption." 14 Dec. 2017. Scientific American, 26 Feb. 2018 <https://www.scientificamerican.com/article/a-year-of-trump-science-is-a-major-casualty-in-the-new-politics-of-disruption/>.

[16] Lindsay, James M. "Ten Most Significant World Events in 2017." 15 Dec. 2017. Council on Foreign Relations, 26 Feb. 2018 <https://www.cfr.org/blog/ten-most-significant-world-events-2017>.

[17] "'Trump looking to loosen Obama limits on drone strikes'." 22 Sep. 2017. Al Jazeera, 26 Feb. 2018 <https://www.aljazeera.com/news/2017/09/loosen-obama-limits-drone-strikes-170922141205178.html>.

[18] "President Trump reestablishes National Space Council." 30 June 2017. SpaceNews, 27 Feb. 2018 <http://spacenews.com/breaking-president-trump-reestablishes-national-space-council/>.

courts[19,20,21] as well as a White House policy banning service in the military by transgender people.[22] And looming above all of this, Trump's presidency remains under the intense spotlight of a Special Counsel inquiry regarding collusion with Russian actors leading to interference in the 2016 election (led by Robert Mueller of the US Department of Justice), with 19 individuals having already been indicted, including 4 from his own campaign team or administration.[23]

Brexit Developments in 2017

The path toward the UK's departure from the European Union, which began in earnest with its June 2016 referendum, saw two major developments in 2017: firstly, the invocation of the Lisbon Treaty's Article 50 on 29 March, being the necessary first step in the withdrawal process and which has given the UK exactly 2 years to negotiate the terms of its departure and secondly, the conclusion of an agreement on 8 December which, in an often-used and apt description, enables Britain and the EU to focus on their *divorce* negotiations.[24] The deal importantly covers, among other negotiation points, the future economic relationship between the two actors, the status of the Irish border, the rights of EU and UK citizens living in each other's territories and the financial settlement of the UK's obligations to the EU, estimated to be as high as 60 billion €.[25,26,27] Up next, preliminary discussions on a future trade deal can begin, although the UK cannot sign on any final document while it is still in

[19]Bradner, Eric. "Resistance to Trump's travel ban mounts." 30 Jan. 2017. CNN, 27 Feb. 2018 <https://edition.cnn.com/2017/01/29/politics/donald-trump-travel-ban-resistance/index.html>.

[20]Siddiqui, Sabrina. "Refugee admissions nearly halved as Supreme Court mulls Trump travel ban." 24 June 2017. The Guardian, 27 Feb. 2018 <https://www.theguardian.com/us-news/2017/jun/24/us-refugees-donald-trump-travel-ban>.

[21]Zapotosky, Matt. "Federal judge blocks Trumps third travel ban." 17 Oct. 2017. Washington Post, 27 Feb. 2018 <https://www.washingtonpost.com/world/national-security/federal-judge-blocks-trumps-third-travel-ban/2017/10/17/e73293fc-ae90-11e7-9e58-e6288544af98_story.html?utm_term=.8beb83a89936>.

[22]Philipps, Dave. "Judge Blocks Trump's Ban on Transgender Troops in Military." 30 Oct. 2017. New York Times, 27 Feb. 2018 <https://www.nytimes.com/2017/10/30/us/military-transgender-ban.html>.

[23]"Trump Russia affair: Key questions answered." Updated regularly. BBC, 27 Feb. 2018 (retrieved) <http://www.bbc.com/news/world-us-canada-42493918>.

[24]*Op. cit.* – "Ten Most Significant World Events in 2017."

[25]Meyer, David. "Here's What the U.K. and EU Just Agreed To in the Brexit Deal." 8 Dec. 2017. Fortune.com, 28 Feb. 2018 <http://fortune.com/2017/12/08/brexit-deal-uk-eu-agree-explainer/>.

[26]Mason, Rowena. "Main points of agreement between UK and EU in Brexit deal." 8 Dec. 2017. The Guardian, 28 Feb. 2018 <https://www.theguardian.com/uk-news/2017/dec/08/main-points-of-agreement-uk-eu-brexit-deal>.

[27]*Op. cit.* – "Ten Most Significant World Events in 2017."

the Union.[28] And at home, British PM Theresa May is now one step closer to the impossible task of pursuing a Brexit arrangement that satisfies all voices; the debate is largely between supporters of so-called "hard" and "soft" Brexits, with their opinions sometimes expressed in very colourful language.[29] A hard Brexit could see the UK leave the EU single market and return to a trade relationship based on World Trade Organization rules. It would also give priority to Britain having full control over its borders. A soft approach would see the British relationship with the EU try to stick to the status quo as much as possible; it would still have access to the single market, but lose its MEPs and European Commissioner.[30,31] Throughout 2017, the Brexit topic continually represented a dramatic source of news, and a final outcome is still not clear. Two big questions linger: Will the resulting relationship look significantly different than that of 2017? And would that make all the effort worth it?

The Conflicts in Iraq and Syria

The conflicts in Iraq and Syria have caused both states' populations to experience horrific conditions and have contributed to a global refugee crisis for which humanitarian systems were not prepared.

In July 2017, Iraqi forces retook Mosul—the city where Abu Bakr al-Baghdadi claimed he was the head of a new caliphate, the Islamic State, 3 years prior—and in December 2017, the Iraqi Prime Minister Haider al-Abadi formally declared an end of fighting against the extremist group.[32,33,34] Despite this landmark in the fight against the Islamic State, it came at a huge cost, and the future remains uncertain for Iraq: Up to possibly 40,000 civilians lost their lives in the fighting for Mosul alone, and the city is in ruins.[35] The 2017 figures also reflect a grave loss of human life

28"Brexit: 'Breakthrough' deal paves way for future trade talks." 8 Dec. 2017. BBC, 28 Feb. 2018 <http://www.bbc.com/news/uk-politics-42277040>.

29*Op. cit.* – "Here's What the U.K. and EU Just Agreed To in the Brexit Deal."

30Sims, Alexandra. "What is the difference between hard and soft Brexit? Everything you need to know." Independent, 28 Feb. 2018 <http://www.independent.co.uk/news/uk/politics/brexit-hard-soft-what-is-the-difference-uk-eu-single-market-freedom-movement-theresa-may-a7342591.html>.

31"Brexit: What are the options?" 12 June 2017. BBC, 28 Feb. 2018 <http://www.bbc.com/news/uk-politics-37507129>.

32"Iraqi Forces Retake City Of Mosul From ISIS Fighters." (Interview transcript, with NPR host Kelly McEvers) 10 July 2017. NPR, 28 Feb. 2018 <https://www.npr.org/2017/07/10/536505269/iraqi-forces-retake-city-of-mosul-from-isis-fighters>.

33Collard, Rebecca. "What We Have Learned Since ISIS Declared a Caliphate One Year Ago." 25 June 2017. Time.com, 28 Feb. 2018 <http://time.com/3933568/isis-caliphate-one-year/>.

34Graham-Harrison, Emma. "Iraq formally declares end to fight against Islamic State." 9 Dec. 2017. The Guardian, 28 Feb. 2018 <https://www.theguardian.com/world/2017/dec/09/iraq-formally-declares-end-to-fight-against-islamic-state>.

35*Op. cit.* – "Ten Most Significant World Events in 2017."

resulting from Syria's civil war—where the Islamic State also claims and has held territory—with over 10,000 civilian deaths recorded.[36,37,38] The UNHCR also reports a total of 5,479,277 registered Syrian refugees that have fled their homes as a result of the conflict from 2013 to the end of 2017.[39] By the fall of 2017, Turkey was hosting over 3 million Syrian refugees, and as of December 2017, 1,015,500 Syrian-citizen asylum applications had been made in Europe.[40,41] Further, host and donor states alike desire for the refugees to return home as soon as possible (due to the costs of supporting the refugees and the extra infrastructural strain), but the fighting in Syria remains intense, and polling among refugees indicates that most want to return, but only once the conflict has ended.[42,43] The UNHCR echoes this attitude, stating that it cannot facilitate returns given the ongoing risks.[44]

Cyber Threats Growing and Growing More Severe

The year 2017 profoundly demonstrated how the cyber domain, existing in the cables and hard drives of society's communications infrastructure, can be both a tool for and a theatre of war and a medium for criminal activity. Cyber-related incidents in 2017 that posed real-world consequences include the WannaCry ransomware attack (effecting 200,000 victims in 150 countries[45] and which

---

[36]El Hilali, Nouran Mohamed and Petkova, Mariya. "MAPPED: The battle against ISIL." 15 Oct. 2017. Al Jazeera, 28 Feb. 2018 <https://www.aljazeera.com/indepth/interactive/2017/04/map-isil-syria-iraq-170413092750456.html>.

[37]"10,204 Civilians Killed In Syria In 2017." 28 Jan. 2018. I AM SYRIA, 28 Feb. 2018 <http://www.iamsyria.org/syrian-conflict-in-2017.html>.

[38]"2017...the year of the military change and the rise of the regime and its allies to head the list of military powers and influences followed by the SDF...and the loss of the opposition and the organization to large spaces of their controlled areas." 28 Dec. 2017. The Syrian Observatory for Human Rights, 28 Feb. 2018 <http://www.syriahr.com/en/?p=81564>.

[39]"Operational Portal, Refugee Situations – Syria Regional Refugee Response." data from 31 Dec. 2017. UNHCR, 1 Mar. 2018 <https://data2.unhcr.org/en/situations/Syria>.

[40]"3RP 2017 Progress Report." (Download page) 17 Oct. 2017. UNHCR, 28 Feb. 2018 <http://data.unhcr.org/syrianrefugees/download.php?id=14381>.

[41]*Op. cit.* – "Operational Portal, Refugee Situations – Syria Regional Refugee Response."

[42]Crisp, Jeff. "Why it's Far Too Early to Talk of Return for Syrian Refugees." 11 Aug. 2017. News Deeply, 28 Feb. 2018 <https://www.newsdeeply.com/refugees/community/2017/08/11/why-its-far-too-early-to-talk-of-return-for-syrian-refugees-2>.

[43]El Gantri, Rim and El Mufti, Karim. "Not Without Dignity: Views of Syrian Refugees in Lebanon on Displacement, Conditions of Return, and Coexistence." 9 June 2017. International Center for Transitional Justice, 28 Feb. 2018 <https://www.ictj.org/publication/syria-refugees-lebanon-displacement-return-coexistence>.

[44]*Op. cit.* – "Why it's Far Too Early to Talk of Return for Syrian Refugees."

[45]Belot, Henry and Borys, Stephanie. "Ransomware attack still looms in Australia as Government warns WannaCry threat not over." 16 May 2017. ABC.net.au, 2 Mar. 2018 <http://www.abc.net.au/news/2017-05-15/ransomware-attack-to-hit-victims-in-australia-government-says/8526346>.

terrifyingly disrupted a third of British NHS locations[46]), the leaking of hacking tools believed to belong to the US NSA (which were later used in the WannaCry attack) and the October announcement of the breach in 2013 of Yahoo's 3 billion customer accounts.[47] The cyber domain has also played a significant role in conflict contexts in recent years, with the Syrian civil war providing many examples: the arrest and torture of activists leading to the access of their social media accounts, the tracking of opposition groups' meeting locations via IP addresses, the spamming of US government websites and news sites by the Syrian Electronic Army as well as a major hacking operation in late 2013 and early 2014 against opposition forces.[48] In the latter incident, "hackers stole a cache of critical documents and Skype conversations revealing the Syrian opposition's strategy, tactical battle plans, supply needs, and troves of personal information and chat sessions", representing data belonging to the opposition and humanitarian aid workers as well as media activists.[49] The hacking group Anonymous has targeted numerous Twitter and Facebook accounts of the Islamic State and has also attacked the Syrian Ministry of Defence.[50] Cyberattacks in other contexts have also been known to shut down major infrastructure and technological systems, including the electrical power supply for over 230,000 residents in Ukraine[51]—an attack later described as "brilliant" by a member of an investigative team—and an American-Israeli "computer worm" known as Stuxnet which targeted an Iranian nuclear enrichment facility, causing its centrifuges to speed up or down in such a way that they destroyed themselves, all while leaving normal indicators on computer readings.[52] Iran is not the only known American target, with President Barack Obama having also ordered the Pentagon to increase its cyber strikes against North Korea's missile programme.[53]

In light of these threats and risks, governments worldwide are reacting through the development of national cybersecurity strategies. According to the ITU's Global

---

46"NHS 'could have prevented' WannaCry ransomware attack." 27 Oct. 2018. BBC, 2 Mar. 2018 <http://www.bbc.com/news/technology-41753022>.

47Larson, Selena. "The hacks that left us exposed in 2017." 20 Dec. 2017. CNN, 2 Mar. 2018 <http://money.cnn.com/2017/12/18/technology/biggest-cyberattacks-of-the-year/index.html>.

48Ruhfus, Juliana. "Syria's Electronic Armies." 18 June 2015. Al Jazeera, 2 Mar. 2018 <https://www.aljazeera.com/programmes/peopleandpower/2015/06/syria-electronic-armies-150617151503360.html>.

49"Behind the Syrian Conflict's Digital Frontlines." 2 Feb. 2015. FireEye Inc., 2 Mar. 2018 <https://www.fireeye.com/blog/threat-research/2015/02/behind_the_syrianco.html>.

50*Op. cit.* – "Syria's Electronic Armies."

51Zetter, Kim. "Inside The Cunning, Unprecedented Hack Of Ukraine's Power Grid." 3 Mar. 2016. Wired, 2 Mar. 2018 <https://www.wired.com/2016/03/inside-cunning-unprecedented-hack-ukraines-power-grid/>.

52Szoldra, Paul. "A new film gives a frightening look at how the US used cyberwarfare to destroy nukes." 7 July 2016. Business Insider, <http://www.businessinsider.com/zero-days-stuxnet-cyber-weapon-2016-7?IR=T>.

53Broad, William J. and Sanger, David E. "Trump Inherits a Secret Cyberwar Against North Korean Missiles." 4 Mar. 2017. New York Times, 2 Mar. 2018 <https://www.nytimes.com/2017/03/04/world/asia/north-korea-missile-program-sabotage.html>.

Cybersecurity Index report of 2017, 38% of countries worldwide have published a cybersecurity strategy, 11% have a standalone and dedicated cybersecurity strategy, and 12% have one in development.[54] There is also discussion on the idea of a digital Geneva Convention, which Microsoft President Brad Smith is calling for.[55]

Given the real-world risks resulting from emails being hacked and released (which American intelligence agencies have concluded Russia authorised against the US Democratic National Committee[56]), from power grid interruptions in the winter in Ukraine and from healthcare appointments being cancelled due to ransomware attacks as in the UK, it is clear that cyber security will continue to grow as a core governance priority.

1.1.2.2 Environment

The 2015 Paris Agreement aims to keep global average temperature increases to below 2 °C above pre-industrial levels, and to make more ambitious efforts to limit temperature increases even further to 1.5 °C, and eliminate the increase of greenhouse gas emissions in the second half of the century.[57] Following its creation in the 21st UN Framework Convention on Climate Change Conference of Parties (UNFCCC/COP), it entered into force on 4 November 2016. This was triggered by the ratification of the European Union on 5 October 2016, which met the threshold that at least 55 parties, accounting for at least an estimated 55% of total global greenhouse emissions, ratify the instrument. China and India, representing 2 of the greatest contributors to greenhouse gas emissions, are among the ratifying states, and 172 in total had ratified it by the end of 2017.[58,59] However, US President Donal Trump's announcement in June to withdraw the US participation was noted

---

[54]"Global Cybersecurity Index 2017." (Download page) 19 July 2017. ITU, 2 Mar. 2018 <https://www.itu.int/pub/D-STR-GCI.01-2017>.

[55]Vanian, Jonathan. "Here's Why Microsoft President Wants a Digital Geneva Convention." 14 Feb. 2017. Fortune, 2 Mar. 2018 <http://fortune.com/2017/02/14/microsoft-president-digital-geneva-convention/>.

[56]*Op. cit.* – "Trump Russia affair: Key questions answered."

[57]"The Paris agreement marks an unprecedented political recognition of the risks of climate change." 12 Dec. 2015. The Economist, 7 Mar. 2018 <http://www.economist.com/node/21683990/>.

[58]"Paris Agreement – Status of Ratification." Updated regularly. UN Climate Change, 7 Mar. 2018 <http://unfccc.int/paris_agreement/items/9444.php>.

[59]"Paris Agreement Reaches 175 Ratifications." 6 Mar. 2018. IISD/SDG Knowledge Hub, 7 Mar. 2018 <http://sdg.iisd.org/news/paris-agreement-reaches-175-ratifications/>.

with disappointment by the global environmental protection community in 2017, and Russia has yet to ratify it.[60,61]

The 23rd UNFCCC/COP took place in Bonn, Germany, from 6 to 17 November 2017.[62] The event likewise served as the 13th session of the Meeting of the Parties to the Kyoto Protocol and was the platform for continued discussions of the Paris Agreement's governing body. In order for the Agreement to be fully operational, its parties' first need to elaborate and adopt decisions on several topics including mitigation (e.g. nationally determined contributions (NDCs)), adaptation communications, finance, transparency, "global stocktake" and market and non-market mechanisms. They aim to do so by the 24th COP in December 2018, ahead of the 2020 timeline from which the agreement was intended to begin, and in Bonn they reaffirmed their commitment to this schedule.[63] Among the main accomplishments of the Bonn Conference was the launch of the InsuResilience Global Partnership, which aims to bring climate and disaster risk insurance solutions to millions who are at risk globally; the German Federal Ministry for Economic Cooperation and Development provided 110 million euros to support this partnership.[64,65]

Looking ahead, the 24th UNFCCC/COP will be held in Katowice, Poland, in December 2018.[66]

1.1.2.3 Energy

With the Paris Agreement's ratification by over 170 countries by the end of 2017[67] and the reports that in 2017 the Arctic saw its second warmest year on record and its overall lowest level of sea ice,[68] the global energy sector is increasingly under the spotlight. Further, this same dynamic between positive developments and alarming

---

[60]Volcovici, Valerie. "U.S. submits formal notice of withdrawal from Paris climate pact." 4 Aug. 2017. Reuters, 7 Mar. 2018 <https://www.reuters.com/article/us-un-climate-usa-paris/u-s-submits-formal-notice-of-withdrawal-from-paris-climate-pact-idUSKBN1AK2FM>.

[61]*Op. Cit.* – "Paris Agreement – Status of Ratification."

[62]"About COP23 – Bonn, 2017." UN Climate Change Conference – COP23 Fiji, 7 Mar. 2018 <https://cop23.com.fj/about-cop-23/about-cop23/>.

[63]"Earth Negotiations Bulletin." (Volume 12 Number 714) 21 Nov. 2017. IISD Reporting Services, 7 Mar. 2018 <http://enb.iisd.org/vol12/enb12714e.html>.

[64]"Key Achievements from COP23." 18 Nov. 2017. UN Climate Change Conference – COP23 Fiji, 7 Mar. 2018 <https://cop23.com.fj/key-achievements-cop23/>.

[65]"InsuResilience Global Partnership." InsuResilience Global Partnership, 7 Mar. 2018 (retrieved) <http://www.insuresilience.org/>.

[66]"Katowice Climate Change Conference (UNFCCC COP 24)." IISD/SDG Knowledge Hub, 7 Mar. 2018 (retrieved) <http://sdg.iisd.org/events/unfccc-cop-24/>.

[67]*Op. cit.* – "Paris Agreement Reaches 175 Ratifications."

[68]"Arctic saw 2nd warmest year, smallest winter sea ice coverage on record in 2017." 12 Dec. 2017. NOAA, 9 Mar. 2018 <http://www.noaa.gov/media-release/arctic-saw-2nd-warmest-year-smallest-winter-sea-ice-coverage-on-record-in-2017>.

statistics can also be seen throughout the International Energy Agency's (IEA) recent reporting; for example, as of 2016, the energy sector accounted for at least two-thirds of greenhouse gas emissions, but in its 2017 World Energy Outlook, it indicates that from the present to 2040, renewable sources will represent 40% of the increase in primary demand.[69] Likewise, although their levels had recently plateaued, energy-related CO_2 emissions will grow slightly in baseline scenario reporting to 2040, and the demand for energy will grow by 30% in the same period. However, global trends broadly indicate that natural gas, the increasing role of electricity and renewables and high-energy efficient technologies are coming to the fore, while coal's dominant years are behind it.[70]

This agrees with the 2016 World Energy Outlook, which indicated that natural gas, as well as oil along with it, will remain the bedrock of the global energy system for many decades to come. The 2017 World Energy Outlook forecasts that through to 2040 the demand for oil will continue to grow and the USA in particular will support this demand by putting out 50% more of oil and gas combined than any other country. In terms of consumption, China's use is expected to exceed the USA by 2030, but demand growth in India is forecasted to exceed that of China after 2025. In terms of coal, China, known for its coal-oriented energy system, is embracing a new energy approach promoting natural gas and high-efficiency and clean technologies, and in India the level of coal in total consumption will decrease to less than half in 2040 from its current three-quarters share. A decrease in the use of coal since 2000 and into the forecasted period up to 2040 is also clearly seen: the capacity of coal-based power generation grew by nearly 900 gigawatts, although net additions as of 2017 into 2040 are less than half.[71]

The role of electricity is also forecasted to grow significantly as an end-use of energy, representing 40% of the growth in final consumption up through 2040. Electric motor systems at the industrial level, in particular, take one-third of the growth in power demand in baseline scenario reporting. Other contributing factors include the spread of electric cars globally, especially in countries which have policy support for reducing the sale of gas and diesel vehicles like the UK and France, the use of electricity for final consumption in heating and mobility and furthermore the growing demand for electrical home appliances as incomes rise. However, to facilitate this growth in this consumption of energy through electricity, major consumers like China and India must significantly expand their infrastructure accordingly. The IEA foresees that to facilitate this rise in consumption, China, for example, will need to expand its existing infrastructural capacity with that of the

[69]"World Energy Outlook 2016 – Executive Summary." 16 Nov. 2016. IAE, 9 Mar. 2018 <http://www.iea.org/Textbase/npsum/WEO2016SUM.pdf>.

[70]"World Energy Outlook 2017 – Executive Summary." 14 Nov. 2017. IAE, 9 Mar. 2018 <http://www.iea.org/publications/freepublications/publication/WEO_2017_Executive_Summary_English_version.pdf>.

[71]Ibid.

USA today. And as further indication of the rise of electricity, consumer electricity spending had already approached parity with that toward oil products in 2016.[72]

1.1.2.4 Resources

With very low global trade growth levels in 2016 at 2.2%, the year 2017 saw significant expansion at a rate of 3.7%, reinforced by the general uptick in global economic growth, which hit 3.0% in 2017. This was the highest since the 3.1% economic growth rate of 2011, and a significant increase on the 2.4% of 2016. Furthermore, global trade growth has also been consistently demonstrating calmer fluctuations, overall, since the dramatic rises and falls during and immediately after the financial crisis. The 2017 growth in trade is also largely attributed to a moderate uptake in investment across certain developing and developed countries, which has also boosted trade of intermediate and capital goods in particular. The year 2017 also saw growth in global industrial output as well as the demand for container shipping and international air freight. And although still moderate, 2018 world trade is forecasted to expand by 3.5% and again in 2019 to 3.6%.[73]

In terms of growth of imports and exports in 2016, the former grew by 2.1% and the latter by 1.7%, representing an average growth of just 1.9%. This figure is furthermore still significantly lower than the 7.2% growth rate seen from 2003 to 2007, before the crisis. And given low global demand levels, the global trade situation overall is not expected to promote growth for any particular set of economies. Although global trade demonstrated some recovery in 2017, associated with significantly increased merchandise imports in Asia and more moderate import growth in Latin America and the USA, the long-term strength of this recovery is certain. China's growth in exports was perfectly flat at 0.0% in 2016, while the volume of imports did grow, however, by 3.1%. In the USA, imports grew at 3.6%, a slight decrease from the 3.7% of 2015, and exports were at −0.2%. Across all developed countries, growth in the volume of imports was 2.7%, while volume of exports was just 1.0% in 2016, the latter largely attributed to weak demand from certain developing countries. In the European Union, imports saw a growth of 2.8% in 2016, while exports declined from 3.3% in 2015 to 1.1%. In the UK, in particular, exports saw a boost in the latter half of 2016 after a drop in the pound sterling's value. African, Latin American and West Asian commodities exporters, specifically, saw declining imports in 2016, the result of decreasing commodity prices. In Africa, for example, growth in the volume of imports was −4.6% in 2016, and in sub-Saharan Africa in particular, the volume of imports growth decreased from −0.3% in 2015 to −6.6% in

[72]Ibid.

[73]*Op. cit.* – "World Economic Situation and Prospects 2018."

2016. In Latin America and the Caribbean, the rate was −4.2% and −2.6% in West Asia.[74]

According to the United Nations Conference on Trade and Development (UNCTAD), while commodity prices experienced declines in all group categories, decreasing by 34.3% in 2015, the largest contraction was in crude oil which decreased by 47.2%.[75] Decreases occurred again in 2016, with commodity prices overall decreasing by 8.7% and crude oil by 15.7%. The year 2017 finally saw positive figures again, with UNCTAD forecasting commodity prices to grow by 14.4% and crude oil by 19.5%. And although this represents a significant turnaround, the rates are still well below respective prices observed during the 2000s commodities boom ending in 2014. For example, in oil prices per barrel, the year started at $52/barrel, dropped to a low of $45 in June and steadily rose again to $62 by December.[76] And with oil and commodity prices still significantly below 2014 levels (when they were just above $100/barrel), recovery is taking place slowly among commodity-exporting economies. Regarding oil in particular, the longevity of the rebound is uncertain, with OPEC taking steps to reduce supply and while there is also a rise in US inventories, largely connected to shale oil production, which appears to be lowering prices.[77]

As stated, UNCTAD forecasted commodity prices to grow by 14.4%. This itself is a jump from the −8.7% percentage change in 2016 and the −34.3% change in 2015. Looking at the specific price growth and contractions figures, all non-fuel commodities were reported to grow by nearly 10% in 2017. Further, all agricultural raw material prices likewise grew by 10%, while mineral, ore and metal prices expanded by 23%, representing a nearly equal turnaround from the −23% contraction of 2015. Iron ore and zinc led the way in 2017, with approximately 27% and 28%, respectively. Precious metals, however, were reported to contract by 0.7% in the year 2017, with gold decreasing at nearly the same rate. Recent metal price declines in 2017 were largely associated with dropping demand in importing economies, the USA and China in particular. Food products prices, though, saw modest growth at 2.5% in 2016 and 1.0% in 2017. In overview, although commodity prices have been returning upwards since the declines seen after the commodities boom, UNCTAD notes that given the most recent declines, the rebound may not necessarily continue for long and that the price movements since the end of the boom are not uniform across all commodity groupings.[78]

---

[74]"Trade and Development Report 2017." 14 Sep. 2017. UNCTAD, 9 Mar. 2018 <http://unctad.org/en/PublicationsLibrary/tdr2017_en.pdf>.

[75]Ibid.

[76]"countryeconomy.com – OPEC Reference Basket (ORB) US Dollars per Barrel." Regularly updated.

Data from those dates mentioned in text. countryeconomy.com, 9 Mar. 2018. <https://countryeconomy.com/raw-materials/opec>.

[77]*Op. cit.* – "Trade and Development Report 2017."

[78]Ibid.

1.1.2.5 Knowledge

By now the advantages of higher education should be seen as worth the effort as employment rates and earnings tend to increase as an adult's level of education and skills increases; moreover, the labour market still regards a diploma or degree as the primary indication of a worker's skills. For Europe, the expansion of its pool of highly skilled and specialised scientists and professionals should be a constant priority if it is to remain a leading actor in the field of space-related scientific and technological R&D. In the year 2016, the percentage of the European[79] population between 25 and 64 years of age with a tertiary degree reached 34%, just below the G20 average of 35% and further below the OECD average of 37%. This does represent however a 2% growth over the previous year. In comparison, 46% of the same age cohort of the USA and 57% in Canada have attained the same level of education in 2016. Among 25–34-year-olds in Europe, Lithuania comes first with 55% having attained a tertiary education in 2016, followed by the UK at 52% and Luxembourg at 51%. These are well above the OECD and European averages of 42% and 40%, respectively, and they are followed by Norway and Switzerland both at 49% and Sweden at 47%. The three largest contributors to ESA's budget in 2017 were Germany, France and Italy,[80] and their tertiary attainment rates among 25–34-year-olds were 31%, 44% and 26%, respectively.[81]

According to the OECD, the enrolment rate of 20–24-year-olds in tertiary education increased on average from 29 to 33% from 2005 to 2014 across OECD countries.[82] For enrolment in educational programmes in general among 20–24-year-olds, the 2017 reporting show that 42% are enrolled across the OECD and 43% across the EU. Further, among 25–65-year-old Europeans studying, 13% are in education; 19% are in the arts, humanities, social sciences, journalism and information; 21% are in business, administration and law; 5% are in natural sciences, mathematics and statistics; 4% are in information and communications technology (ICT); 18% are in engineering, manufacturing and construction; and 12% are in health and welfare. These data align nearly perfectly with the OECD averages. And focusing on the upcoming generation of European students focusing on science, technology, engineering and mathematics (STEM), Germany is leading with nearly 40% of new tertiary-level entrants in STEM. Following are Estonia and Finland with both having nearly 35%, and Slovenia and Austria with each just over 30%. The respective OECD and EU averages are at just over 25%.[83]

---

[79]Not including Bulgaria, Croatia, Cyprus, Lithuania, Malta and Romania

[80]"ESA Budget for 2017." 16 Jan. 2017. ESA, 28 Mar. 2018 <http://www.esa.int/spaceinimages/Images/2017/01/ESA_budget_2017>.

[81]"Education at a Glance 2017." (Download page) 12 Sep. 2017. OECD iLibrary, 28 Mar. 2018 <https://www.oecd-ilibrary.org/education/education-at-a-glance-2017_eag-2017-en>.

[82]"Education at a Glance 2016." (Download page) 15 Sep. 2016. OECD iLibrary, 28 Mar. 2018 <https://www.oecd-ilibrary.org/education/education-at-a-glance-2016_eag-2016-en>.

[83]*Op. cit.* – "Education at a Glance 2017."

In considering gender, the percentages of female students entering STEM tertiary programmes in those same countries are as follows: In Germany, 28% of new STEM entrants are women, 34% in Estonia, 23% in Finland, 29% in Slovenia and 28% in Austria. The respective OECD and EU averages are both 30%, and, further, the UK and Iceland are leading the way with 37% of women entering STEM programmes. These recent figures reconfirm the progress, but continuing work to be done in reducing the gender disparity in STEM. More broadly speaking, women entering tertiary studies are most represented in education, where the OECD average is 78%, followed by health and welfare (76%) and thirdly in social sciences, journalism and information (at 64%). In engineering, they represent only 24% of students entering tertiary programmes, 50% of students in natural sciences, mathematics and statistics and 19% in ICT. Across many disciplines, Latvia and Iceland are leading the way among European countries. For example, Iceland has the largest percentage of women in OECD countries entering engineering, at 37%.[84]

1.1.2.6 Mobility

Maritime transport remains a critical element in the infrastructure of the global economy, carrying over 80% of the world's trade by volume and over 70% of its value. From 2015 to 2016, world seaborne trade grew from 1.8% up to 2.6%, which, although representing growth, is still below the 3% historical average observed during the past 40 years. The total volume of seaborne trade in 2016 reached nearly 10.3 billion tons, up from just over 10 billion tons in 2015. In more precise terms, this represented an increase of over 260 million cargo tons over this time period, with more than half of it in tanker trade. A significant influence on global trade levels came from China's strong import demand in 2016, but growth was held back by limited export demand and especially limited import demand from other developing economies. However, and despite this, developing regions are still the major basis for world shipments in seaborne cargo in 2016. Specifically, developing countries represent 59% of the world goods exported and about two-thirds imported. Developed economies, for their part, represent about 35% in both regards. Regionally, Europe represents about nearly 20% of both goods exported and imported in 2016. The most dominant region, however, remains Asia, which loaded 40% of goods in 2016 and unloaded just over 60%. These seaborne shipment levels are significantly higher than the second most active region, being the Americas, which imported 14% of the world trade and exported 22%. Furthermore, and looking ahead, UN reporting indicates that seaborne trade is expected to grow 2.8% and reach 10.6 billion tons of volume in 2017, and over the following 5 years expanding by 3.2% annually.

[84]Ibid.

Looking at the cross section of the cargo to be shipped, containerised cargo and major dry bulk commodities are anticipated to grow the most.[85]

Security of the ships and crew active in international maritime transport remained a major issue in 2017. Between 1984 and the end of 2015, the number of incidents of piracy and armed robbery against ships totalled 7346 worldwide.[86] In 2015, specifically, the Maritime Safety Committee reported 303 incidents, which decreased by 27% to a total of 221 incidents in 2016. However, despite this overall reduction, West Africa saw a 77% increase during the same reporting time period. Furthermore, piracy is not the only threat, with cybersecurity representing a growing threat to the maritime sector. At the root, seaborne shipping heavily utilises computer-based and ICT technologies which are exposed to the same risks as those referred to in the Geopolitics section of this chapter. Essentially, on-board computers or computer systems in other facilities, such as at ports, can be the objects of cyberattacks such as phishing, so-called Trojan horses, hacking, malware, denials of service and other forms. Results of such attacks can disrupt container inventory and tracking operations or, more dramatically, the navigation systems and control of propulsion on the actual ships. They may also lead to the theft of company data, used for diverse purposes. Ultimately, such threats have very direct and real consequences for individuals working in various applications in the maritime industry and for safety, the environment, in addition to consequences to the economy. Concerns have also been expressed that the industry culture is not sufficiently aware of such risks. However, a major recent regulatory development came in the form of the International Maritime Organization's adoption in 2017 of guidelines on maritime cybersecurity risk management. Several real examples from 2011 to 2017 have occurred, including one targeting Maersk which involved the use of ransomware. Assuming such attacks may occur more frequently, the recent guidelines will hopefully lead to further regulatory activity.

1.1.3 Main Science and Technology Indicators Relevant for Space Activities

1.1.3.1 Research and Development Inputs: EU and International Expenditures

Science and technology inputs at the European level are measured against the benefits they generate; in 2017, R&D investment continued to push for increased effectiveness and stronger R&D integration in broader industrial and macro-

[85]"Review of Maritime Transport 2017." (Download page) 25 Oct. 2017. UNCTAD, 29 Mar. 2018 <http://unctad.org/en/Pages/Publications/Review-of-Maritime-Transport-(Series).aspx>.

[86]"Review of Maritime Transport 2016." (Download page) 7 Nov. 2016. UNCTAD, 29 Mar. 2018 <http://unctad.org/en/Pages/Publications/Review-of-Maritime-Transport-(Series).aspx>.

economic policies. The reporting period results are measured by gross domestic expenditure on R&D (GERD), a statistical measure showing nominal changes in these expenditures.

According to Eurostat, GERD as a percentage of GDP spent by the 28 EU countries was provisionally assessed at 2.03% in 2016, compared to the 2.04% of 2015.[87] The EU's GDP in 2016 stood at 14.921 trillion € and reached 15.336 trillion € in 2017[88]; as such, the EU's total 2016 GERD spending reached 302.896 billion €.

Within the EU in 2016, the five leading members in terms of GERD were Sweden (3.25%), Austria (3.09%), Germany (2.94%), Denmark (2.87%) and Finland (2.75%). In terms of absolute spending in euros, Germany spent 92.897 billion € (up from 87.188 billion € in 2015), followed by France spending 50,143 billion € (up from 48.463 billion €) and the UK at 40,617 billion € (down from 43.878 billion €). At the global level, the latest available GERD data is from 2015, when the EU performance stood at 2.04%, compared with the following states: South Korea at 4.23% (a decrease from 2014), Japan at 3.29% (a decrease), the USA at 2.79% (an increase) and Russia at 1.1% (an increase).

Further, countries with higher GERD in general tend to match those with high investment in innovation within the private sphere: according to Eurostat, R&D expenditure in 2016 as a percentage of GDP, under the business enterprise sector, was at 1.32% in the EU overall and at 2.26% in Sweden, 2.2% in Austria, 2% in Germany, 1.89% in Denmark and 1.81% in Finland, which represent the leading five nations.[89]

1.1.3.2 Research and Development Inputs: Corporate Activity and Policy Guidance

Based on the European Commission's 2017 EU Industrial R&D Investment Scoreboard, over 2016–2017, 2500 companies worldwide each invested at least 24 million € in R&D, with a global total coming to 741.6 billion €.[90] This figure represents 6.6% growth over the previous reporting period's 696.0 billion €. This growth was mainly driven by activity in the ICT services industry, which expanded 11.7%, although R&D growth is generally led also by the health and automotive industries.

---

[87]"Eurostat – Gross domestic expenditure on R&D (GERD)." Data from years 2015 & 2016. Eurostat, 22 Aug. 2018 <http://ec.europa.eu/eurostat/tgm/table.do?tab=table&plugin=1&language=en&pcode=t2020_20>.

[88]"Eurostat – Gross domestic product at market prices." Data from years 2016 & 2017. Eurostat, 22 Aug. 2018 <http://ec.europa.eu/eurostat/tgm/table.do?tab=table&plugin=1&language=en&pcode=tec00001>.

[89]"Eurostat – Research and development expenditure, by sectors of performance." Data from year 2016. Eurostat, 22 Aug. 2018 <http://ec.europa.eu/eurostat/tgm/table.do?tab=table&plugin=1&language=en&pcode=tsc00001>.

[90]"The 2017 EU Industrial R&D Investment Scoreboard." (Download page) 4 Dec. 2017. European Commission, 22 Aug. 2018 <http://iri.jrc.ec.europa.eu/scoreboard17.html>.

In the EU in particular, R&D growth over 567 companies grew by 7%, representing 192.5 billion € (26% of the global 741.6 billion €). The EU growth was largely based in ICT producers (up 14.4% over the previous year) and health (up 7.9%), although aerospace R&D investment contracted 5.4%, while globally it grew 2.2%. Within the EU, almost two-thirds of the 1000 highest investing companies are based in Germany, France and the UK, aligning with the GERD reporting of the previous section above. Further, 16 of the 50 largest investing companies are based in the EU. Internationally, the USA experienced 7.2% R&D growth over the previous reporting period, while China saw an 18.8% increase and Japan contracted by 3%.

Moreover, the European Commission's bi-annual Conferences on Corporate R&D and Innovation (CONCORDi) "provide a platform for discussion on how research can support evidence-based policy making in the area of private R&D and innovation investments".[91] Conference documentation states that "Corporate R&D, innovation and technological development are crucial" industry drivers. Moreover, the 6th CONCORDi meeting, held in September 2017, had a stated aim to provide input "to the EU industrial research and innovation policy agenda".[92] The conference's resulting policy brief—"Innovation and Industry: Policy for the Next Decade"—assessed key challenges resulting from industrial change and innovation and provides four major policy considerations for the future. These include (1) a focus on the goals of supporting "science for science", "science for society" and "science for industry", (2) greater coordination of relevant actors across governance levels combined with simplification of the existing policy instruments (which are considered to be too numerous), (3) greater "tailoring" of research and innovation policy with regard to the particular conditions in the EU and (4) "embrace experimentation on a larger scale".[93]

---

[91] "Conferences on Corporate R&D and Innovation (CONCORD)." Regularly updated. European Commission, 22 Aug. 2018 (retrieved) <http://iri.jrc.ec.europa.eu/concord.html>.

[92] "CONCORDi 2017." 14 Nov. 2016. European Commission, 22 Aug. 2018 <http://iri.jrc.ec.europa.eu/concord/2017/index.html>.

[93] Moncada-Paternò-Castello P., Grassano N., & Vezzani A. "Innovation and Industry: Policy for the Next Decade." (Download page) December 2017. European Commission, 22 Aug. 2018 <http://iri.jrc.ec.europa.eu/concord/2017/index.html but direct to the Policy Brief link>.

1.2 Worldwide Space Policies and Strategies

1.2.1 The United Nations System: UN Entities Relevant to Space Activity

1.2.1.1 United Nations General Assembly

In 2017, the United Nations General Assembly (UNGA) passed seven resolutions relating to the use and exploration of outer space during in its 72nd session. These include:

On 4 December 2017

- *Resolution 72/26: Prevention of an arms race in outer space,*[94] which reaffirms the positions of the General Assembly in previous years' resolutions and emphasises "the paramount importance of strict compliance with existing arms limitation and disarmament agreements relevant to outer space" and raises the need for "further measures with appropriate and effective provisions for verification to prevent an arms race in outer space".
- *Resolution 72/27: No first placement of weapons in outer space,*[95] which reiterates the resolution of the same name from the previous session. It emphasises "the paramount importance of strict compliance with the existing legal regime providing for the peaceful use of outer space" and states "that practical measures should be examined and taken in the search for agreements to prevent an arms race in outer space in a common effort".
- *Resolution 72/56: Transparency and confidence-building measures in outer space activities,*[96] which builds on the content of the previous session's resolution of the same name and reaffirms "the right of all countries to explore and use outer space in accordance with international law". It "Encourages Member States to continue to review and implement, to the greatest extent practicable, the proposed transparency and confidence–building measures contained in the report" from the group of governmental experts on transparency and confidence-building measures in outer space activities.

---

[94]United Nations General Assembly. 72nd Session. Resolution adopted by the General Assembly on 4 Dec. 2017. Resolution 72/26. Prevention of an arms race in outer space, 23 Aug. 2018 (retrieved) <http://undocs.org/A/RES/72/26>.

[95]United Nations General Assembly. 72nd Session. Resolution adopted by the General Assembly on 4 Dec. 2017. Resolution 72/27. No first placement of weapons in outer space, 23 Aug. 2018 (retrieved) <http://undocs.org/A/RES/72/27>.

[96]United Nations General Assembly. 72nd Session. Resolution adopted by the General Assembly on 4 Dec. 2017. Resolution 72/56. Transparency and confidence-building measures in outer space activities, 23 Aug. 2018 (retrieved) <http://undocs.org/A/RES/72/56>.

On 7 December 2017

- *Resolution 72/77: International cooperation in the peaceful uses of outer space,*[97] which recalls the previous session's resolution of the same name and raises the General Assembly's serious concern regarding a possible arms race in space and the harm caused by disasters. It further refers to the role of space technology and science "in areas such as tele-health, tele-education, disaster management, environmental protection, natural resources management and ocean and climate monitoring" and "Considers that it is essential that Member States pay more attention to the problem of the gradually increasing probability of collisions of space objects, especially those with nuclear power sources, with space debris".
- *Resolution 72/78: Adoption of the Declaration on the 50th anniversary of the Treaty on Principles Governing the Activities of States in the Exploration and Use of Outer Space, including the Moon and Other Celestial Bodies,*[98] which notes a total of 105 states that had become parties to the Outer Space Treaty as of January 2017 and highlights "the role of the Treaty as the cornerstone of the international legal regime governing outer space activities and that the Treaty manifests the fundamental principles of international space law".
- *Resolution 72/79: Consideration of the 50th anniversary of the United Nations Conference on the Exploration and Peaceful Uses of Outer Space,*[99] where the General Assembly decided "to consider in plenary meeting at its seventy-third session an agenda item entitled 'Space as a driver of sustainable development' in the context of the fiftieth anniversary of the United Nations Conference on the Exploration and Peaceful Uses of Outer Space (UNISPACE+50)". It furthermore requested COPUOS "to submit a draft resolution on the outcomes of UNISPACE+50 for the consideration of the General Assembly, in plenary meeting, at its seventy-third session".

On 24 December 2017

[97] United Nations General Assembly. 72nd Session. Resolution adopted by the General Assembly on 7 Dec. 2017. Resolution 72/77. International cooperation in the peaceful uses of outer space, 23 Aug. 2018 (retrieved) <http://www.unoosa.org/res/oosadoc/data/resolutions/2016/general_assembly_71st_session/ares7277_html/N1742901.pdf>.

[98] United Nations General Assembly. 72nd Session. Resolution adopted by the General Assembly on 7 Dec. 2017. Resolution 72/78. Adoption of the Declaration on the 50th anniversary of the Treaty on Principles Governing the Activities of States in the Exploration and Use of Outer Space, including the Moon and Other Celestial Bodies, 23 Aug. 2018 (retrieved) <http://www.unoosa.org/res/oosadoc/data/resolutions/2017/general_assembly_72nd_session/ares7278_html/N1742908.pdf>.

[99] United Nations General Assembly. 72nd Session. Resolution adopted by the General Assembly on 7 Dec. 2017. Resolution 72/79. Consideration of the 50th anniversary of the United Nations Conference on the Exploration and Peaceful Uses of Outer Space, 23 Aug. 2018 (retrieved) <http://www.unoosa.org/res/oosadoc/data/resolutions/2017/general_assembly_72nd_session/ares7279_html/N1742915.pdf>.

- *Resolution 72/250: Further practical measures for the prevention of an arms race in outer space,*[100] which importantly "Urges the Conference on Disarmament to agree on and implement at its earliest opportunity a balanced and comprehensive programme of work that includes the immediate commencement of negotiations on an international legally binding instrument on the prevention of an arms race in outer space". It further "Requests the Secretary-General to establish a United Nations Group of Governmental Experts, with a membership of up to 25 Member States, chosen on the basis of fair and equitable geographical representation, to consider and make recommendations on substantial elements of an international legally binding instrument on the prevention of an arms race in outer space, including, inter alia, on the prevention of the placement of weapons in outer space". The group is to meet for 2-week sessions in 2018 and 2019.

1.2.1.2 UNGA Committees Including Space Within Their Purview

The Disarmament and International Security Committee

The Disarmament and International Security Committee (First Committee) works in close cooperation with the United Nations Disarmament Commission and the Geneva-based Conference on Disarmament to deal with disarmament, global challenges and threats to peace that affect the international community and seeks solutions to the challenges in the international security regime, including the prevention of an arms race in outer space. On 2 November 2017, the First Committee forwarded 58 draft resolutions and decisions to the UNGA for adoption, created out of a session which discussed in particular nuclear weapons and other weapons of mass destruction.[101] On 4 December 2017, the General Assembly went on to adopt 53 draft resolutions and 4 draft decisions of these First Committee texts.[102] Among these, three draft resolutions directly concern activity in space, addressing in particular:

- *Transparency and confidence-building measures in outer space activities* (adopted without votation), encouraging "Member States to continue to review and implement the proposed measures through the relevant national mechanisms".

[100]United Nations General Assembly. 72nd Session. Resolution adopted by the General Assembly on 24 Dec. 2017. Resolution 72/250. Further practical measures for the prevention of an arms race in outer space, 23 Aug. 2018 (retrieved) <https://undocs.org/A/RES/72/250>.

[101]"Closing Session, First Committee Approves Draft on Chemical Weapons Convention, Sending Total of 58 Texts to General Assembly" 2 Nov. 2017. United Nations, 24 Aug. 2018 <https://www.un.org/press/en/2017/gadis3594.doc.htm>.

[102]"Adopting 57 First Committee Texts, General Assembly Addresses Myriad Security Issues, from Use of Chemical Weapons in Syria to Nuclear-Arms-Ban Treaty" 4 Dec. 2017. United Nations, 24 Aug. 2018 <https://www.un.org/press/en/2017/ga11984.doc.htm>.

- *Prevention of an arms race in outer space* (182 in favour, none against and 3 abstentions (Israel, Palau, the USA)), which calls upon states "in particular those with major space capabilities, to contribute actively to the objective of the peaceful use of outer space and of the prevention of an arms race in that domain".
- *No first placement of weapons in outer space* (131 in favour, 4 against (Georgia, Israel, Ukraine, the USA) and 48 abstentions), "by which the Assembly reaffirmed the importance and urgency of that objective and the willingness of States to contribute to reaching that common goal".

The UNGA also addressed the report entitled "Joint panel discussion of the First and Fourth Committees on possible challenges to space security and sustainability" and deferred action on a draft resolution following up on the 2013 high-level General Assembly meeting concerning further practical measures for the prevention of an arms race in outer space.

The Committee on the Peaceful Uses of Outer Space (COPUOS)

The UN COPUOS reports to the UNGA's Fourth Committee. Its activities in 2017 included:

- The 54th session of its *Scientific and Technical Subcommittee* from 30 January to 10 February
- The 56th session of its *Legal Subcommittee* from 27 March to 7 April
- The 60th *plenary session of COPUOS* from 7 to 16 June

The Committee addressed a new topic during its 60th plenary session, being a general exchange of views on potential legal models for activities in exploration, exploitation and utilisation of space resources. In addition, the Committee provided its endorsement to the draft declaration on the 50th anniversary of the Treaty on Principles Governing the Activities of States in the Exploration and Use of Outer Space, including the Moon and Other Celestial Bodies. This declaration was subsequently adopted on 7 December 2017 by the UN General Assembly as Resolution 72/78.

The Committee likewise endorsed the findings and recommendations of its subcommittees from their preceding sessions, considering:

- Space and sustainable development
- The current status of the spin-off benefits of space technology
- Space and water
- Space and climate change
- The use of space technology in the United Nations system
- The future role of the Committee
- Other related matters

At the 2017 Scientific and Technical Subcommittee meeting, discussions and presentations expanded on a number of agenda items, including space technology for socio-economic development; matters relating to remote sensing of the Earth by satellite, including applications for developing countries and monitoring of the Earth's environment; space debris; space-system-based disaster management; recent developments in global navigation satellite systems; space weather; near-Earth objects; the use of nuclear power sources in outer space; and the long-term sustainability of outer space activities. The Subcommittee also examined the physical nature and technical attributes of the geostationary orbit and its utilisation and applications, including the field of space communications with particular note of the needs and interests of developing countries.

As in previous years, the agenda items of the 2017 Legal Subcommittee addressed information on the activities of international governmental and non-governmental organisations related to space law, the status and application of the five United Nations treaties on outer space, matters relating to the definition and delimitation of outer space and the utilisation of the geostationary orbit, national legislation relevant to the peaceful exploration and use of outer space, capacity building in space law and review of the Principles Relevant to the Use of Nuclear Power Sources in Outer Space. Agenda items also included general exchanges of information on legal mechanisms relating to space debris mitigation measures, on non-legally binding United Nations instruments on outer space, on the legal aspects of space traffic management and on the application of international law to small satellite activities. A new agenda item was also included in 2017, including a general exchange of views on potential legal models for activities in exploration, exploitation and utilisation of space resources. The Subcommittee also held a review of international mechanisms for cooperation in the peaceful exploration and use of outer space and welcomed proposals for new items to be considered by the Legal Subcommittee at its following session.

1.2.1.3 Other UN Bodies and Organs Monitoring Outer Space Activities

International Telecommunication Union (ITU)

The ITU International Satellite Symposium 2017—on Space Development and Sustainability in the Americas Region—was co-organised by the International Telecommunications Union and the Ministry of Communications of Argentina and held in Argentina on 29–31 May 2017. The Symposium was organised for industry, satellite operators, space agencies and regulators and discussed, among other topics, national policies and space plans, space regulations and licensing, satellite markets, the results of the World Radiocommunication Conference (WRC) 2015 and the agenda of the upcoming WRC 2019.[103] Further, on 8–12 October 2017, a regional

[103] "ITU International Satellite Symposium 2017 – S.C.Bariloche, Argentina – 29–31 May 2017." ITU, 29 Aug. 2018 (retrieved) <https://www.itu.int/en/ITU-R/space/workshops/2017-Bariloche/Pages/default.aspx>.

training workshop was held in Egypt on Satellite Communications: Technical and Regulatory considerations, focusing in particular on "VSAT and Satellite Systems"; the training was led by the ITU and the International Telecommunications Satellite Organization.[104,105]

Regional Radiocommunication Seminars (RRS) inform ITU member states on the development of general issues of spectrum management and international regulations for terrestrial and space services, including the relevant coordination, examination and registration procedures. In 2017, four RRS events took place: in Dakar, Senegal, in March for Africa; in Lima, Peru, in September for the Americas; in Phnom Penh, Cambodia, in December for Asia and Pacific; and in Muscat, Oman, in December for the Arab countries.[106] Moreover, the first ITU Inter-regional Workshop on WRC-19 Preparation was held in Geneva, Switzerland, from 21 to 22 November 2017. The purpose of this Workshop was to "provide participants with the opportunity to exchange views and have a better understanding of the most challenging WRC-19 agenda items and issues, with draft preliminary common views and/or positions of the concerned entities".[107] Two further Workshops will take place in 2018 and 2019.[108]

Programmes Under the United Nations Office for Outer Space Affairs

UN-SPIDER

Set up by the United Nations General Assembly in 2006, the United Nations Platform for Space-based Information for Disaster Management and Emergency Response (UN-SPIDER) aims to provide universal access to all types of space-

[104]"ITU-ITSO Regional Training Workshop on "Satellite Communications": Technical and Regulatory considerations." ITU, 29 Aug. 2018 (retrieved) <https://academy.itu.int/index.php?option=com_joomdle&view=coursecategoryextended&cat_id=:&course_id=1177:itu-itso-regional-training-workshop-on-satellite-communications-technical-and-regulatory-considerations&lang=en>.

[105]"ITU-ITSO Regional Training on "Satellite Communications": Technical and Regulatory Considerations, Giza – Egypt, 8–12 October 2017, INFORMATION NOTE." ITU, 29 Aug. 2018 (retrieved) <https://academy.itu.int/courses//auth/joomdle/pluginfile_joomdle.php?file=/115909/course/summary/20300-Info%20Note_ITSO_ITU_VSAT.pdf>.

[106]"Radiocommunication Seminars and Workshops." ITU, 29 Aug. 2018 (retrieved) <https://www.itu.int/en/ITU-R/seminars/Pages/default.aspx>.

[107]"1st ITU Inter-regional Workshop on WRC-19 Preparation – Geneva, Switzerland, 21–22 November 2017." ITU, 29 Aug. 2018 (retrieved) <https://www.itu.int/en/ITU-R/conferences/wrc/2019/irwsp/Pages/2017.aspx>.

[108]"ITU Inter-regional Workshops on WRC-19 Preparation." ITU, 29 Aug. 2018 (retrieved) <https://www.itu.int/en/ITU-R/conferences/wrc/2019/irwsp/Pages/default.aspx>.

based information and services relevant to disaster management support. In 2017, it organised several workshops and regional meetings, including[109]:

- A training course on synthetic aperture radar satellite imagery use, Libreville, Gabon, 20–24 February 2017
- A regional expert meeting on enhancing the use of space-based information in multi-hazard early warning systems, Mexico City, Mexico, 11–13 July 2017
- The United Nations International Conference on Space-based Technologies for Disaster Risk Reduction: Building Resilience through Integrated Applications, Beijing, China, 23–25 October 2017
- A training course on the integration of multisource Earth observation data for disaster damage assessment, Beijing, China, 25–31 October 2017
- The United Nations/Germany International Conference on International Cooperation toward Low-Emission and Resilient Societies, Bonn, Germany, 22–24 November 2017
- A regional workshop on Building drought resilience in agriculture: partnerships and outreach, Bangkok, Thailand, 4 December 2017 and a regional training course on Earth observation-based tools for drought monitoring, Si Racha, Thailand, 5–8 December 2017

UN-SPIDER also conducted a technical advisory mission to Nepal from 31 July to 4 August 2017 "to evaluate the current and potential use of space-derived information in all aspects of disaster management and to offer recommendations to strengthen the disaster risk management and emergency response efforts in" Nepal. It additionally conducted five follow-ups to previous missions in El Salvador, Guatemala, Myanmar, the Solomon Islands and Sri Lanka.[110]

UN-SPIDER also provides disaster support via the Charter on Cooperation to Achieve the Coordinated Use of Space Facilities in the Event of Natural or Technological Disasters (the International Charter on Space and Major Disasters). UN-SPIDER activated the Charter or supported its activation on three occasions in 2017: in Sri Lanka in May 2017, in Nepal in August 2017 and in Sierra Leone in August 2017.[111]

UN Programme on Space Applications (PSA)

The UN Programme on Space Applications (PSA) is concerned with cooperation in space science and technology, and its mission is "to enhance the understanding and subsequent use of space technology for peaceful purposes in general, and for

[109]"Report on activities carried out in 2017 in the framework of the United Nations Platform for Space-based Information for Disaster Management and Emergency Response." 22 Nov. 2017. UN Committee on the Peaceful Uses of Outer Space, 30 Aug. 2018 <http://www.un-spider.org/sites/default/files/V1708225.pdf>.

[110]Ibid.

[111]Ibid.

national development, in particular, in response to expressed needs in different geographic regions of the world". Several activities were carried out under the PSA's auspices in 2017, including workshops (on the Government of Italy's "Open Universe" initiative, human capacity building in space science and technology, on space technology for socio-economic benefits and on the International Space Weather Initiative) and other events.[112] In chronological order these include:

- Expert meeting on preparation of the United Nations/Italy Workshop on the Open Universe Initiative, Rome, Italy, 11–12 April 2017
- COSPAR-UNOOSA Coordination Meeting on COSPAR's contributions to UNISPACE+50, Vienna, Austria, 22–23 May 2017
- UN/USA Workshop on the International Space Weather Initiative: The Decade after the International Heliophysical Year 2007, Boston, Massachusetts, USA, 31 July–4 August 2017
- UN/WHO/Switzerland Conference on Strengthening Space Cooperation for Global Health, Geneva, Switzerland, 23–25 August 2017
- UN/Austria Symposium on Access to Space: Holistic Capacity Building for the 21st Century, Graz, Austria, 4–7 September 2017
- UN/Russian Federation Workshop on Human Capacity-Building in Space Science and Technology for Sustainable Social and Economic Development, 30 October–2 November 2017
- United Nations/Italy Workshop on the Open Universe Initiative, Vienna, Austria, 20–22 November 2017
- United Nations/South Africa Symposium on Basic Space Technology, Stellenbosch, South Africa, 11–15 December 2017

International Committee on Global Navigation Satellite Systems (ICG)

The International Committee on Global Navigation Satellite Systems (ICG) was established in 2005 to promote voluntary cooperation on matters of mutual interest related to civil satellite-based positioning, navigation, timing and value-added services. It aims to make positioning, navigation and timing available globally for societal benefits at a lower cost through encouraging coordination among providers GNSS, regional systems and augmentation services to ensure greater compatibility, interoperability and transparency. It also aims to promote the introduction and utilisation of navigation services and their future enhancements through providing assistance for integration into their infrastructures and by existing as a focal point for information exchange to assist GNSS users with their development plans and applications. The UNOOSA serves as the Executive Secretariat of the ICG and the associated Providers' Forum. The 12th Meeting of the ICG was organised by the

[112] "Events." (Events tagged with PSA) UN Office for Outer Space Affairs, 30 Aug. 2018 (retrieved) <http://www.unoosa.org/oosa/events/search-events-ng.jspx#?c=%7B%22filters%22:%5B%7B%22fieldName%22:%22event.tags.tag_s%22,%22value%22:%22PSA%22%7D%5D,%22sortings%22:%5B%5D%7D>.

Japanese Cabinet Office and the Ministry of Foreign Affairs on behalf of the Government of Japan and took place on 2–7 December 2017 in Kyoto.[113] The meeting brought together stakeholders in industry, government, non-governmental officials and academia to review and discuss developments in global navigation systems.

The meeting consisted of three plenary sessions and a series of working group meetings and included presentations to update ICG members, associate members and observers on various matters of interest.[114] An applications and experts seminar discussed, among other topics:

- Water vapour analysis over the ocean using shipborne GNSS measurements for the mitigation of weather disaster
- Variations of water vapour observed with a hyper-dense GNSS network
- How the real-time GNSS can contribute to near-field tsunami early warning?
- Developments of GNSS buoy for a synthetic geohazard monitoring system

The work activities of the ICG are organised via its four working groups, which also met during the meeting, and focus on:

- Systems, signals and services
- Enhancement of GNSS performance, new services and capabilities
- Information dissemination and capacity building
- Reference frames, timing and applications

In addition to discussing cross-cutting topics concerning open service information-sharing and service performance monitoring, the working groups reviewed a report on the international GNSS monitoring and assessment (IGMA) task force and space service information, as well as a report concerning GNSS space service volume (SSV).[115]

The 13th Meeting of the ICG is to take place in Xi'an, China, on 4–9 November 2018.[116]

[113] "International Committee on Global Navigation Satellite Systems (ICG): Annual Meeting." UN Office for Outer Space Affairs, 31 Aug. 2018 (retrieved) <http://www.unoosa.org/oosa/en/ourwork/icg/meetings/ICG-2017.html>.

[114] "AGENDA, 12th Meeting of the International Committee on Global Navigation Satellite Systems." ICG, 31 Aug. 2018 (retrieved) <http://www.unoosa.org/documents/pdf/icg/2017/icg12-agenda.pdf>.

[115] "International Committee on Global Navigation Satellite Systems (ICG): Past Annual Meetings." ICG, 31 Aug. 2018 (retrieved) <http://www.unoosa.org/oosa/en/ourwork/icg/meetings/past.html>.

[116] "International Committee on Global Navigation Satellite Systems (ICG): Annual Meeting." ICG, 31 Aug. 2018 (retrieved) <http://www.unoosa.org/oosa/en/ourwork/icg/meetings/ICG-2018.html>.

United Nations Geographic Information Working Group (UNGIWG)

The United Nations Geographic Information Working Group (UNGIWG) is a comprehensive, decentralised geospatial information network of UN cartography and geospatial information management science professionals that lay the foundations for the United Nations Spatial Data Initiative (UNSDI) and facilitate decision-making mechanisms between member states, regional organisations and partners that advance social, economic development, environmental and humanitarian interests.

The UNGIWG held its 15th Plenary Meeting in New York from 3 to 5 August 2015. This latest meeting was co-chaired by representatives from the United Nations Department of Safety and Security (UNDSS) and the United Nations Office for Outer Space Affairs (UNOOSA).[117] Its plenary conclusions confirmed the need to move the voluntary best-effort approach of the group to the next level and to formalise this geospatial coordination mechanism in the UN system, building on models such as UN-Water, UN-Space or the EMG. It also called for a permanent, dedicated Secretariat for the Working Group and highlighted that coordination was required both on the policy level and the technical level. It also found that challenges remained in the availability of core datasets and of commercial very high-resolution satellite imagery.[118]

Conference on Disarmament (CD)

The Conference on Disarmament (CD) is the key multilateral disarmament and arms control negotiating forum within the international community. The CD meets in Geneva, and the 2017 session occurred in three parts: from 23 January to 31 March, from 15 May to 30 June and from 31 July to 15 September. In 2017, 32 formal plenary meetings occurred, during which member states and non-member states were invited to participate and present their positions on the Conference's discussions. Six informal plenary meetings also took place.[119]

Moreover, on 17 February 2017 the CD adopted decision CD/2090, thereby establishing "a working group on the 'way ahead' to identify common ground for a programme of work with a negotiating mandate".[120] Included among the working

[117]"15th UNGIWG Plenary Meeting." UNGIWG 20 Apr. 2016 <http://www.ungiwg.org/meetings/15th-ungiwg-plenary-meeting>.

[118]"United Nations Geographic Information Working Group | Report to the UN Committee of Experts on Global Geospatial Information Management (GGIM)." 25 Aug. 2015. UNGIWG 20 Apr. 2016 <http://ggim.un.org/docs/meetings/GGIM5/UNGIWG%20Introduction.pdf>.

[119]"Report of the Conference on Disarmament to the General Assembly of the United Nations." 22 Sep. 2017. Conference on Disarmament, 3 Sep. 2018 <https://www.unog.ch/80256EDD006B8954/(httpAssets)/4C940C4CB4AF2D5AC12581AA004B7826/$file/2112E.pdf>.

[120]Ibid.

group's meeting reports, and with regard to the Conference's space-related work in 2017, there was a "*note verbale* dated 7 September 2017 from the Permanent Mission of Chile addressed to the Secretary-General of the Conference on Disarmament transmitting the report on the informal discussions on the prevention of an arms race in outer space" (CD/2100). In summarising the CD's meetings on the prevention of an arms race in outer space (PAROS), this document found that "it was broadly recognized that the existing legal framework, with the Outer Space Treaty (1967) at its core, was no longer sufficient as it left legal gaps" and additionally that "some delegations called for a comprehensive approach to deal with the prevention of an arms race in outer space".[121]

The CD furthermore took note of UNGA Resolutions from its 71st session in 2016, including three which pertained to the CD and space activities:

- 71/31 Prevention of an arms race in outer space
- 71/32 No first placement of weapons in outer space
- 71/42 Transparency and confidence-building measures in outer space activities

And with regard to the CD's agenda item C. Prevention of an Arms Race in Outer Space, the aforementioned *note verbale* CD/2100 was submitted to the Conference as well as a "letter dated 9 August 2017 from the Permanent Representative of the Russian Federation, addressed to the Secretary General of the Conference on Disarmament, transmitting the Joint Statement by President of the Russian Federation Vladimir Putin and President of the Socialist Republic of Vietnam Tran Dai Quang of 29 June 2017, with regard to the no first placement of weapons of any kind in Outer Space" (CD/2098); the Joint Statement contained within this document reads that "the presidents of the Russian Federation and Viet Nam declared their countries' adherence to the policy of no first placement of weapons in outer space and called upon all countries possessing space capabilities to follow their example".[122]

United Nations Institute for Disarmament Research (UNIDIR)

The United Nations Institute for Disarmament Research (UNIDIR) aims to produce "ideas and promotes action on disarmament and security"; among other topics, some

[121]"Note verbale dated 7 September 2017 from the Permanent Mission of Chile addressed to the Secretary-General of the Conference on Disarmament transmitting the report on the informal discussions on the prevention of an arms race in outer space." 14 Sep. 2017. Conference on Disarmament, 3 Sep. 2018 <http://undocs.org/cd/2100>.

[122]"Letter dated 9 August 2017 from the Permanent Representative of the Russian Federation, addressed to the Secretary General of the Conference on Disarmament, transmitting the Joint Statement by President of the Russian Federation Vladimir Putin and President of the Socialist Republic of Vietnam Tran Dai Quang of 29 June, 2017, with regard to the no first placement of weapons of any kind in Outer Space." 6 Sep. 2017. Conference on Disarmament, 3 Sep. 2018 <http://undocs.org/cd/2098>.

of its projects cover, both directly and indirectly, issues of space security. It seeks to review former proposals and to propose new options for breaking the deadlock in space weaponisation and other emerging security matters at the CD. On 25 July 2018, the UNGA received the report of the Director of UNIDIR on its activities over January 2017–May 2018 as well as the proposed work and financial plan for 2018 and 2019. This report simultaneously represents the UNIDIR Annual Report for the respective period.[123]

In 2017, UNIDIR conducted a wide range of space-related activities, all carrying a focus on security. This work takes place in particular under UNIDIR's "emerging security issues" programme, where it focuses, among other substantive topics, on security in outer space. In 2017, this programme's work saw the launch of a report series on outer space security, referred to as the UNIDIR "space dossier" and which "contains specific ideas for the progressive development of space security norms". The first publication of the series deals with strategic stability in outer space. Further, UNIDIR was also chosen for a consultant role to the recently formed Group of Governmental Experts on the prevention of an arms race in outer space (see UNGA Res. 72/250 in the previous section on the UNGA). Lastly, UNIDIR held its annual space security conference in April 2017, focusing on "The Outer Space Treaty's 50th Anniversary: Reviewing the Regime". The conference panels focused on, among other topics, "National and Regional Approaches to Space Security", "Multilateral International Initiatives: Status Update and Ways Forward" and "The Outer Space Regime in an Evolving Space Environment".[124] The event was co-organised with the Secure World Foundation and the Simons Foundation Canada and aimed to review "the main multilateral international initiatives to safeguard access to, and use of, outer space and assess the adequacy of the outer space regime in an evolving space environment, where new threats to space systems and developments in space activities are believed to be potentially destabilizing".[125]

1.2.2 The Group on Earth Observations

The Group on Earth Observation (GEO), with its system-of-systems cooperation platform (GEOSS), is a voluntary partnership of governments and international

[123]"United Nations Institute for Disarmament Research, Note by the Secretary-General." 25 July 2018. UN General Assembly, 4 Sep. 2018 <http://www.unidir.org/files/files/Director's%20Report%202018.pdf>.

[124]"2017 Space Security Conference, 'Celebrating the Outer Space Treaty: 50 Years of Space Governance and Stability'." UNIDIR, 4 Sep. 2018 (retrieved) <http://www.unidir.org/files/medias/pdfs/agenda-eng-0-721.pdf>.

[125]"Space Security 2017—Celebrating the Outer Space Treaty: 50 Years of Space Governance and Stability." UNIDIR, 4 Sep. 2018 (retrieved) <http://www.unidir.org/programmes/emerging-security-issues/annual-outer-space-security-conference/space-security-2017-celebrating-the-outer-space-treaty-50-years-of-space-governance-and-stability>.

organisations that facilitates the coordination of current and future Earth observing systems while also providing structured and sustained data worldwide as a decision support tool for its users.

GEO remained active in 2017, participating in or conducting 52 workshops, symposia, webinars and other events, covering a range of topics including GEOSS, climate change, water management, disaster risk reduction and other Earth observation and environmental monitoring topics.[126] Major events included, in particular, the 9th and 10th GEOSS Asia-Pacific Symposia, the 2nd AfriGEOSS Symposium, the 11th GEO European Projects Workshop, the first meeting of the GEO Human Planet Initiative as well as GEO's 14th Plenary Session, which was held over 25–26 October 2017 in Washington, D.C.[127] Topics raised during the Plenary included, among others, Earth Observations in Public Policy, Earth Observations in the Commercial Sector, Earth Observations in International Development, the 2017–2019 GEO Work Programme and National Earth Observations. Moreover, the Plenary took place in the context of GEO Week 2017, held over 23–27 October 2017, also in Washington.

1.2.3 Europe

1.2.3.1 European Space Agency

The European Space Agency consists of 22 member states, including Austria, Belgium, Czech Republic, Denmark, Estonia, Finland, France, Germany, Greece, Hungary, Ireland, Italy, Luxembourg, the Netherlands, Norway, Poland, Portugal, Romania, Spain, Sweden, Switzerland and the UK. Canada participates in certain projects via a Cooperation Agreement and sits on the ESA Council, while Slovenia is an Associate Member, and Bulgaria, Croatia, Cyprus, Latvia, Lithuania, Malta and Slovakia also hold Cooperation Agreements with the Agency.[128]

ESA's budget increased by 9.52% to 5.75 billion € in 2017 from 5.25 billion € in the previous year. Direct funding from ESA member states increased by about 1% to 3.78 billion €, while funding from the EU and EUMETSAT grew by just over 30% to 1.97 billion € for the year.[129] *Navigation* saw the largest boost across all budgetary

[126]"Past events." GEO, 4 Sep. 2018 (retrieved) <http://www.earthobservations.org/meetings.php#past>.

[127]"DRAFT AGENDA, Group on Earth Observations, 14th Plenary Session – GEO-XIV." GEO, 4 Sep. 2018 (retrieved) <http://www.earthobservations.org/documents/geo_xiv/GEO-XIV-1-3(Rev1)_Draft%20Agenda.pdf>.

[128]"New Member States." ESA, 10 Apr. 2018 (retrieved) <https://www.esa.int/About_Us/Welcome_to_ESA/New_Member_States>.

[129]"ESA Budget 2017." 16 Jan. 2017. ESA, 10 Apr. 2018 <http://m.esa.int/spaceinimages/Images/2017/01/ESA_budget_2017>. *See also*: "ESA Budget for 2016." 20 Jan. 2016. ESA, 10 Apr. 2018 <http://www.esa.int/spaceinimages/Images/2016/01/ESA_budget_2016>.

items, growing nearly 66% to 1010.8 million euros, reflecting an increase of 401.3 million euros. This boost occurs in line with the implementation in 2016 of the new Navigation Innovation and Support Programme—of the Directorate of the Galileo Programme and Navigation-related Activities—which is an optional programme and reflects ESA's large and ongoing focus on navigation capabilities through the EGNOS and GALILEO programmes.[130] Funding for both *Earth Observation* and *Telecommunications and Integrated Applications* decreased in 2017, the former by almost 4% (dropping to 1543 million €) and the latter by approximately 11% (reaching 319 million €). Space Science saw a 1% growth, reaching 513.1 million euros in 2017 from 507.9 the year before. And Space Situational Awareness grew by 17% (a value of 2.2 million) to 15.1 million euros in 2017. The remaining 8.1% of funding was allocated to ESA's *Basic Activities*, activities *Associated with the General Budget* and the *European Cooperating States Agreements*. In 2017, two budgetary label changes occurred: specifically, *Human Spaceflight* and *Robotic Exploration and PRODEX* were reorganised as *Human Spaceflight and Robotic Exploration* (HRE) and *PRODEX*. In other words, Human Spaceflight was coupled with Robotic Exploration in the budgetary breakdown for 2017, while PRODEX was separated. HRE in 2017 received 633 million euros, and PRODEX received 47.2 million.[131]

During the reporting period, several formal steps were taken to promote cooperation between ESA and other partners, including Croatia, Japan, Luxembourg and the satellite industry, indicating significant collaboration around ongoing and new projects into the future. In early 2018, ESA signed a Cooperation Agreement with the Republic of Croatia, a process which had begun already in 2014. This Agreement will lead to the two creating a framework for further and more intensive activity in ESA programmes.[132] The signing of this Agreement also marks the establishment of formal cooperation between ESA and all EU member states that are not already ESA member states. The Japan Aerospace Exploration Agency JAXA and ESA also prepared a joint statement regarding their collaboration to date and their future cooperation. As stipulated in the joint statement, lunar exploration as well as cooperation around data concerning terrestrial greenhouse gases will be the objects of continued collaboration.[133] In 2017, at the Paris Air Show, ESA and Luxembourg

[130]"Green Light for ESA's Advanced SATNAV Technology and Innovation Programme." 5 Dec. 2016. ESA – Navigation, 10 Apr. 2018 <https://www.esa.int/Our_Activities/Navigation/Green_light_for_ESA_s_advanced_satnav_technology_and_innovation_programme>.

[131]"ESA budget 2017: by domain." 16 Jan. 2017. ESA, 10 Apr. 2018 <http://www.esa.int/spaceinimages/Images/2017/01/ESA_budget_2017_by_domain>. *See also*: "ESA budget 2016 by domain." 14 Jan. 2016. ESA, 10 Apr. 2018 <http://m.esa.int/spaceinimages/Images/2016/01/ESA_budget_2016_by_domain>.

[132]"Croatia Signs Cooperation Agreement." 19 Feb. 2018. ESA, 25 Apr. 2018 <https://www.esa.int/About_Us/Welcome_to_ESA/Croatia_signs_Cooperation_Agreement>.

[133]"ESA and JAXA Confirm Further Cooperation in Space." 8 Mar. 2018. ESA, 25 Apr. 2018 <https://www.esa.int/Our_Activities/Human_Spaceflight/ESA_and_JAXA_confirm_further_cooperation_in_space>.

signed a joint statement concerning the study of technical, scientific and other aspects of the exploration and use of space resources. With this step, Luxembourg has further demonstrated its proactive approach toward the utilisation of space resources, and ESA will gain experience and expertise across its related technical and regulatory aspects.[134] Also taking place at the 2017 Paris Air Show, 16 leaders of the European satellite industry and the European Space Agency signed a joint statement regarding cooperation on Satellite for 5G, in which the signatories express their intention to support satellite and 5G convergence efforts, and thus further support efforts already in place by the European Commission, such as through Horizon 2020 and the 5G Infrastructure Public Private Partnership. As part of ESA's *Satellite for 5G initiative* over 2018–2020, the agency will work with industry toward 5G service trials, in line with the EC Action Plan, and incorporating terrestrial telecommunication infrastructure, with a focus on public safety, transport, entertainment and media. Other activities will include interoperability demonstration, standardisation and applications development.[135] Days after the announcement of this joint statement, ESA and ESPI co-hosted a conference on "Space and SATCOM for 5G: European Transport and Connected Mobility".[136]

With the last ESA Council at Ministerial Level having taken place in Switzerland in 2016, the next is scheduled for late 2019 in Spain. Discussion will likely include the progress thus far of Director General Jan Wörner's strategy proposal "Space 4.0 for a United Space in Europe" as well as its further implementation.[137]

1.2.3.2 European Union

With the EU's three flagship space programmes (Copernicus, Galileo and EGNOS) being well advanced, Europe has shifted its focus from building space infrastructure toward ensuring strong market uptake of their space data and services by the public and private sectors. Building on this, the European Commission released its Space Strategy in late 2016, comprising four strategic goals, with the overarching aim of building a sustainable space economy. In addition to this, it aims to advance Europe's Space Situational Awareness capability and Governmental Satellite

[134]"Luxembourg and ESA Announce Enhanced Cooperation on Space Resources." 21 June 2017. Spaceresources.lu, 25 Apr. 2018 <https://spaceresources.public.lu/en/actualites/2017/Luxembourg-ESA-Cooperation-signed.html>.

[135]"Signing Up For Satellite-5g Convergence." 24 July 2017. ESA – ARTES, 25 Apr. 2018 <https://artes.esa.int/news/signing-satellite-5g-convergence>.

[136]"ESA/ESPI Conference on 'Space and SATCOM for 5G: European Transport and Connected Mobility'." ESA, 25 Apr. 2018 (retrieved) <https://artes.esa.int/esaespi-conference-%E2%80%98space-and-satcom-5g-european-transport-and-connected-mobility>.

[137]"European Ministers Ready ESA for a United Space in Europe in the Era of Space 4.0." 2 Dec. 2016. ESA, 25 Apr. 2018 <https://www.esa.int/About_Us/Ministerial_Council_2016/European_ministers_ready_ESA_for_a_United_Space_in_Europe_in_the_era_of_Space_4.0>.

Communications (GOVSATCOM) programme.[138] It furthermore aims to boost investment in start-ups through its Investment Plan for Europe and via the Pan-European Venture Capital Fund-of-Funds, supporting European venture capital.

In overview of the Strategy, its first goal aims to "maximise the benefits of space for society and the EU economy" by encouraging the uptake of space services and data and advancing the EU space programmes and meeting new user needs. Its second goal is to "foster a globally competitive and innovative European space sector" by supporting research and innovation and development of skills and fostering entrepreneurship and new business opportunities. Its third goal is to "reinforce Europe's autonomy in accessing space in a safe and secure environment", ensuring access to radio frequency spectrum, ensuring the protection and resilience of critical European space infrastructure and reinforcing synergies between civil and security space activities. And, fourthly, the Commission aims to "strengthen Europe's role as a global actor and promoting international cooperation" by pursuing space dialogues with strategic international partners, by taking space policy into account in EU export control dialogues with third countries, and it will use economic diplomacy and trade policy instruments to assist European companies active in global markets and to address societal challenges.[139]

A major component of the Strategy is the Investment Plan for Europe, which has a stated aim of investing 315 billion euros as a programme total, of which 256 billion had been approved by the end of 2017.[140] This financing is split up across numerous sectors, including Agriculture; Digital; Energy; Environment and Resource Efficiency; Research, Development and Innovation; Social; and Transport. Furthermore, through the Pan-European Venture Capital Fund-of-Funds, which forms part of the aforementioned Investment Plan, the European Commission will provide up to 300 million € in cornerstone investments to one or multiple venture capital fund-of-funds, with a total budget of 400 million €. With EU investments limited to 25%, selected promotors are then to raise the remaining funds, leading to a total at around 1.6 billion €. The fund-of-funds programme aims to enlarge European venture capital funds, increase the amount of private investment relative to public funding and thirdly expand the operational scope across multiple European states.[141]

138"Space." Regularly updated. European Commission, Growth, Sectors, Space, 18 July 2018 (retrieved) <http://ec.europa.eu/growth/sectors/space_en>.

139"Communication from the Commission to the European Parliament, the Council, the European Economic and Social Committee and the Committee of the Regions – Space Strategy for Europe." 26 Oct. 2016. European Commission, 3 Apr. 2018 <https://ec.europa.eu/transparency/regdoc/rep/1/2016/EN/COM-2016-705-F1-EN-MAIN.PDF>.

140"Investment Plan results – Breakdown of results by country and sector from the European Fund for Strategic Investments (EFSI)." 12 Dec. 2017. European Commission webpage archived via the Internet Archive Wayback Machine, 3 Apr. 2018 <https://web.archive.org/web/20180124070047/https://ec.europa.eu/commission/priorities/jobs-growth-and-investment/investment-plan-europe-juncker-plan/investment-plan-results_en#infographic>.

141"Commission and EIF start selecting Pan-European Venture Capital Fund-of-Funds promoters." 7 Feb. 2017. European Commission, 4 Apr. 2018 <http://ec.europa.eu/research/index.cfm?pg=newsalert&year=2017&na=na-070217>.

The Governmental Satellite Communications initiative was defined in 2013 by the European Council as one of the four capability development programmes and is planned to “provide by 2018 EDA Member States and European CSDP actors with access to a GOVSATCOM capability based on existing, pooled, governmental SATCOM resources”. Furthermore, “GOVSATCOM’s objective is to ensure in both the civil and military environment reliable, secure and cost-effective satellite communication services for EU and national public authorities managing security critical missions and operations”.[142] Ultimately, it is envisaged as a “capability that is placed in between the commercial satellite communication market and the highly protected military satellite communication capability”.[143] Its inclusion within the European Space Policy underlines its importance for the European Commission, and important steps were taken in 2017: In March, both a GOVSATCOM Common Staff Requirements and a connected Business Case were approved by the European Defence Agency (EDA) Steering Board, and in June the Outline Description for the Governmental Satellite Communications Pooling and Sharing demonstration project was additionally accepted, with the project led by Spain and comprising 14 EDA member states as well as Norway. This demonstration project aims to begin offering services to governmental users by the summer of 2018. It will provide the platform of the project ultimately to be led by the European Commission, which is anticipated to be operational in 2020.[144] Lastly, near the end of 2017, an implementing arrangement was signed between the EDA and ESA toward supporting the European Commission in the GOVSATCOM initiative.[145]

Furthermore, in December of 2017, the Competitiveness Council of the Council of the European Union convened and, among other topics, focused on the implementation of the Space Strategy for Europe, introduced approximately 1 year prior. Ministers present pressed for closer ties between the space sector and the digital economy. Lastly, the Council of the European Union also adopted late in the year the

[142]“Governmental Satellite Communications (GovSatcom).” 15 June 2017. European Defence Agency, 19 July 2018 <https://www.eda.europa.eu/what-we-do/activities/activities-search/governmental-satellite-communications-(govsatcom)>.

[143]“Latest News, Future European GOVSATCOM programme takes next step.” 12 Sep. 2017. European Defence Agency, 19 July 2018 <https://eda.europa.eu/info-hub/press-centre/latest-news/2017/09/12/future-european-govsatcom-programme-takes-next-step>.

[144]Pultarova, Tereza. “Govsatcom demonstration aims to start service next summer.” 8 Nov. 2017. SpaceNews, 19 July 2018 <http://spacenews.com/govsatcom-demonstration-aims-at-start-service-next-summer/>.

[145]“Chief Executive Domecq at ESA.” 5 Feb. 2018. European Defence Agency, 9 April 2018 <https://eda.europa.eu/info-hub/press-centre/latest-news/2018/02/05/chief-executive-domecq-at-esa>.

conclusions of the midterm reviews of Galileo, EGNOS and Copernicus.[146,147] Conclusions from the reporting place a clear emphasis on the facilitation and creation of downstream applications, built around these programmes' services, including through their integration.[148,149]

1.2.3.3 EUMETSAT

The European Organisation for the Exploitation of Meteorological Satellites (EUMETSAT) is an intergovernmental organisation that supplies weather and climate-related satellite data to the National Meteorological Services of its Member and Cooperating States in Europe, and other users worldwide.

EUMETSAT's 87th Council meeting took place in Darmstadt, Germany, on 27 June 2017. At the meeting, the Council reached three major decisions.[150] Firstly, it formally approved to extend operations of the Jason-2 programme, an ocean surface topography mission supporting oceanography, originally launched in 2008. The extended period will cover 1 January 2018 to the end of 2019. This programme marks a transition from research activities to operations and involves collaboration between Europe's EUMETSAT and CNES and the US NASA and NOAA.[151] The extension supports the Jason-3 high-precision ocean altimetry mission, launched in 2016. Secondly, contracts were awarded to Arianespace and Telespazio around both launch services as well as Launch and Early Operations services for the first two Metop-SG satellites (including a spare unit), forecasted for launch in late 2021 and 2022. These units will be employed in the second generation of the EUMETSAT Polar System, a EUMETSAT mandatory programme which will contribute to the

---

[146]"Competitiveness Council, 30/11–01/12/2017." 1 Dec. 2017. European Council |Council of the European Union, 4 Apr. 2018 <http://www.consilium.europa.eu/en/meetings/compet/2017/11/30-01/>.

[147]"Transport, Telecommunications and Energy Council, 04–05/12/2017." 5 Dec. 2017. European Council |Council of the European Union, 4 Apr. 2018 <http://www.consilium.europa.eu/en/meetings/tte/2017/12/04-05/>.

[148]"Council conclusions on 'The Mid-term Evaluation of the Galileo and EGNOS programmes and of the performance of the European GNSS Agency' – Council conclusions (adopted on 5 December 2017)." 5 Dec. 2017. Council of the European Union, 4 Apr. 2018 <http://data.consilium.europa.eu/doc/document/ST-15435-2017-INIT/en/pdf>.

[149]"Copernicus space programme: conclusions on mid-term review." 1 Dec. 2017. European Council |Council of the European Union, 4 Apr. 2018 <http://www.consilium.europa.eu/en/press/press-releases/2017/12/01/the-mid-term-evaluation-of-the-copernicus-programme-council-adopts-conclusions/>.

[150]"EUMETSAT Extends Jason-2 Operations and Approves Contracts for the Deployment of EPS-Second Generation and the Improvement of Data Access Services." 28 June 2017. EUMETSAT, 4 May 2018 <https://www.eumetsat.int/website/home/News/DAT_3536769.html?lang=EN&pState=1>.

[151]"JASON-2." EUMETSAT, 4 May 2018 (retrieved) <https://www.eumetsat.int/website/home/Satellites/CurrentSatellites/Jason2/index.html>.

Joint Polar System, planned with the US NOAA.[152] Thirdly, a contract was approved which extends at higher bandwidth the EUMETCast Africa data service and includes arrangements for deploying a distributed Copernicus Data and Information Access Service platform.

The Council's 88th meeting followed in early December 2017, when it (1) decided on a 7-year extension of Meteosat Second Generation and EUMETSAT Polar System exploitation as well as (2) approved ground segment contracts for the (mandatory) Meteosat Third Generation programme and (optional) Jason-CS (Continuity of Service) systems.[153] The Sentinel-6 mission will be carried out with the Jason-CS systems, and the mission overall is referred to as Jason-CS/Sentinel-6.[154] As such, it contributes to the Copernicus programme and will employ two Jason-CS satellites maintaining a continuity of ocean surface topography observations with and following Jason-3.[155] In Europe, ESA is also participating in the mission as well as NASA and NOAA in the USA.[156] The first of the two units is forecasted for launch in 2020. The Meteosat Third Generation's first satellite is forecasted for launch in 2021, and the mission is a cooperative one with ESA. The first unit is forecasted for launch in 2023.

1.2.3.4 National Governments

France

During the 2017–2018 reporting period, France, through the French space agency (CNES), took a number of diverse cooperative steps with foreign authorities, including in Australia, China, Ethiopia, Germany, India, Malta, Mexico, Saudi Arabia, the UAE, the UK and the USA. With China, in particular, CNES President Jean-Yves Le Gall and his counterpart Wu Yanhua of the China National Space Administration (CNSA) signed in January 2018 a Memorandum of Understanding on climate actions and space exploration. The MoU will place a priority on the dissemination of data from the CFOSat (Chinese-French Oceonography Satellite)

---

[152] "EUMETSAT Polar System – Second Generation." EUMETSAT, 4 May 2018 (retrieved) <https://www.eumetsat.int/website/home/Satellites/FutureSatellites/EUMETSATPolarSystemSecondGeneration/index.html>.

[153] "EUMETSAT takes steps to extend Meteosat and EPS Operations and approves Ground Segment Contracts for Next Generation Systems." 6 Dec. 2017. EUMETSAT, 4 May 2018 <https://www.eumetsat.int/website/home/News/DAT_3745326.html?lang=EN&pState=1>.

[154] Kramer, Herbert J. "Sentinel-6/Jason-CS." eoPortal Directory, 11 Sep. 2018 (retrieved) <https://directory.eoportal.org/web/eoportal/satellite-missions/content/-/article/jason-cs>.

[155] "Copernicus Satellites." EUMETSAT, 4 May 2018 (retrieved) <https://www.eumetsat.int/website/home/Satellites/FutureSatellites/CopernicusSatellites/index.html>.

[156] "JASON-CS Programme enters into force – Two Months ahead of COP 21." 11 Sep. 2015. EUMETSAT, 4 May 2018 <https://www.eumetsat.int/website/home/News/DAT_2779482.html?lang=EN&pState=1>.

mission, which is due to launch in mid-2018 and represents a main element in joint Chinese/French activities concerning climate change. The signing took place in the presence of Chinese President Xi Jinping and French President Emmanuel Macron.[157] Likewise, in March 2018, Jean-Yves Le Gall and ISRO Chairman K. Sivan signed an agreement in the presence of President Macron and Indian Prime Minister Narendra Modi for the joint development and operation of a micro-satellite constellation for remote sensing and telecommunications. President Macron and Prime Minister Modi additionally adopted a general policy statement affirming the strategic value of the French and Indian cooperation in space activities.[158] In April 2018, during President Macron's US state visit, CNES and the US National Aeronautics and Space Administration (NASA) signed a joint statement around the two agencies' continuing partnership in ocean and atmospheric observation as well as space exploration.[159]

On 7 December 2017, the CNES Board of Directors came to two key decisions at its 353rd session. Firstly, the French contribution to the JUICE mission—a large (L-class) mission within the ESA Cosmic Vision 2015–2025 programme—was approved to move forward. The mission will send a satellite to the Jovian system (the planet Jupiter and its magnetosphere and moons) to make observations on the system's habitability and mechanisms. The Board of Directors additionally approved a resolution on the renewal of draft industrial contracts for the Guiana Space Centre; the contracts concern, notably, launch operations support services from 2018 to 2024.[160]

Germany

On 24 July 2017, the German space agency (DLR) introduced, along with Germany's Federal Ministry for Economic Affairs and Energy, the DLR's new Strategy 2030. Implementation of the strategy began in 2018, and it has for an overarching goal to benefit both the society and economy, with focus areas on space

[157]"State Visit of President Emmanuel Macron to the People's Republic of China: France and China Step Up Space Cooperation in the Fields of Climate And Exploration." 10 Jan. 2018. CNES, 9 May 2018 <https://presse.cnes.fr/en/state-visit-president-emmanuel-macron-peoples-republic-china-france-and-china-step-space-cooperation>.

[158]"French President Emmanuel Macron and Indian Prime Minister Narendra Modi Adopt General Policy Statement Extending Scope of Space Cooperation between France and India." 10 Mar. 2018. CNES, 9 May 2018 <https://presse.cnes.fr/en/french-president-emmanuel-macron-and-indian-prime-minister-narendra-modi-adopt-general-policy>.

[159]"State Visit of President Macron to United States CNES and Nasa Step Up their Cooperation." 25 Apr. 2018. CNES, 9 May 2018 <https://presse.cnes.fr/en/state-visit-president-macron-united-states-cnes-and-nasa-step-their-cooperation>.

[160]"353rd Session of CNES Board of Directors." 7 Dec. 2017. CNES, 25 Apr. 2018 <https://presse.cnes.fr/en/353rd-session-cnes-board-directors>.

and aeronautics, digitalisation, energy, transport and security.[161] The strategy introduces ten cross-sectoral projects on cyber security, broadband rollout and future fuels, among others, and it will see the DLR establish seven new research institutes.[162] These were approved at the DRL Senate's June 2017 meeting, and the German federal government will provide 42 million € annually in funding.[163]

Further, in 2017 and in the first months of 2018, the DLR formed and reinforced several collaborative relationships with partners in Australia, Canada, France, Israel, Japan and South Africa and additionally signed a memorandum of understanding with the UN Human Settlements Programme. With Japan, in particular, several milestones occurred in 2017. In March, Chair of the DLR Executive Board Pascale Ehrenfreund, Programme Director of DLR Energy Research Bernhard Milow and President of the Japanese National Institute of Advanced Industrial Science and Technology (AIST) Ryoji Chubachi signed a Memorandum of Understanding concerning scientific collaboration around energy research.[164] In April 2017, two agreements were signed with JAXA and one was with Japan's Electronic Navigation Research Institute, all concerning aeronautics research generally and specifically dealing with air traffic management, electric aviation and research on reduction of the sonic boom.[165] In September 2017, the DLR, JAXA and AIST signed two cooperation agreements again around energy research, focusing in particular on high-performance batteries for space missions. And in December 2017, the DLR, Japan's National Institute for Environmental Studies (NIES) and JAXA signed a collaboration agreement on validating satellite systems used for greenhouse gases measurements.[166]

---

161 "DLR in 2018." 1 Feb. 2018. DLR, 16 May 2018 <http://www.dlr.de/dlr/en/desktopdefault.aspx/tabid-10081/151_read-25907/year-2018/151_page-2/#/gallery/14331>.

162 "DLR and the German Federal Ministry for Economic Affairs and Energy present their new strategy." 24 July 2017. DLR, 16 May 2018 <http://www.dlr.de/dlr/en/desktopdefault.aspx/tabid-10081/151_read-23429/year-2017/151_page-3/#/gallery/27704>.

163 "DLR Senate gives the go-ahead for the foundation of seven DLR institutes." 29 June 2017. DLR, 16 May 2018 <http://www.dlr.de/dlr/en/desktopdefault.aspx/tabid-10081/151_read-23037/year-2017/151_page-4/#/gallery/27334>.

164 "DLR and Japanese research organisation AIST agree on cooperation on energy research." 23 Mar. 2017. DLR, 16 May 2018 <http://www.dlr.de/dlr/en/desktopdefault.aspx/tabid-10081/151_read-21784/year-2017/151_page-6/#/gallery/26642>.

165 "DLR expands cooperation with Japanese partners in aeronautics research." 5 Apr. 2017. DLR, 16 May 2018 <http://www.dlr.de/dlr/en/desktopdefault.aspx/tabid-10081/151_read-21980/year-2017/151_page-6/#/gallery/26764>.

166 "DLR and Japan sign collaboration agreement on climate research." 12 Dec. 2017. DLR, 16 May 2018 <http://www.dlr.de/dlr/en/desktopdefault.aspx/tabid-10081/151_read-25401/year-2017/151_page-1/#/gallery/29335>.

Italy

In 2017, Italy, through the Italian space agency (ASI), established and/or reinforced several collaborative relationships with institutional partners from Argentina, China, Kenya and the USA. In late 2016, Italy and Kenya renewed their agreement around ASI's Luigi Broglio Space Center, based in Malindi, Kenya, and in January 2017, ASI President Roberto Battiston, Italian Ambassador Mauro Massone and Kenya's Defence Secretary Kirimi Kaberia prepared in the context of the first Italy-Kenya Joint Steering Committee the operational and technical aspects of this renewed agreement.[167] The agreement covers, in particular, the development of a regional Earth observation centre; accessibility of Earth observation data, telemedicine as well as educational and training activities; and support generally to the Kenyan National Space Agency.[168] In Beijing in February 2017 during Italian President Sergio Mattarella's state visit to China, an accord was signed by ASI President Battiston and Wang Zhaoyao, Director General of the China Manned Space Agency (CMSA). The activities covered by the agreement will concern scientific experimentation on long-duration astronaut missions in the context of the Chinese space station programme. It additionally will lead to the creation of a Joint Cooperation Committee by ASI and CMSA which will meet yearly and address various aspects of the collaboration.[169] In May 2017, ASI and the Argentinian Comisión Nacional de Actividades Espaciales (CONAE) signed an agreement around commercial exploitation of satellite data from the SIASGE radar technology programme, which includes the Argentinian SAOCOM and Italian COSMO-SkyMed satellite constellations. SIASGE is the Spanish acronym for the Italian-Argentine System of Satellites for the Management of Environmental Emergencies and Economic Development and is an Earth observation/natural disaster prevention system.[170] And in June 2017, ASI furthered its collaboration with NASA, firstly, through the signing of an implementing arrangement providing ASI associate membership in the Solar System Exploration Research Virtual Institute managed by NASA Ames.[171] Additionally, NASA's then acting administrator Robert Lightfoot and ASI President Roberto Battiston signed an agreement around ASI's cooperation in the IXPE mission which will take measurements on the polarisation of cosmic X-rays.[172]

---

[167]Zanini, Andrea. "First meeting of the Italy-Kenya Joint Steering Committee." 23 Jan. 2017. ASI, 16 May 2018 <https://www.asi.it/en/news/first-meeting-italy-kenya-joint-steering-committee>.

[168]"ASI in Kenya for the first Joint Steering Committee." 27 Jan. 2017. Research Italy, 16 May 2018 <https://www.researchitaly.it/en/news/asi-in-kenya-for-the-first-joint-steering-committee/>.

[169]"Agreement Italy-China." 22 Feb. 2017. ASI, 16 May 2018 <https://www.asi.it/en/news/agreement-italy-china>.

[170]"Italy-Argentina: el espacio clásico." 10 May 2017. ASI, 16 May 2018 <https://www.asi.it/en/news/italia-argentina-el-espacio-clasico>.

[171]"A new covenant for planetary exploration." 15 June 2017. ASI, 16 May 2018 <https://www.asi.it/en/news/a-new-covenant-planetary-exploration>.

[172]"IXPE mission, NASA teams with ASI." 20 June 2017. ASI, 16 May 2018 <https://www.asi.it/en/news/ixpe-mission-nasa-teams-asi>.

The UK

The UK aims to further develop its space sector through both the government's new Industrial Strategy published in November 2017 as well as a series of initiatives directed specifically at space activity.[173] The latter focus in particular around commercial launch in the UK and include the passage in March 2018 of the Space Industry Act. This legislation provides for the operation of spaceports within the UK and launching of spacecraft from its territory, and it will be complemented by later legislation.[174,175] Additionally, a campaign led by the UK Space Agency and grouped under the LaunchUK title aims to promote the national commercial launch industry. At the campaign's March 2017 launch, issues relevant to space launch in the UK were discussed, and nine prospective spaceport sites were presented, from sites across England, Wales and Scotland.[176] Further, a programme worth £50 million was also announced, at the same time as the UK's Industrial Strategy, which will support satellite launch activities from British spaceports.[177] This was also preceded in February 2017 by a £10 million grant programme to support the national commercial launch sector. The British government has furthermore published a series of guidance resources on regulatory aspects of operating spaceports in the UK, launching from the UK, and on spaceplane and suborbital flight activities.[178]

Beyond the activities described above, the UK Space Agency and the French space agency CNES signed a joint statement in January 2018 to enhance space activities cooperation, in particular around projects on climate action and Mars exploration, as well as space applications and the European space programme.[179] The Agency's Corporate Plan for 2017–2018 was also released in August 2017, and

[173]"Industrial Strategy: building a Britain fit for the future." (Download page) 27 Nov. 2017. Government of the United Kingdom, 16 May 2018 <https://www.gov.uk/government/publications/industrial-strategy-building-a-britain-fit-for-the-future>.

[174]Pultarova, Tereza. "U.K. Space Industry Act to future-proof against Brexit." 10 Apr. 2018. SpaceNews, 16 May 2018 <http://spacenews.com/u-k-space-industry-act-to-future-proof-against-brexit/>.

[175]Hirst, Paul & Payne, Lauren. "The sky's no longer the limit for the UK." 20 Mar. 2018. Addleshaw Goddard LLP, 16 May 2018 <https://www.lexology.com/library/detail.aspx?g=66d366b2-b49b-4d7d-88c9-6c4330df8982>.

[176]"LaunchUK: campaign launch." 7 Mar. 2017. Government of the United Kingdom, 16 May 2018 <https://www.gov.uk/government/publications/launch-uk-event>.

[177]"UK space launch programme receives £50 million boost in Government's Industrial Strategy." 27 Nov. 2017. Government of the United Kingdom, 16 May 2018 <https://www.gov.uk/government/news/uk-space-launch-programme-receives-50-million-boost-in-governments-industrial-strategy>.

[178]"Commercial spaceflight." 8 Feb. 2017. Government of the United Kingdom, 16 May 2018 <https://www.gov.uk/government/collections/commercial-spaceflight>.

[179]"UK-France space co-operation at Sandhurst Summit." 18 Jan. 2018. Government of the United Kingdom, 16 May 2018 <https://www.gov.uk/government/news/uk-france-space-co-operation-at-sandhurst-summit>.

it details the Agency's four priorities for the period, including the publication of a Space Strategy, enabling UK commercial space access, oversight of key national and international programmes and reform of the British space regulatory regime.[180]

1.2.4 The USA

On June 30, 2017, US President Trump signed an executive order which reestablished the National Space Council. The Council had been previously disbanded in 1993 and in its new form was given an ambitious mandate of maintaining American space leadership.[181] It held its first meeting on 5 October 2017, a second on 21 February 2018 and a third on 18 June 2018. At the first, US Vice President Mike Pence directed NASA to create plans for human lunar missions which would lead on to future Mars missions. The Vice President additionally referred the US Office of Management and Budget Director and the Secretaries of Commerce and Transportation to make a review of the US commercial space regulatory framework, with an eye to making operations more efficient.[182] At the National Space Council's second meeting, four recommendations were put forward to reform commercial space regulations, concerning launch licensing, consolidation of the Office of Space Commerce and the Commercial Remote Sensing Regulatory Affairs Office, protection of the radiofrequency spectrum in commercial space activity and reform of export control.[183] At the third meeting, President Trump signed a space policy directive on space traffic management (known also as Space Policy Directive-3), largely based on a draft policy revealed by Vice President Pence 2 months prior. The Directive focuses on a range of space traffic management topics and relocates core responsibilities from the Department of Defense to the Commerce Department. Media quote Vice President Pence referring to the original draft as the US' "first comprehensive space traffic management policy".[184] Moreover, on 11 December 2017, President Trump signed Space Policy Directive-1, which builds

---

[180]"UK Space Agency Corporate Plan 2017–18." 24 Aug. 2017. Government of the United Kingdom, 16 May 2018 <https://www.gov.uk/government/publications/uk-space-agency-corporate-plan-2017-18>.

[181]"President Trump reestablishes National Space Council." 30 June 2017. SpaceNews, 18 May 2018 <http://spacenews.com/breaking-president-trump-reestablishes-national-space-council/>.

[182]Foust, Jeff. "National Space Council calls for human return to the moon." 5 Oct. 2017. SpaceNews, 18 May 2018 <http://spacenews.com/national-space-council-calls-for-human-return-to-the-moon/>.

[183]Foust, Jeff. "National Space Council backs incremental space regulatory reform." 21 Feb. 2018. SpaceNews, 18 May 2018 <http://spacenews.com/national-space-council-backs-incremental-space-regulatory-reform/>.

[184]Foust, Jeff. "President to sign space traffic management policy." 28 June 2018. SpaceNews, 18 May 2018 <http://spacenews.com/president-to-sign-space-traffic-management-policy/>.

on the first National Space Council meeting in instructing NASA to again conduct human missions to the Moon.[185]

Regarding NASA financing, US President Trump signed the NASA Transition Authorization Act of 2017 on 21 March 2017, which authorised a $19.5 billion budget for NASA in fiscal year 2017 and provided support to the Orion spacecraft (a human spaceflight capsule) and the heavy-lift Space Launch System which will launch it, both under development.[186]

Further, the White House released an initial budget blueprint for the fiscal year 2018 on 16 March 2017, with a complete budget request released later on 23 May.[187] Under this final budget request, NASA was to receive $19.1 billion in 2018, representing a drop from the $19.653 billion the agency received for fiscal year 2017.[188] Notably, the White House budget request proposed to cancel the Asteroid Redirect Mission, strengthen NASA's cybersecurity capabilities, allocate $3.7 billion for the Orion capsule and Space Launch System heavy-lift rocket and reduce earth science activity support by $102 million, thereby cancelling four missions.[189] The final outcome, as seen in the omnibus spending bill passed by Congress and signed by President Trump on 23 March 2018, resulted in broad budgetary increases for science and research activities generally, including for NASA and other science agencies.[190] For NASA's part, it received $20.74 billion for the 2018 fiscal year, which represents a 6% increase over its fiscal year 2017 budget. Human exploration received an 11% increase to $4.79 billion in 2018; space technology also received an 11% increase, reaching $0.76 billion; the Science Mission Directorate received an 8% increase, leading to $6.22 billion in funding; and Aeronautics received a 4% increase to $0.69 billion, and Space operations dropped 4% to $4.75 billion. And whereas the White House budget request would have eliminated the NASA Office of Education, it received unchanged funding into 2018 at $0.10 billion. Within the Science Mission Directorate, the Planetary Science Division's budget in particular

[185]Foust, Jeff. "Trump formally establishes lunar landing goal, but without details." 11 Dec. 2017. SpaceNews, 18 May 2018 <https://spacenews.com/trump-formally-establishes-lunar-landing-goal-but-without-details/>.

[186]Cofield, Calla. "President Trump Signs NASA Authorization Bill." 21 Mar. 2017. Space.com, 18 May 2018 <https://www.space.com/36154-president-trump-signs-nasa-authorization-bill.html>.

[187]Smith, Marcia. "NASA's FY2018 Budget Request." 26 Mar. 2018. SpacePolicyOnline.com, 18 May 2018 <https://spacepolicyonline.com/wp-content/uploads/2017/03/NASA-FY2018-budget-request-Mar-24-2018.pdf>.

[188]Smith, Marcia. "NASA Budget to soar over $20 Billion in Final FY2018 Appropriations." 21 Mar. 2018. SpacePolicyOnline.com, 18 May 2018 <https://spacepolicyonline.com/news/nasa-budget-to-soar-over-20-billion-in-final-fy2018-appropriations/>.

[189]*Op. Cit.* – "NASA's FY2018 Budget Request."

[190]Koren, Marina. "Congress Ignores Trump's Priorities for Science Funding." 23 Mar. 2018. The Atlantic, 18 May 2018 <https://www.theatlantic.com/science/archive/2018/03/trump-science-budget/556229/>.

increased to $2.23 billion, a 21% gain.[191] Further, the Orion capsule and heavy-lift Space Launch System programmes received the same funding as in the 2017 fiscal year.[192]

Moreover, US President Donald Trump signed on 18 April 2017 the Weather Research and Forecasting Innovation Act, which provides new direction from the US Congress for NOAA with regard to weather research and forecasting as well as observation programmes, and it is seen as the first comprehensive policy on weather since previous legislation from 1992. The third of its five titles deals entirely with "Weather Satellite and Data Innovation" and tasks NOAA to, among other steps, "complete the COSMIC-2 microsatellite system, including deployment of constellations in both the equatorial and polar orbits, and to integrate the resulting data into 'all national operational and research weather forecast models'".[193]

1.2.5 Canada

On 18 April 2017, Canada's Innovation, Science and Economic Development (ISED) Minister Navdeep Bains announced the renewal of the Government of Canada's Space Advisory Board. The process of renewing the Board had been announced previously in October of 2016, and its members were chosen via a merit-based application procedure. The Space Advisory Board is composed of a Chair person and ten members with space sector expertise and is mandated to engage with the Canadian public toward the development of a "new vision for Canada's space sector", to advise the government on space objectives and to identify the core aspects of a new space strategy.[194]

As part of its mandate, the Board conducted a series of roundtable discussions throughout early 2017 with Canadian space sector stakeholders.[195] The major findings from these discussions were published in a final report released in August 2017 and are ultimately intended to assist in defining the new space strategy. The Board's findings can be outlined in two main recommendations: (1) that space should be designated "as a national strategic asset to ensure that" a platform of priorities can be achieved, with a focus on Canada's ability to enjoy the benefits of

---

[191]Thomas, Will. "Final FY18 Appropriations: NASA." 30 Mar. 2018. American Institute of Physics, 18 May 2018 <https://www.aip.org/fyi/2018/final-fy18-appropriations-nasa>.

[192]"American Institute of Physics – Federal Science Budget Tracker." Data for NASA from FY2018. American Institute of Physics, 18 May 2018 <https://www.aip.org/fyi/federal-science-budget-tracker/FY2018#tabs-section-nasa>.

[193]Henry, Mike. "First Comprehensive Weather Policy Update Since 1992 Now Law." 21 Apr. 2017. American Institute of Physics, 18 May 2018 <https://www.aip.org/fyi/2017/first-comprehensive-weather-policy-update-1992-now-law>.

[194]"Space Advisory Board." Government of Canada, 22 May 2018 (retrieved) <https://www.ic.gc.ca/eic/site/ad-ad.nsf/eng/h_ad03983.html>.

[195]Ibid.

space activities and (2) to further mandate the Space Advisory Board in supporting the implementation of the eventual Space Strategy and the evaluation thereof. A third central commentary strongly expressed "the need for a reinvigorated and fully funded set of space activities and supporting policies".[196] However, although it was previously announced that the strategy would be published in June of 2017, the final publication date has been extended to allow for revision after comments were received following the circulation of a draft version.[197,198]

1.2.6 Russia

As made known in government documents and announcements throughout 2017, Russia is continuing to pursue construction and operations of it spaceports with significant investments. As announced by Russian Prime Minister Dmitry Medvedev, the federal programme for the development of spaceports and their infrastructure over 2017–2025 is to be more than $5.9 billion.[199] For the Vostochny Cosmodrome in particular, which has been under construction since 2010, the Russian Finance Ministry is to provide $39 million, which is from funds classified in the budget for Outer Space Exploration and Use. This funding will be transferred specifically to Roscosmos for the operation of the spaceport's ground infrastructure and maintenance of supporting infrastructure.[200] A further $630 million will be used for the construction at Vostochny of the future Angara rocket's launch infrastructure, forecasted for completion in 2022.[201]

The Russian Space Agency Roscosmos also strengthened its ties with its counterpart agencies in the USA and China in 2017 and 2018. On 27 September 2017, at the 68th International Astronautical Congress in Australia, Roscosmos and NASA signed a joint statement "on Researching, Exploring Deep Space" and which focuses in particular on research which may lead to development of NASA's concept Deep

---

[196]"Consultations on Canada's future in space: What we heard." 30 Oct. 2017. Government of Canada, 22 May 2018 <https://www.ic.gc.ca/eic/site/ad-ad.nsf/eng/ad03996.html>.

[197]Boucher, Marc. "New Canadian Space Strategy Not Ready – Exclusive." 19 Mar. 2018. SpaceQ, 22 May 2018 <http://spaceq.ca/new-canadian-space-strategy-not-ready/>.

[198]Boucher, Marc. "New Canadian Space Strategy Delayed." 21 June 2017. SpaceQ, 22 May 2018 <https://spaceq.ca/new-canadian-space-strategy-delayed/>.

[199]"More than $5.9 billion to be invested in spaceports development by 2025 — Russian PM." 13 Sept. 2017. TASS, 11 July 2018 <http://tass.com/science/965379>.

[200]"Russian government allocates $39 mln for Vostochny spaceport operation." 18 Aug. 2017. TASS, 11 July 2018 <http://tass.com/science/960961>.

[201]"Russia to spend $630 mln on launch pad for Angara carrier rocket at Vostochny spaceport." 11 Aug. 2017. TASS, 11 July 2018 <http://tass.com/science/959991>.

Space Gateway.[202,203] And on 3 March 2018, The Chinese National Space Administration with Roscosmos signed an agreement around cooperation in lunar and deep space exploration as well as the creation of a Russian-Chinese Data Center for lunar projects. Two possible avenues of implementing this cooperation include the Russian Luna-Resurs-1 (Luna-26) and Chinese lunar south pole missions.[204]

1.2.7 Japan

Important developments occurred in 2017 related to Japan's industrial, legal and other efforts to increase domestic launch activity. Regarding the H3 rocket, the government of Japan through JAXA awarded a contract to Mitsubishi Heavy Industries in 2014 for the construction of the H3, Japan's next-generation launcher which will follow its H-2A and H-2B rockets. In this contract, JAXA stipulated that the launch price per kilogramme should be cut by 50% as compared with the H-2A, and it was announced in 2018 that Japan plans to adapt the Tanegashima spaceport's infrastructure to enable an increase in its launch rate with the new launcher.[205] Currently, Tanegashima hosts one pad for the H2-A and one for the H2-B, both of which MHI will adapt for the H3. The H3 is scheduled to start launching in 2020 and is intended to boost the overall launch rate to ten per year from the current approximately four per year.[206] Furthermore, in November 2017, the Japanese Cabinet Office began to take applications for launching and operating satellites under new space activities legislation passed in November 2016, opening up private sector involvement in access to space activities.[207,208]

---

[202]Warner, Cheryl. "NASA, Roscosmos Sign Joint Statement on Researching, Exploring Deep Space." 27 Sept. 2017. NASA, 11 July 2018 <https://www.nasa.gov/feature/nasa-roscosmos-sign-joint-statement-on-researching-exploring-deep-space>.

[203]Foust, Jeff. "NASA and Roscosmos to study Deep Space Gateway." 28 Sept. 2017. SpaceNews, 11 July 2018 <http://spacenews.com/nasa-and-roscosmos-to-study-deep-space-gateway/>.

[204]Messier, Doug. "Russia, China Sign Lunar Cooperation Agreement." 7 Mar. 2018. Parabolic Arc, 11 July 2018 <http://www.parabolicarc.com/2018/03/07/russia-china-sign-lunar-cooperation-agreement/>.

[205]Henry, Caleb. "MHI says H3 rocket development on track for 2020." 26 June 2017. SpaceNews, 12 July 2018 <http://spacenews.com/mhi-says-h3-rocket-development-on-track-for-2020/>.

[206]Henry, Caleb. "Japan to modernize spaceport launch pads to support H3 rocket." 23 Mar. 2018. SpaceNews, 12 July 2018 <http://spacenews.com/japan-to-modernize-spaceport-launch-pads-to-support-h3-rocket/>.

[207]"Japan Legal Update – Volume 32." Jones Day, 12 May 2018 (retrieved) <http://www.jonesday.com/files/Publication/d3cb154e-014a-4014-9b83-60baf57dbcf4/Presentation/PublicationAttachment/c150f476-2611-4188-8823-cda77404824d/Japan%20Legal%20Update%20Dec%202017%20Jan%202018.pdf>.

[208]Setsuko, Aoki. "New Law Aims to Expand Japan's Space Business." 3 Mar. 2017. Nippon.com, 12 July 2018 <https://www.nippon.com/en/currents/d00294/>.

In further support to the private sector, Prime Minister Shinzo Abe announced an initiative on 20 March 2018 to provide venture capital from a fund worth $940 million to support Japan's space startups. The programme will run during a period of 5 years, with funding from the Development Bank of Japan and the Industrial Innovation Organization, among others, and will also match investors with startups.[209] Furthermore, the Japanese Cabinet Office's Committee of National Space Policy has also released a Space Industry Vision 2030, aiming to double the Japanese space industry's market size from the current 1.2 trillion yen to 2.4 trillion yen ($21 billion as of May 2017).[210,211,212]

In terms of space exploration, it was announced in June 2017 that JAXA aims to send an astronaut to the Moon by 2030. JAXA intends to join the lunar-orbiting Deep Space Gateway mission led by NASA and use it as a platform for access to the Moon's surface. JAXA intends to develop water and air purifiers in its cooperation in the Deep Space Gateway, and further programme descriptions were announced for 2018.[213,214]

With a series of diverse cooperative instruments, JAXA also continued to pursue its cooperation with external partners throughout 2017 and 2018, including:

- An implementing arrangement with the French space agency CNES[215]
- Two joint statements on bilateral cooperation with ESA[216,217]

---

[209] Foust, Jeff. "New fund to boost Japanese space startups." 21 Mar. 2018. SpaceNews, 12 July 2018 <http://spacenews.com/new-fund-to-boost-japanese-space-startups/>.

[210] Messier, Doug. "NewSpace Business Book Published in Japan." 27 May 2018. Parabolic Arc, 12 July 2018 <http://www.parabolicarc.com/2018/05/27/space-business-book/>.

[211] "Japan aims to double its space market to $21bn by 2030s." 12 May 2017. Nikkei Asian Review, 12 July 2018 <https://asia.nikkei.com/Economy/Japan-aims-to-double-its-space-market-to-21bn-by-2030s>.

[212] Tsuruda, Masanori. "Japanese Space Industry Policy Overview – Space industry in the big data era." 6 Sept. 2017. Japanese Ministry of Economy, Trade and Industry, 12 July 2018 <https://www.eu-japan.eu/sites/default/files/imce/1._meti_tsuruda_1.pdf>.

[213] Foust, Jeff. "Japan has plans to land astronauts on the moon by 2030 — with a little help from the United States." 29 June 2017. SpaceNews, 12 July 2018 <http://spacenews.com/mda-establishes-company-to-commercialize-satellite-servicing-technology/>.

[214] "JAXA reveals plans to put a Japanese on the moon by 2030." 1 July 2017. The Japan Times, 12 July 2018 <https://www.japantimes.co.jp/news/2017/07/01/national/science-health/jaxa-reveals-plans-put-japanese-moon-2030/#.Wwa7tu6FO70>. *If page expired, see:* <https://web.archive.org/web/20171223144400/https://www.japantimes.co.jp/news/2017/07/01/national/science-health/jaxa-reveals-plans-put-japanese-moon-2030/#.W2hSZ9Uza70>.

[215] "JAXA and CNES Make and Sign Implementing Arrangement on Martian Moons Exploration (MMX)." 10 Apr. 2017. JAXA, 12 July 2018 <http://global.jaxa.jp/press/2017/04/20170410_cnes.html>.

[216] "JAXA-ESA Joint Statement concerning the bilateral cooperation." 15 May 2017. JAXA, 12 July 2018 <http://global.jaxa.jp/press/2017/05/20170515_esa.html>.

[217] "JAXA-ESA Joint Statement concerning the bilateral cooperation." 3 Mar. 2018. JAXA, 12 July 2018 <http://global.jaxa.jp/press/2018/03/20180303_esa.html>.

- A joint statement on bilateral cooperation with the German space agency DLR[218]
- A cooperation agreement with the Vietnam National Space Center[219]
- An implementation arrangement with the Indian Space Research Organisation ISRO[220]
- A collaboration agreement mutually with the Japanese National Institute for Environmental Studies NIES, ESA and the DLR[221]
- A joint statement on space exploration with NASA[222]
- And a partnership agreement with the University of Tokyo[223]

Lastly, the Cabinet Office of Japan released an Implementation Plan of the Basic Plan on Space Policy, revised for fiscal year 2017 and dated 12 December 2017. Its first objective is to "Ensure space security" as well as "Promote the use of civil space" and "Maintain and strengthen the science and technology industrial base". To pursue these objectives, the Implementation Plan leverages satellite positioning, satellite remote sensing, satellite communications and broadcasting, space transportation, Space Situational Awareness, Maritime Domain Awareness, early-warning functions, "strengthening of guarantee for functions of the overall space systems", and space science, exploration and manned space activity.[224]

1.2.8 China

China released the fourth version of its White Paper on space activities on 27 December 2016, detailing its major objectives for the following 5 years as well as development measures and policies, with a core principal being that China's space

---

[218]"DLR-JAXA Joint Statement concerning the bilateral cooperation." 20 Sept. 2017. JAXA, 12 July 2018 <http://global.jaxa.jp/press/2017/09/20170920_dlr.html>.

[219]"Topics 2017." JAXA, 12 July 2018 (retrieved) <http://global.jaxa.jp/news/2017/>.

[220]"Japan Aerospace Exploration Agency (JAXA) Signs Implementation Arrangement (IA) with Indian Space Research Organization (ISRO) concerning joint study of Joint Lunar Polar Exploration." 6 Dec. 2017. JAXA, 12 July 2018 <http://global.jaxa.jp/press/2017/12/20171206_isro.html>.

[221]*Op. cit.* – "Topics 2017."

[222]"NASA-JAXA Joint Statement on Space Exploration." 26 Jan. 2018. JAXA, 12 July 2018 <http://global.jaxa.jp/press/2018/01/20180126_nasa.html>.

[223]"UTokyo and JAXA Partnership Agreement to Promote Kibo Use." 26 Apr. 2018. JAXA, 12 July 2018 <http://global.jaxa.jp/press/2018/04/20180426_kibo.html>.

[224]"Implementation Plan of the Basic Plan on Space Policy." (Download page) 12 Dec. 2017. Japanese Cabinet Office, 12 July 2018 <http://www8.cao.go.jp/space/english/basicplan/basicplan.html>.

activity be "subject to and serves the national overall development strategy".[225] Key objectives include[226]:

- The launch of the Tianzhou-1 cargo vessel and its docking with the Tiangong-2 space laboratory.
- Assemble and operate the space station.
- Launch both the Chang'e-5 lunar probe (sample return mission) by the end of 2017 (ultimately delayed to 2019 by the failed launch of a Long March 5 in July 2017[227]).
- The launch of the Chang'e-4 lunar probe in 2018, which is to make a soft landing on the far side of the Moon (its relay satellite having been launched in May 2018 to the Earth—Moon L2 Lagrange Point[228]).
- As well as a wide-reaching policy approach to promote the space sector across many levels, including enhancing innovation and improving the industry's capacity.

An additional component of the White Paper is the promotion of international cooperation in Chinese space activities. As of 2017, the People's Republic of China has signed 43 memoranda of understanding or cooperation agreements with 29 international organisations, space agencies or other countries since 2011.[229] Most recently, this has included a memorandum of understanding on climate actions and space exploration with the French space agency CNES. CNES President Jean-Yves Le Gall and his counterpart Wu Yanhua of the China National Space Administration (CNSA) signed the memorandum in January 2018 in the presence of Chinese President Xi Jinping and French President Emmanuel Macron.[230] China also signed a series of cooperation documents with Cambodia on 11 January 2018. Among these is a framework agreement between the Royal Group of Cambodia and the China

---

[225]"China's Space Activities in 2016 – the 4th version of white paper." 30 Jan. 2017. China National Space Administration (powerpoint hosted online by unoosa.org), 16 July 2018 <http://www.unoosa.org/documents/pdf/copuos/stsc/2017/tech-01E.pdf>.

[226]"Full text of white paper on China's space activities in 2016." 28 Dec. 2016. The State Council – english.gov.cn, 16 July 2018 <http://english.gov.cn/archive/white_paper/2016/12/28/content_281475527159496.htm>.

[227]Jones, Andrew. "China reveals cause of Long March 5 failure; lunar sample mission to follow return-to-flight." 16 Apr. 2018. SpaceNews, 16 July 2018 <http://spacenews.com/china-reveals-cause-of-long-march-5-failure-lunar-sample-mission-to-follow-return-to-flight/>.

[228]Barbosa, Rui C. & Bergin, Chris. "Queqiao relay satellite launched ahead of Chang'e-4 lunar mission." 20 May 2018. NASASPACEFLIGHT.com, 16 July 2018 <https://www.nasaspaceflight.com/2018/05/queqiao-relay-satellite-launched-change-4-lunar-mission/>.

[229]*Op. cit.* – "China's Space Activities in 2016 – the 4th version of white paper."

[230]"State Visit of President Emmanuel Macron to the People's Republic of China: France and China Step up Space Cooperation in the Fields of Climate and Exploration." 10 Jan. 2018. CNES, 16 July 2018 <https://presse.cnes.fr/en/state-visit-president-emmanuel-macron-peoples-republic-china-france-and-china-step-space-cooperation>.

Great Wall Industry Corp. around the build, launch, insuring and ground systems of a new geostationary telecommunications satellite.[231]

And in further pursuit of international cooperation, China (through the China Manned Space Agency) and the United Nations Office for Outer Space Affairs signed a Memorandum of Understanding in 2016 around experiment opportunities for UN member states aboard the China Space Station. An Announcement of Opportunity was made jointly on 28 May 2018 by UNOOSA and the Permanent Mission of China to the UN and Other International Organizations in Vienna, and application instructions were made available. Operations on the future space station are forecasted from 2022; this cooperation aims to "promote international cooperation in human space flight and activities related to space exploration", as well as "provide flight experiment and space application opportunities on-board the CSS for United Nations Member States", among other objectives.[232]

1.2.9 India

Although heavily oriented toward civilian space activities, India's space sector is increasingly emphasising defence and security, especially through international cooperation, and in particular with Japan, the USA and France.[233]

The Indian Space Research Organisation formed and/or reinforced a number of collaborative relationships in the reporting period over 2017–2018, showing a continuing emphasis on international cooperation in its activities. These include 15 implementing arrangements, Memoranda of Understanding or other agreements with 19 actors, including[234]:

- The California Institute of Technology (Caltech)
- The Canadian Space Agency CSA
- The French space agency CNES[235]
- Geoscience Australia (a national agency of Australia)

---

[231]Henry, Caleb. "Cambodia to buy Chinese satellite as relations tighten on Belt and Road Initiative." 12 Jan. 2018. SpaceNews, 16 July 2018 <http://spacenews.com/cambodia-to-buy-chinese-satellite-as-relations-tighten-on-belt-and-road-initiative/>.

[232]"United Nations and China invite applications to conduct experiments on-board China's Space Station." 28 May 2018. United Nations Office for Outer Space Affairs, 16 July 2018 <http://www.unoosa.org/oosa/en/informationfor/media/2018-unis-os-496.html>.

[233]Rajagopalan, Rajeswari Pillai. "India changing tack on space policy." 3 Apr. 2018. ORF, 13 July 2018 <https://www.orfonline.org/research/india-changing-tack-on-space-policy/>.

[234]"Government of India – Department of Space Annual Report 2017–2018." 12 Jan. 2018. ISRO, 13 July 2018 <https://www.isro.gov.in/sites/default/files/flipping_book/AnnualReport2017-18-en/files/assets/common/downloads/annualreport2017-18-en.pdf>.

[235]"France-India Space Cooperation – Agreements Signed on Future Launchers and Lunar Exploration." 9 Jan. 2017. CNES, 13 July 2018 <https://presse.cnes.fr/en/france-india-space-cooperation-agreements-signed-future-launchers-and-lunar-exploration>.

- The Israel Space Agency ISA
- The Japan Aerospace Exploration Agency JAXA
- The National Remote Sensing Department of Vietnam[236]
- The US National Aeronautics and Space Administration NASA

Beyond the scope of ISRO, India has also entered into cooperative agreements with several actors, including between:

- The Indian Department of Space and the European Commission, in a Cooperation Agreement around the sharing of Earth observation data[237]
- The Indian Department of Science and Technology and the Israeli National Technological Innovation Authority, in a Memorandum of Understanding[238]
- The Indian Institute of Space Science and Space Technology, of the Indian Science Ministry, and the Technion—Israel Institute of Technology, in a Memorandum of Understanding[239]
- The governments of India and France through the India-France Joint Vision for Space Cooperation, which lays out future areas of cooperation between ISRO and CNES[240,241]
- The Governments of Armenia, Bangladesh, the Netherlands and Portugal (four Memoranda of Understanding)[242]

[236]"India, Vietnam sign two agreements." 24 Jan. 2018. The Economic Times, 13 July 2018 <https://economictimes.indiatimes.com/news/politics-and-nation/india-vietnam-sign-two-agreements/articleshow/62639822.cms>.

[237]"European Commission and India to share satellite data." 17 Apr. 2018. UN-SPIDER, 13 July 2018 <http://www.un-spider.org/news-and-events/news/european-commission-and-india-share-satellite-data>.

[238]Goh, Deyana. "Modi, Netanyahu sign space cooperation agreements between ISRO and ISA." 6 July 2017. SpacetechAsia, 13 July 2018 <http://www.spacetechasia.com/modi-netanyahu-sign-space-cooperation-agreements-between-isro-and-isa/>.

[239]Siegel-Itzkovich, Judy. "Space Exploration Cooperation Agreement Signed While Netanyahu in India." 18 Jan. 2018. The Jerusalem Post, 13 July 2018 <https://www.jpost.com/HEALTH-SCIENCE/Space-exploration-cooperation-agreement-signed-while-Netanyahu-in-India-538147>.

[240]Datta, Anusuya. "India, France to collaborate on satellite technology to tackle Climate Change." 12 Mar. 2018. Geospatial World, 13 July 2018 <https://www.geospatialworld.net/blogs/india-france-collaboration-on-satellite-technology/>.

[241]"India-France Joint Vision for Space Cooperation (New Delhi, 10 March 2018)." 10 Mar. 2018. Government of India, Ministry of External Affairs, 13 July 2018 <http://mea.gov.in/bilateral-documents.htm?dtl/29597/IndiaFrance+Joint+Vision+for+Space+Cooperation+New+Delhi+10+March+2018>.

[242]*Op. cit.* – "Government of India – Department of Space Annual Report 2017–2018."

1.3 Worldwide Space Budgets and Revenues

In 2017, the global government space actors figure at close to 50, most of them having an agency or an institute related to space activity and at least 9 having a budget exceeding $1 billion (being China, the USA, France, Germany, Italy, Russia, India, Japan and also the European Commission).

Global space activity accounted for $383.51 billion, and the total revenues from the commercial space sector were the 80.1% of the global economic activity in space, or $307.32 billion, including private and public activities.[243] In terms of commercial revenues, the Space Report 2018 indicates that the commercial infrastructure and support industries generated $95.87 billion (+7.5% from 2016), including manufacturing of commercial satellites, provision of launch services and provision of space asset insurance to satellite-launching and satellite-operating companies. The main sources of revenues were from ground stations and equipment and satellite manufacturing, representing $85.84 billion and 6.82 billion, respectively, in revenue in 2017. In the navigation sector for 2017, the European Global Navigation Satellite System Agency (GSA) estimated that devices for Global Navigation Satellite Systems (GNSS) generated $57 billion in revenue, a healthy increase (+7.8%) from $52.9 million of 2016.[244]

1.3.1 Overview of Institutional Space Budgets

The total institutional spending on space in 2017, including that of intergovernmental organisations, is estimated to be approximately $76.197 billion, an increase of 4.8% compared to 2016.[245] While the US institutional spending decreased slightly to $43.344 billion from $44.444 billion of 2016, the USA spent $22.7 billion on civil space activities in 2017—up 0.9% from the previous year. Approximately half of the US government spending went toward national security space activities within the Department of Defense (DoD) budget, the National Reconnaissance Office (NRO) and the National Geospatial-Intelligence Agency (NGA). According to the Space Report 2018, DoD programmes made up 47.7% of the total US space budget in FY 2017, representing $20.7 billion or 6.0% lower than $22.0 billion estimated in 2016.

Figure 1.1 and the following figure (Fig. 1.2) present a relative picture of the space sector. They should be viewed holistically, and compared to each other may not reflect the real output of spacefaring countries.

The state hierarchy is not far from 2016 one, even considering China's recent growth. While spending among space actors saw significant changes in 2017, they

[243]The Space Report 2018. Colorado Springs: The Space Foundation, 2018: 4.

[244]Ibid. p. 6.

[245]The Space Report 2018. Colorado Springs: The Space Foundation, 2018: 8.

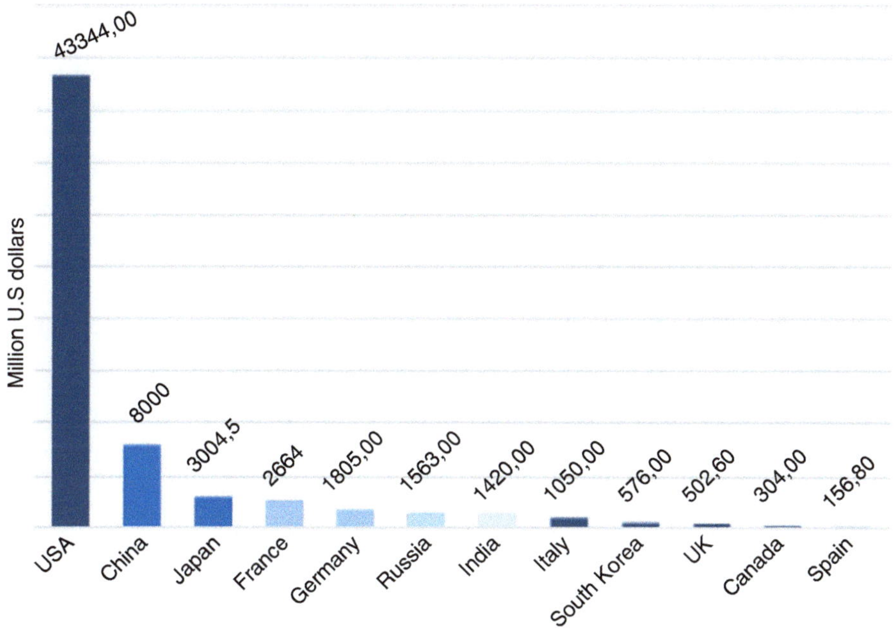

Fig. 1.1 Public space budgets of major space power in 2017 (Source: The Space Report 2018)—Military space budgets excluded

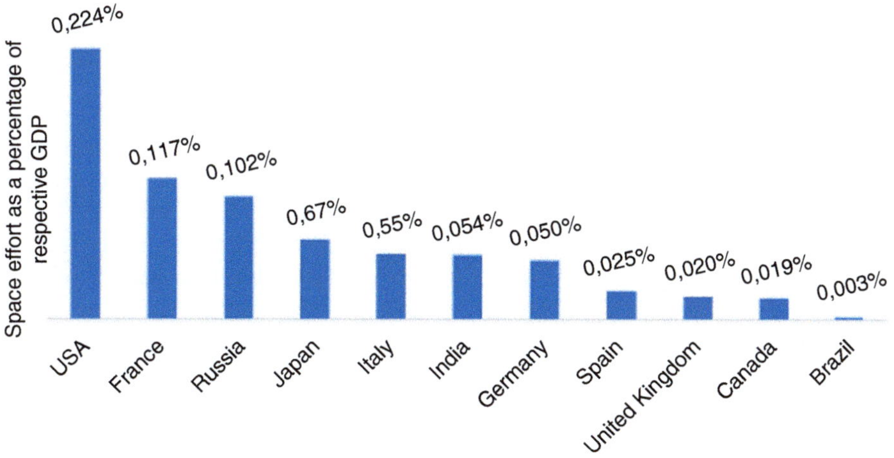

Fig. 1.2 Public Space Budgets as a share of nom. GDP in 2017 (Source Space Report 2018)

should not be ranked against each other given the fluctuations in currency exchange rates as these budgets are converted to US dollars (in Fig. 1.1).[246]

The USA column of Fig. 1.1 shows that it continues to have the largest space budget, remaining around $43.344 billion (2.5% less than 2016), while it decreased its military expenditure 6.0 percentage points[247] and increased its civilian space spending to $22.656 billion in 2017. Confirming the data analysed for 2016, China's budget is still the second largest, doubling in 2017 to $8.006 billion[248] in comparison to the $4.317 billion of 2016—attesting a growth of 85.4%. Japan's space budget for 2017 is $30.446 billion (¥342.1 billion), the result of an increase of JAXA's allocated supplemental budget.

France (at 2.334 billion €, which reflects +13.2% growth), Russia ($1.563 billion, −11.5%), India ($1.417 billion, +21.9%) and Germany ($1.805 billion, +11.6%) all follow Japan while also witnessing a Russian budget decrease that began in 2015. India had an increase of 870% in the number of commercial spacecraft launched given that in 2017 a total of 138 spacecraft were launched, 113 (82%) of which were commercial.[249] The Italian budget was $1.05 billion (+19.4% from 2016), while the UK budget grew by 4% reaching $502.6 million, including ESA budget.

Moreover, national contributions to ESA's budget were included within these figures. In 2017, ESA operated with 5.75 billion € ($6.56 billion), increasing from 2016 by 12.8% (5.25 billion € or $5.82 billion). The members' contributions grew up to 3.78 billion € ($4.32 billion), and the majority of funds came from optional contributions as determined by the interest of the member states. Among ESA members, the five biggest contributors are Germany and France both at 22.7%, Italy at 14.6%, the UK at 7.9% and Belgium at 5.5%. Spain was the next highest contributor at 4%, followed by Switzerland at 3.8% for seventh position in the 2017 budget.[250]

In order to measure the concrete effort provided by a country in the space sector, it is necessary to measure the investments with regard to GDP generated in 2017 (Fig. 1.2).

The US position confirms its strong engagement in space activities, although its spending decreased to 0.224% from the 0.239% of 2016. France enjoyed a slight increase from 0.113 to 0.117%, whereas Russia dropped to 0.102% from 0.127% in 2016. Japan came next with 0.067%, followed by Italy with 0.055%. Other leading space countries in Europe and the rest of the world invested less than 0.050% of their GDP in space activities. An additional perspective is furnished by the space budget per capita (Fig. 1.3) where the USA is holding undisputedly the first position, although its spending decreased 2.5%.

---

[246]Ibid. at page 8.

[247]Ibid. at page 15.

[248]Chinese space budget is not publicly available. The calculation is a Space Foundation's estimate.

[249]The Space Report 2018. Space Foundation 2018: 8.

[250]Italy, the UK, Belgium and Spain contribute the same % of amount as 2016. "ESA Budget 2017". 16th January 2017. ESA 6th August 2018:<http://www.esa.int/spaceinimages/Images/2018/01/ESA_budget_2018>.

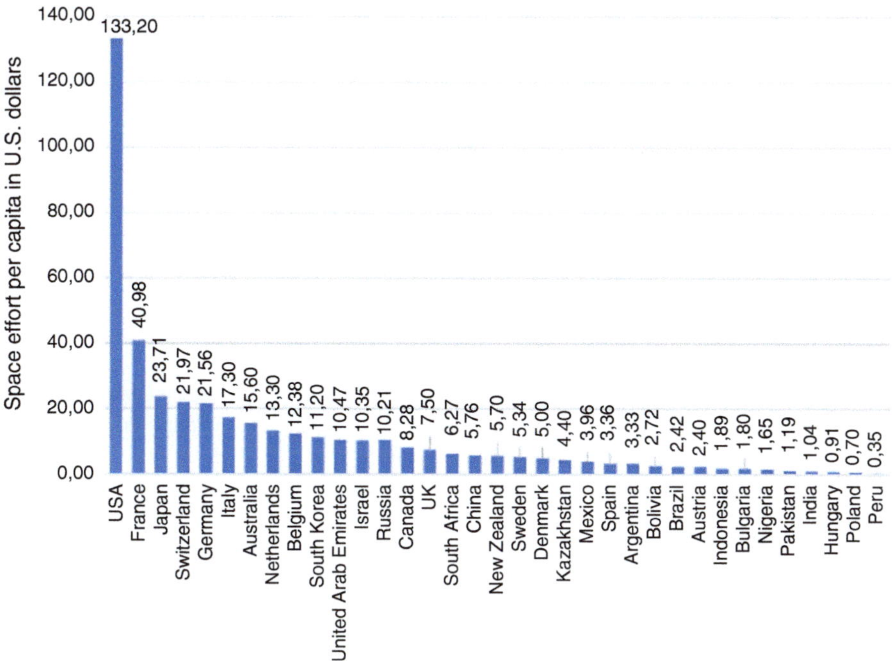

Fig. 1.3 Public space budgets per capita (selection) in 2017 (Source: The Space Report and Population Reference Bureau)

The second in ranking is France at $40.98 (a decrease from the $43.22 of 2016), followed Japan ($23.71), Switzerland ($21.97), and looking to China ($5.7) its relatively lower figure is clear. The situation is similar for India (at $1.04), a unique position with a large part of its GDP spent on space although having a budget per capita well behind the other spacefaring nations. Some caution is needed when considering these figures, though, due to fluctuating exchange rates and the uncertainty of reported values.

1.3.2 Overview of Commercial Space Markets

The Satellite Industry Association (SIA) reported that global industry revenues from satellites services, satellite manufacturing, launch industry and ground equipment segments grew by 3% in 2017, reaching $269 billion in 2017, up from the $261 billion in 2016 (representing 79% of the space economy in 2017).[251]

---

[251]"2018 State of the Satellite Industry Report." 13th June 2018. Satellite Industry Association and Bryce Space Technology. 6th August <https://www.sia.org/wp-content/uploads/2018/06/2018-SSIR-2-Pager-.pdf>.

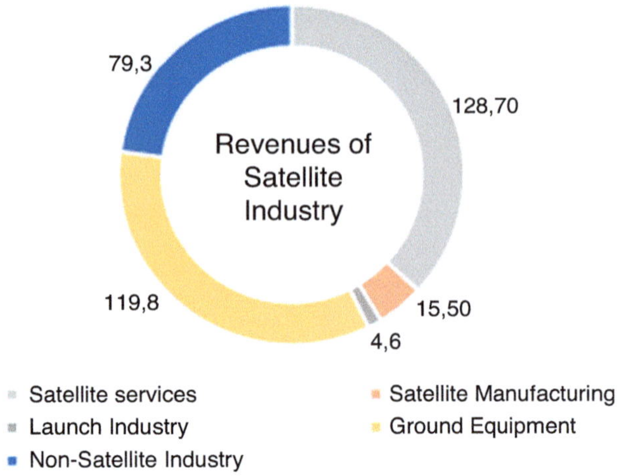

Fig. 1.4 Status of the satellite industry (Source: SIA)

1.3.2.1 Satellite Services

According to the SIA, the revenue earned from satellite services hovered around the same value of 2016, growing 1% and reaching $128.7 billion in 2017—up from $127.7 billion in 2016.[252] Considering satellites services subgroups, the highest revenue comes from direct-to-home television at $97 billion, while Earth observation was $2.2 billion.[253] Satellite services can further be deconstructed into their component parts, including consumer services (i.e. satellite TV (DBS/DTH), satellite radio and satellite broadband services), fixed satellite services (e.g. transponder agreements and managed network services) as well as mobile services (voice and data) and Earth observation. The following is a breakdown of the industry's key developments and trends, according to the nature of the services provided (Fig. 1.4).

1.3.2.2 Consumer Services

As mentioned above, consumer services include satellite television, radio and broadband services. The segment's downstream services revenue experienced a slight decrease, from $104.7 billion in 2016 to $104.5 billion in 2017. The television revenues were $97 billion, $5.4 billion for radio and $2.1 billion for broadband. The general consumer equipment (including GNSS devices like Garmin and TomTom, satellite TV dishes, etc.) increased by 5.6% to $108 billion. The network equipment (VSATs, gateways, etc.) revenues were $11.8 billion, and according to SIA, the total revenues from ground equipment were $119.8 billion. With around 220 million satellite television subscribers worldwide, DBS/DTH customers, increasingly in

[252]Ibid.

[253]Ibid.

emerging markets, are a key driver in consumer services revenue; however, there is the potential for a further slowdown in demand growth for DBS/DHT services as customers opt for Internet-based video services.[254]

Fixed Satellite Services (FSS) refer to the use of spacecraft that utilise land terminals in fixed positions to broadcast (such as through Intelsat, SES, Inmarsat, etc.). Whereas consumer services cover satellite broadband Internet, communications and network television and radio broadcasts, FSS relates to commercial signal agreements, such as transponder agreements and managed network services. The FSS segment saw figures comparable to 2015, earning $17.9 billion of revenues in 2017, likened to the $17.4 billion earned in 2016.

Mobile Satellite Services (MSS) offer both mobile data service and mobile voice service (including satellite phones). The MSS revenue grew by 17.6% in 2017, earning $4 billion from the $3.4 billion of 2016.

Earth Observation Services (offered, e.g. by Airbus and Digital Globe) refer to commercial companies that provide optical and radar images to the open market; however, demand for such services is mostly driven by government entities. Moreover, this sector is rapidly growing and changing; new entrants such as Terra Bella and Planet (formerly Planet Labs) have continued to raise capital and have begun deploying constellations. The majority of the revenue is generated through the development of value-added services, information products and big data-enabled analytics. The latter includes the processing and analysis of multiple images and a variety of datasets in order to extrapolate statistical information not in the base data itself.[255] Earth observation services saw a robust growth of 11.1%, increasing to $2.2 billion in 2017 from the $2.0 billion in 2016.[256]

1.3.2.3 Satellite Manufacturing

The total revenue of satellite manufacturers (such Thales Alenia Space, Airbus, OHB, Boeing, etc.), serving both governmental and commercial customers, reached $15.5 billion (up +10%) from the $13.9 billion of 2016, although still did not return to the $16.0 billion of 2015 (Fig. 1.5). It should be noted that the manufacturing sector, while mostly driven by the telecommunications sector, remains sensitive to downturns which took place despite the emergence of new players driving an increase in competition. US manufacturers' revenues were static from 2016, passing from $8.9 billion in 2016 to $8.8 billion in 2017; however, this activity nevertheless generated half of the overall revenues. In contrast, non-US manufacturers' revenues grew up to $6.7 billion from $5.0 billion in 2016. The overall market growth is

[254]Ibid.

[255]The Space Report 2018. Colorado Springs: The Space Foundation, 2018: 7.

[256]"2018 State of the Satellite Industry Report." 13th June 2018. Satellite Industry Association and Bryce Space Technology. 6th August <https://www.sia.org/wp-content/uploads/2018/06/2018-SSIR-2-Pager-.pdf>.

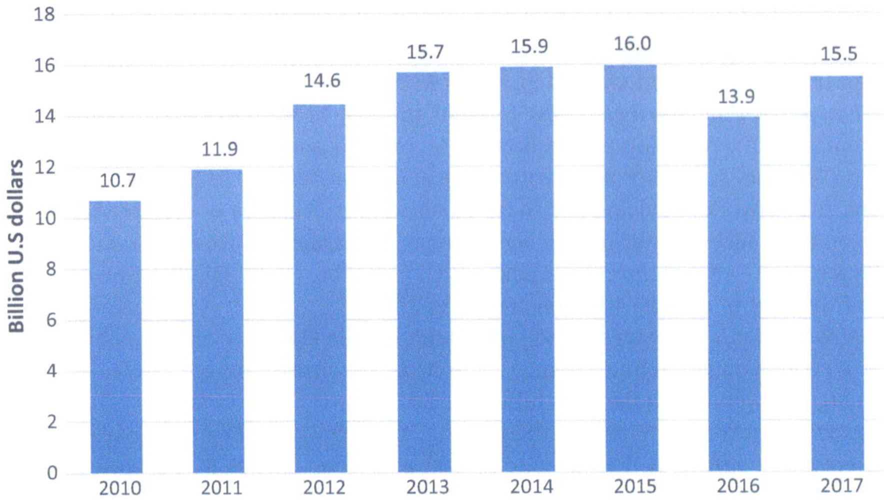

Fig. 1.5 Worldwide satellite manufacturing revenue (Source: SIA)

calculated at 10%. Furthermore, Earth observation services accounted for 49% of the revenues generated for the year, followed by commercial communications, which earned 18% of the total revenue. Next, meteorology services represented 15% of revenues, R&D and military surveillance each accounted for 6%, civil/military communications stood at 3%, and navigation and scientific satellites represented 2% and 1%, respectively.[257]

1.3.2.4 Launch Sector

In 2017, a total of 90 orbital launches were conducted, 33 of which were commercial (Fig. 1.6). Five of the 90 launches failed and in these are included four government launches. Specifically, these include a Soyuz 2.1b carrying the Meteor-M 2-1 satellite and 18 secondary payloads, a PSLV-XL carrying the IRNSS-1H satellite, a Long March 5A carrying the Shijan 18-01 satellite, an Electron carrying the Humanity Star and a SS-520 Upgrade carrying the TRICOM-1 satellite. China's Long March 3B launch in June was a partial success when ChinaSat 9A (Sinosat 4) was located in an incorrect orbit. For comparison, in 2016 there were 85 launches, including 22 commercial launches.[258] Revenues from the 33 commercial orbital

[257]"2018 State of the Satellite Industry Report." 13th June 2018. Satellite Industry Association and Bryce Space Technology. 6th August <https://www.sia.org/wp-content/uploads/2018/06/2018-SSIR-2-Pager-.pdf>.

[258]Federal Aviation Administration. The Annual Compendium of Commercial Space Transportation: 2018. Washington DC: FAA, Jan. 2018: 39.

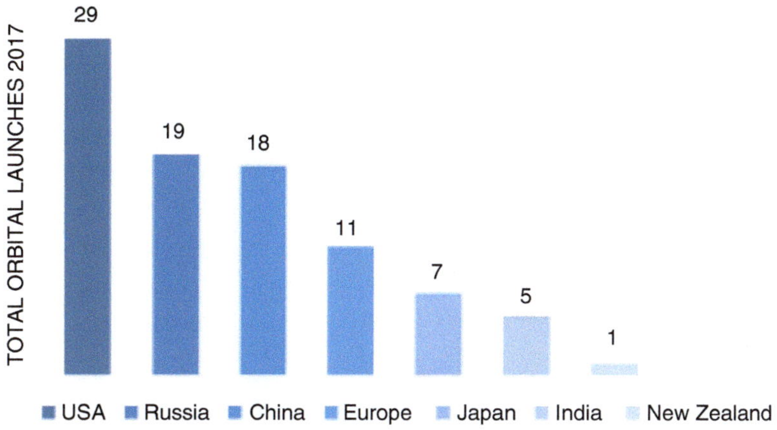

Fig. 1.6 Total orbital launches in 2017 (Source: ESPI database)

launches in 2017 are estimated at roughly over $3 billion, growing from the $2.5 billion in 2016.

For US providers, the commercial orbital launch revenues are estimated at $1.7 billion, compared to $1.2 billion in 2016. In 2017, US launch providers conducted 22 commercial orbital launches, compared to 11 licensed launches in 2016, with a share of 64% of the commercial launches. SpaceX's Falcon 9 conducted 18 launches, of which 10 were conducted for commercial purposes, including for Iridium, EchoStar, SES, Inmarsat, Bulsatcom, Intelsat and KT Sat.[259]

China conducted 18 non-commercial launches, but did not have any commercial launches in 2017. Russia was third in terms of launches, with only 3 out of 19 launches for commercial purposes. Moreover, its share of total commercial launches lowered to 9% from 9.5% in 2016. Europe conducted 8 commercial launches and 3 non-commercial launches in 2017 (the same figures of 2016), representing a share of commercial-launch activity of 24%. Japan conducted seven non-commercial launches, India five non-commercial launches and New Zealand a single commercial launch (Fig. 1.7).[260]

US launch commercial revenue decreased to $1.731 million from $1.185 million in 2016, when it had nearly doubled its commercial launch revenue from $617 million in 2015 (Fig. 1.8). Europe generated the second highest revenue for 2017, despite decreasing to $1.092 million in 2017 from $1.152 million in 2016. Russia held the third position earning an estimated $195 million, increasing from the $130 million earned in 2016.[261] Those results come from a combination of factors; due to the arrival of SpaceX's Falcon 9 vehicle on the global market in 2014, US providers

---

[259]ESPI database.

[260]Ibid.

[261]Federal Aviation Administration "The Annual Compendium of Commercial Space Transportation: 2018". Washington DC: FAA, Jan. 2018: 40, figure 7.

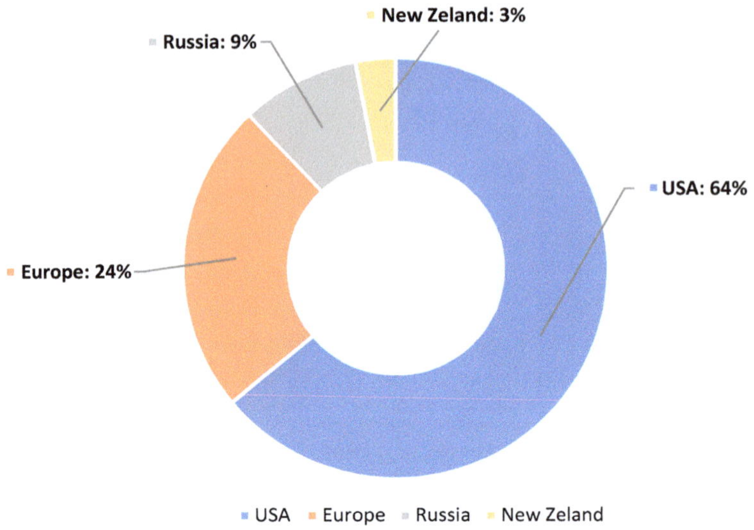

Fig. 1.7 Commercial launch activity by country in 2017 (Source: FAA)

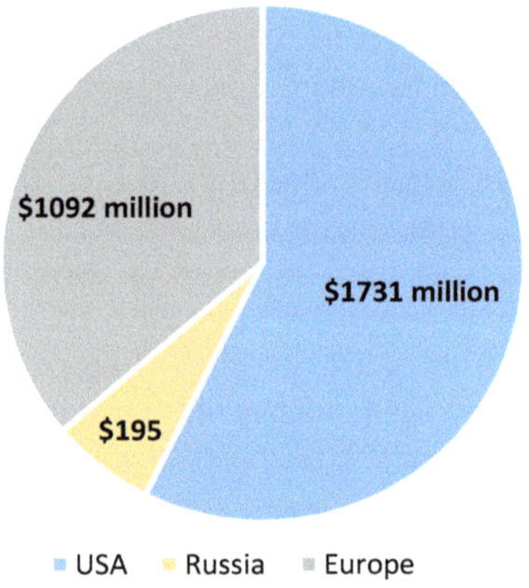

Fig. 1.8 Commercial launch revenues by country in 2017 (Source: FAA)

have begun to cut into the existing share of commercial launches that before were covered by Russian providers. Over the past few years, the Russian space industry was affected by issues that led some customers to find alternative providers.[262] Europe remains a steadfast option, however, with Arianespace offering services though the Ariane 5, Soyuz and Vega launchers. Arianespace conducted a total of 11 launches from its French Guiana spaceport in 2017. Its Ariane 5 ECA launcher had 5 launches, lifting 9 commercial payloads to orbit (Intelsat 32e, Telkom 03s,

[262]Ibid. p.1.

Koreasat 07, SDGC 1, Eutelsat 172B, ViaSat 02, Hellas-Sat 3/Inmarsat-S-EAS, BSat 4A and Intelsat 37e) and a single civil governmental telecommunication satellite to LEO (GSat 17). The Ariane 5 ES launcher had one launch, placing four Galileo navigation satellites (Galileo FOC −15, −16, −17, −18) to medium Earth orbit (MEO) for the European Commission. The Europeanised Soyuz had two launches, lifting Hispasat 36W-1 into GEO and the second that lifted SES 15 into GEO. The Vega launcher conducted three launches, lifting four civil governmental Earth observation satellites to GEO, including Sentinel 2B, OPTSAT-3000, VENμS and Mohammed VI A.

1.3.2.5 Ground Equipment

Ground equipment revenue includes infrastructure elements, such as mobile terminals, gateways and control stations and consumer equipment, such as very small aperture terminals (VSAT), ultra-small aperture terminals (USAT), DTH broadcast dishes, satellite phones and digital audio radio satellite (DARS) equipment. Portable navigation devices (PND) form one of the subsegments of end-user electronics that incorporate GNSS chip sets.

Ground equipment revenues increased by 5.6% to $119.8 billion in 2017 from $113.4 billion in 2016, driven by growth in consumer equipment for satellite navigation including standalone devices and embedded chipsets for smartphones, traffic information systems and transport vehicles and by increasing demand for network equipment for managed network services. Consumer equipment for satellite TV, satellite radio and satellite broadband saw a growth from $84.6 billion in 2016 to $108 billion in 2017, while network equipment (VSATs, gateways) saw a flatter growth to $11.8 billion in 2017 from $10.3 billion.[263]

The two companies leading the PND market, Garmin and TomTom, present a static income situation for 2017 (Table 1.1). TomTom earned 903 million € ($1.082 billion) in revenue for the year ending 31 December 2017, a 9% decrease from 2016. Its Automotive & Enterprise segment reached 328.8 million € (+22%) and +4% for Telematics (162.1 million €). Both segments were offset by a −27% decrease in revenue in the consumer segment (412.5 million €).[264] Conversely, Garmin's total revenue was $3.087 million, growing 2% over the prior year, with outdoor, fitness, marine and aviation segments collectively growing 9% over the prior year and contributing 76% of total revenue.[265]

---

[263]"2018 State of the Satellite Industry Report." 13th June 2018. Satellite Industry Association and Bryce Space Technology. 28th August.

[264]"TomTom Annual Report 2017." 31 December 2017. TomTom 20th July 2018 p.34: <http://files.shareholder.com/downloads/TOMTOM/0x0x971824/D69A3581-1EBD-4174-A248-0791015F088C/Annual_Report_2017.pdf>.

[265]Garmin Reports Solid Fiscal 2017 Revenue and Operating Income Growth" 21 February 2018. Garmin 20th July 2018 <https://www8.garmin.com/aboutGarmin/invRelations/releases/2017_Q4_Press_Release.pdf>.

Table 1.1 TomTom and Garmin variables 2017

Companies	Total revenues	2017	2016
TomTom		**903 million € ($1.084 billion)**	987.239 million € ($1.040 billion)
Garmin		**$3.087 billion**	$3.019 billion
	Geographical sales		
TomTom	Europe	704.34 million € ($838.164 million)	773.235 million € ($814.649 million)
	North America	144.48 million € ($171.931 million)	167.361 million € ($176.325 million)
	Rest of the world	54.18 million € ($64.474 million)	46.733 million € ($49.236 million)
Garmin	Europe/Middle East/ Africa	$1175.155 billion	$1.111 billion
	Americas	$1475.661 billion	$1.521 billion
	Asia-Pacific region	$436.188 million	$386.549 million

1.3.2.6 Insurance Sector

Roughly 60% of orbital launches were insured in 2017, including half of commercial GEO satellites. In 2017, net premiums reached $712 million (+12.7%). Losses of $451 million are estimated for the same year, while profits are estimated at $261 million.[266] The global insurance market is roughly $5 trillion dollars per year, and space insurance represents about $500 million to $1 billion per year, or approximately $750 million. At the moment, there are around 30 insurers; they can compete on coverage terms and capacity, but most are only competing on price. Typically packaged together in most insurance products, there is a distinction between property (first party) and liability insurance (third party). Property insurance insures against the failure of a satellite during launch or operation and will typically cover the cost of the satellite, while excluding lost future revenue. Conversely, liability insurance of a satellite insures against damage caused to a third party by the operator's satellite. On-orbit liability insurance is required by a small number of countries; the vast majority of all satellite ventures carry property insurance, and it is typically their third largest expenditure after launch and manufacturing.[267] Considering the events of 2017, in April ExactEarth received a nearly $2.7 million insurance pay-out for EV5, an AIS satellite from Fairfax, Virginia-based SpaceQuest that ceased communicating in February. The satellite had been launched in

[266] The Space Report 2018. Colorado Springs: The Space Foundation, 2018: 6.

[267] Stimson Centre and the Secure World Foundation (SWF) "Insurance and Responsible Behaviour in Space Event Report", 3rd April 2018. Stimson, 29th August 2018. <https://www.stimson.org/sites/default/files/fileattachments/2018_stimson_swf_insurance_event_report.pd>.

November 2013 on a Kosmotras Dnepr rocket.[268] The same occurred for ViaSat, which expects to file an insurance claim for an antenna glitch that reduced the performance of its newest satellite by around 15%, according to statements from CEO Mark Dankberg. The antenna malfunction on ViaSat-2, revealed 7 months after the satellite's launch, specifically affects a set of antennas designed to target pockets of high-demand customers. As a consequence, some areas are already experiencing bandwidth shortages.[269] ViaSat does not expect to be compensated in the existing quarter, and as a result the company will focus on "near-term" cost reduction for 2018.[270]

Further, in 2017 the Angolan communication satellite Angosat-1 was delivered into orbit by the Russian rocket Zenit-3F. Despite there having been no problem at launch, on 23 April 2018 the Angolan government declared the satellite a loss. The Angola government accepted from Roscosmos to build a replacement for the satellite (Angosat-2), and the cost of the construction will be partially covered by an insurance payment of $121 million.[271]

1.3.3 Developments in the Space Industry

1.3.3.1 Industrial Developments in Europe

ArianeGroup

As of 1 July 2017, the corporate name of Airbus Safran Launchers was changed to ArianeGroup. As stated by the group, the new name "strengthens the coherence of the image with its Arianespace subsidiary".[272] Under its new name, ArianeGroup continued its development of the Ariane 6 launcher in 2017, which it is building as

[268]Caleb Henry "Spire, 40 cubesats in orbit, competing more directly in space-based ship-tracking market", Space News 31st August 2017. <https://spacenews.com/spire-40-cubesats-in-orbit-competing-more-directly-in-space-based-ship-tracking-market/> 28th August 2018.

[269]Caleb Henry "Viasat preps big insurance claim for ViaSat-2 antenna anomaly", 30th May 2018 SpaceNews. <https://spacenews.com/viasat-preps-big-insurance-claim-for-viasat-2-antenna-anomaly/> 29th August 2018.

[270]Craig Barner "Soaring inflight connectivity growth boosts ViaSat as insurance claim and JV questions linger", 29th May 2018 Satellite Finance. <https://www.satellitefinance.com/insights/soaring-inflight-connectivity-growth-boosts-viasat-insurance-claim-and-jv-questions-linger> 30th August 2018.

[271]Jarowslav Adamowski "Angola eyes new satellite as African space race accelerates", 12th June 2018. SpaceNews 30th August 2018.<https://spacenews.com/angola-eyes-new-satellite-as-african-space-race-acceleratcs>.

[272]"Airbus Safran Launchers to Become ArianeGroup." 17 May 2017. Safran, 5 June 2018 <https://www.safran-group.com/media/airbus-safran-launchers-become-arianegroup-20170517>.

the lead contractor and under the oversight of ESA.[273,274] As reported, the new launcher is envisioned to begin operations in 2020 and cost about half the price of its predecessor, the Ariane 5.[275] In January 2018, it was announced that ArianeGroup and the Arianespace subsidiary had an order for 10 further Ariane 5 launch vehicles, all to be launched from the Guiana Space Centre. The order comes to a value of over 1 billion euros for the European space industry, and it leads to a total of 23 Ariane 5 launchers to be built or under production.[276] These will be launched over 2020–2022, creating a 3-year overlap with Ariane 6 operations.[277] Further, ArianeGroup also announced in December 2017 that it will start production of the first Ariane 6 launcher; the first unit will be an Ariane 62, which is one of the two configurations of the new launch vehicle.[278] And in September of 2017, ESA signed the first launch contract for the Ariane 6, to launch a total of four Galileo satellites on two Ariane 62 launchers near the end of 2020 and in the middle of 2021.[279] Eutelsat had also expressed interest in June of 2017 in launching with the new launch vehicle.[280] And looking further ahead, ArianeGroup agreed on a contract with ESA in July 2017 around the development of the Prometheus reusable engine demonstrator, envisioned for powering European launch vehicles from 2030.[281]

[273]"CNES And ArianeGroup Step Up Cooperation In The Launch Vehicles Sector." 20 Sep. 2017. ArianeGroup, 5 June 2018 <https://www.ariane.group/en/news/cnes-and-arianegroup-step-up-cooperation-in-the-launch-vehicles-sector/>.

[274]Henry, Caleb. "ESA signs up as Ariane 6 inaugural customer with two Galileo missions." 14 Sep. 2017. SpaceNews, 5 June 2018 <http://spacenews.com/esa-signs-up-as-ariane-6-inaugural-customer-with-two-galileo-missions/>.

[275]Henry, Caleb. "Eutelsat awards three launches to Arianespace, hints at Ariane 6 interest." 2 June 2017. SpaceNews, 5 June 0218 <http://spacenews.com/eutelsat-awards-three-launches-to-arianespace-hints-at-ariane-6-interest/>.

[276]"Arianespace And ArianeGroup Kick Off Production For The Final 10 Ariane 5 Launchers Across Europe's Space Industry In Parallel To The Operational Debut Of Ariane 6." 9 Jan. 2018. ArianeGroup, 5 June 2018 <https://www.ariane.group/en/news/arianespace-and-arianegroup-kick-off-production-for-the-final-10-ariane-5-launchers-across-europes-space-industry-in-parallel-to-the-operational-debut-of-ariane-6/>.

[277]Henry, Caleb. "Ariane 5 down to two dozen launches before Ariane 6 takes over." 16 Jan. 2018. SpaceNews, 5 June 2018 <http://spacenews.com/ariane-5-down-to-two-dozen-launches-before-ariane-6-takes-over/>.

[278]"ArianeGroup To Start Production Of The First Ariane 62." 18 Dec. 2017. ArianeGroup, 5 June 2018 <https://www.ariane.group/en/news/ariane6-production-en/>.

[279]*Op. cit.* – "ESA signs up as Ariane 6 inaugural customer with two Galileo missions."

[280]*Op. cit.* – "Eutelsat awards three launches to Arianespace, hints at Ariane 6 interest."

[281]"ArianeGroup Signs A First Contract With ESA To Develop The Future Prometheus Engine." 21 June 2017. ArianeGroup, 5 June 2018 <https://www.ariane.group/en/news/arianegroup-signs-a-first-contract-with-esa-to-develop-the-future-prometheus-engine/>.

Arianespace

In 2017, Arianespace put 20 satellites in orbit over 11 launches for 18 commercial and institutional customers and from this activity generated a turnover of 1.3 billion €. These included six launches of the Ariane 5, two Soyuz launches and three Vega launches. Breaking these satellites down by categories, 12 were destined to geostationary orbit, 4 to medium Earth orbit and 4 to Sun-synchronous orbit. Further, 14 were commercial and 6 were institutional. And 12 were for telecommunications, 4 for Earth observation and 4 for navigation. Arianespace won 19 new launch contracts over the year, which account for 27 satellites. These will include 8 on Ariane 5, 1 on Vega and the first 3 to use Vega C, 5 for Soyuz and the first 2 employing Ariane 6 with ESA (for the European Commission). On the order book, Arianespace currently has 29 customers employing 58 launches, and with a total coming to 5 billion €. 66% will be commercial and 34% will be governmental. In terms of application, 7% are in science and technology, 7% in navigation, 24% in Earth observation and 62% in telecommunications. Looking to 2018, Arianespace is aiming to hit a record of 14 launches.[282] Concerning the Ariane 6 launch vehicle, it is built for approximately five institutional missions annually, and according to the company, these missions are there up to and after 2022, coming in particular from ESA, EUMETSAT, governments and the European Commission.[283]

Eutelsat

In the year ending 30 June 2017, Eutelsat earned a revenue of just over 1.477 billion €, representing a 3.3% decrease from the previous reporting period of 1.529 billion €. Its EBITDA at mid-2017 was 1133.6 billion €, down 2.7% from 1.165 billion € in mid-2016. Eutelsat's operating profit also decreased 7.1% from 662 million € in mid-2016 to 614.8 million € over the following year. Its net income grew by nearly 1% to 351.8 million €.[284] Eutelsat's share of revenue from the European region dropped to 52.3% by mid-2017 from the 56.1% from 1 year prior. Aside from Asia and the "Other" region, whose shares both dropped 0.1% during the reporting period, the share of revenue from all other regions increased, with values at the

[282]"After meeting its commitments in 2017, Arianespace prepares for an intense 2018 and looks to the future with Ariane 6 and Vega C." 9 Jan. 2018. Arianespace, 5 June 2018 <http://www.arianespace.com/press-release/after-meeting-its-commitments-in-2017-arianespace-prepares-for-an-intense-2018-and-looks-to-the-future-with-ariane-6-and-vega-c/>.

[283]*Op. cit.* – "Ariane 5 down to two dozen launches before Ariane 6 takes over."

[284]"Eutelsat Communications Full Year 2016–17 Results." 28 July 2017. Eutelsat, 6 June 2018 <http://news.eutelsat.com/pressreleases/eutelsat-communications-full-year-2016-17-results-2085832>.

Americas (22.3%), the Middle East (15.8%) and Africa (7.2%).[285] Eutelsat's backlog decreased 8% to 5.2 billion € by mid-2017.[286] However, notable contract activity came in the form of a multi-year agreement with NTV-Plus of Russia as well as continuation of capacity with Digitürk and Arqiva.[287] On 15 June 2016 and 2 June 2017, Eutelsat launched EUTELSAT 117WB (entering service in January of 2017) and EUTELSAT 172B, respectively.[288] The latter employs electric propulsion, as will two future units that Eutelsat will launch, including the African Broadcast Satellite and EUTELSAT 7C.[289] During the reporting period, Eutelsat also pursued a new launch contract with Blue Origin for its New Glenn rocket as well as a contract for multiple launches with Arianespace, which will cover the two aforementioned electric propulsion satellites. The company's EUTELSAT 5 West B unit was also selected "by the European Global Navigation Satellite Systems Agency (GSA) for the next-generation EGNOS payload" worth approximately 100 million euros for a period of 15 years.[290]

SES

In the year ending 31 December 2017, SES, with headquarters in Luxembourg, earned 2.035 billion € in revenue, a drop of 1.6% from the previous year's 2.069 billion €. EBITDA decreased 8.8% from 1.452 billion € in 2016 to 1.324 billion € in 2017, and the EBITDA margin likewise dropped from 70.2 to 65.1% in the year. SES' operating profit stood at 610.6 million €, down 54% from the 1.315 billion € in 2016. This is the value of the operating profit "excluding the reported gain on deemed disposal of equity interest of EUR 495.2 million which was recognised directly after the consolidation of O3b (August 2016) and consequently not repeated in 2017". The company's backlog of "fully protected" contracts reached 7.5 billion € down from the 8.1 billion € of 2016.[291] As announced in its Annual Report 2017,

[285]"2016–17 Reference Document – eutelsat communications." Eutelsat, 6 June 2018 (retrieved): p. 162. <https://www.eutelsat.com/files/contributed/investors/pdf/Eutelsat_Communications_Reference_Document_2016-17.pdf>.

[286]*Op. cit.* – "Eutelsat Communications Full Year 2016–17 Results."

[287]*Op. cit.* – "2016–17 Reference Document – eutelsat communications." at p. 02.

[288]*Op. cit.* – "2016–17 Reference Document – eutelsat communications." at p. 138.

[289]*Op. cit.* – "2016–17 Reference Document – eutelsat communications." at p. 02.

[290]*Op. cit.* – "2016–17 Reference Document – eutelsat communications." at p. 06.

[291]"SES Annual Report 2017 – New Frontiers." 5 April 2018. SES, 7 June 2018 <https://www.ses.com/sites/default/files/2018-04/SES_AR_2017_web.pdf>.

SES planned to deploy a total of seven satellites in 2018, all of which have been launched.[292,293,294,295]

Intelsat

Intelsat earned $2.149 billion in revenue for the year ending 31 December 2017, representing a 1.8% decrease from the 2016 result of $2.188 billion. EBITDA stood at $1.629 billion and adjusted EBITDA at $1.665 billion, representing 77% of revenue as well as 0.85% growth from the previous year's $1.651 billion. Operating profit dropped to $914.6 million from the $920.6 million of the previous year, and the company experienced a net loss of $178.7 million, down from the $990.2 million net profit in 2016. The company's contracted backlog—future revenue from current contracts—furthermore dropped to $7.8 billion from the $8.7 billion reported at the end of 2016. In terms of launch activity, the satellites Intelsat 32e (an Epic Next-Generation Ku-band unit), Intelsat 35e and Intelsat 37e were all launched in February, July and September of 2017, respectively.[296,297,298] Satellites to be launched in 2018 include Intelsat 38e in Q2 2018 and Horizons 3e in the latter half of 2018.

HISPASAT

HISPASAT earned 235.1 million € in revenue for the year 2017, which equals a 2.7% increase over last year's 228.9 million €. Space capacity revenues dropped to 218.8 million € in 2017, though, from the approximately 225 million € the year before. As reported by the company, this was a consequence of a satellite market experiencing significant evolution. HISPASAT obtained a net profit of 80.5 million €—a 220% rise

[292]"SES-14: Now launched." 18 Jan. 2018. SES, 7 June 2018 <https://www.ses.com/newsroom/ses-14-now-launched>.

[293]Clark, Stephen. "SpaceX rocket flies on 60th anniversary of first U.S. satellite launch." 31 Jan. 2018. SPACEFLIGHT NOW, 7 June 2018 <https://spaceflightnow.com/2018/01/31/spacex-rocket-flies-on-60th-anniversary-of-first-u-s-satellite-launch/>.

[294]"Seamlessly Scaling our O3b Fleet to Meet Exponential Demand for Connectivity." 26 Feb. 2018. SES, 7 June 2018 <https://www.ses.com/newsroom/seamlessly-scaling-our-o3b-fleet-meet-exponential-demand-connectivity>.

[295]"SES-12 Mission." 3 June 2018. SpaceX, 7 June 2018 <http://www.spacex.com/news/2018/06/03/ses-12-mission>.

[296]"Intelsat Announces Fourth Quarter and Full-Year 2017 Results." 26 Feb. 2018. INTELSAT, 7 June 2018 <http://www.intelsat.com/wp-content/uploads/2018/02/4Q17ERFINAL.pdf>.

[297]"Intelsat Announces Fourth Quarter and Full-Year 2016 Results." 28 Feb. 2017. INTELSAT, 7 June 2018 <http://www.intelsat.com/wp-content/uploads/2017/03/Intelsat4Q2016EarningsRelease2.29.17.pdf>.

[298]"Capture New Growth with the World's Most Advanced Satellite Platform." INTELSAT, 7 June 2018 (retrieved) <http://www.intelsat.com/global-network/satellites/fleet/intelsat-35e/>.

over the last fiscal year, during which the accounts were faced with extraordinary expenditures. Concerning EBITDA, it grew 9.4% to 192.1 million €, attributed to "strict control over managing the operating revenues which, during last fiscal year, dropped by more than 10 million Euros, representing more than 19% of consolidated savings". Its backlog of guaranteed long-term satellite capacity contracts is reported at 4.4 times its revenues, which comes to just above 1 billion €. Geographically, Europe represents 36.4% and the Americas 62.7% of its revenues from leasing its space capacity; just under 1% belongs to "other regions". HISPASAT has furthermore reduced by 2.4 million € its recurring operating costs and launched the satellites Amazonas 5 and Hispasat 36W-1 during the year 2017. 115.1 million € of investment was also directed to innovation projects and the aforementioned satellite programmes, as well as Hispasat 30W-6, whose construction was also finalised in 2017. This investment is "to provide a response to the challenges and opportunities present in the market".[299]

Telenor

Telenor Satellite, based near Oslo, Norway, is entirely owned by Telenor Broadcast Holding AS, itself owned by Telenor Group.[300] Among its broadcast revenues breakdown, Telenor Group's reported 2017 revenue for satellite services came to 892 million kroner (90.6 million € as of the 1 January 2018 exchange rate[301]), down 6.6% from the 955 million kroner of 2016. EBITDA for satellite services dropped as well from 650 million kroner (66 million €) for the year 2016 to 585 million kroner in 2017, representing a negative 10% decrease. Total broadcast revenues represent globally satellite services, Canal Digital DTH (revenues from subscriptions to Nordic DTH as well as households in SMATV networks) and Norkring (terrestrial radio and TV transmission). For the year 2017, total broadcasting revenue stood at 6.071 billion kroner (614 million €), down about 5% from the 6.366 billion kroner of 2016.[302] Telenor attributes this drop to the "shut-down of FM broadcasting in Norway and a settlement in Norkring in 2016".[303]

299"HISPASAT reaches a net profit of 80.5 million Euros." 8 Mar. 2018. HISPASAT, 6 June 2018 <https://www.hispasat.com/en/press-room/press-releases/archivo-2018/311/hispasat-reaches-a-net-profit-of-805-million-euros>.

300"Telenor Satellite – One of Europe's leading satellite operators." Telenor – Satellite, 8 June 2018 (retrieved) <http://www.telenorsat.com/our-business/>.

301"xe.com – Current and Historical Rate Tables." data for the Norwegian Krone from 1 Jan. 2018. xe.com, 8 June 2018 <https://www.xe.com/currencytables/?from=NOK&date=2018-01-01>.

302"telenor group Q4–2017 Interim report January – December 2017." 30 Jan. 2018. Telenor Group, 8 June 2018: p. 9 <https://www.telenor.com/wp-content/uploads/2018/01/Telenor-Group-Q4-2017-report-182c40290db6af1b736fccdd6cd644d6.pdf>.

303"telenor group Annual Report 2017." 20 Mar. 2018. Telenor Group, 8 June 2018: p. 9 <https://www.telenor.com/wp-content/uploads/2018/04/iPDF-Annual-Report-2017-Q-3e7f8e04515b822ddfe375bf65839757.pdf>.

Inmarsat

Inmarsat earned $1.400 billion in revenue for the year ending 31 December 2017, an increase of 5.4% from the $1.329 billion of 2015. The breakdown of this revenue according to its business groups, including percentage change from 2016, is as follows: Maritime (−1.8% to $564.7 M), Government (+11% to $366.7 M), Aviation (+36.7% to $195 M), Enterprise (−8.3% to $132.6 M) and Other (Central Services and Ligado Networks, +3.8% to $141.2 M). The strong growth on the part of the Aviation section is in part explained by revenues coming from In-Flight Connectivity installation. Inmarsat's EBITDA dropped however 8% from $794.8 million in 2016 to $731.5 million in 2017. Furthermore, adjusted EBITDA dropped $43.4 million or by 5.5% to $751.4 million.[304] Operating profit decreased 28.1% from $447.1 million to $321.5 million in 2017, which is partly attributed to an increase in depreciation and amortisation related to two new satellites entering service.[305] This included, in particular, the launch of Inmarsat-5 F4 in May as well as Inmarsat S EAN in June. The former is Inmarsat's fourth satellite in its Global Xpress constellation and the latter an "S-band satellite for the European Aviation Network".[306,307]

Airbus

As initially announced in late 2016, Airbus implemented a significant merger in 2017 between its Group structure and the biggest of its divisions, Commercial Aircraft. The Helicopters as well as Defence and Space segments remain as divisions under Airbus.[308,309] Focusing on the latter, the Airbus Defence and Space (D&S) comprises four programmatic sections: Military Aircraft; Communications, Intelligence & Security; Unmanned Aerial Systems; and Space Systems. The D&S

[304] "Inmarsat plc reports Preliminary Full Year Results 2017." 9 Mar. 2018. Inmarsat, 7 June 2018 <https://investors.inmarsat.com/wp-content/uploads/2018/03/Inmarsat-plc-Preliminary-Full-Year-Results-2017-Final.pdf>.

[305] "Inmarsat PLC Annual Report and Accounts 2017." 9 Mar. 2018. Inmarsat, 7 June 2018 <https://investors.inmarsat.com/wp-content/uploads/2018/04/2017-Annual-Report.pdf>.

[306] "Inmarsat confirms successful launch of the fourth Global Xpress satellite."16 May 2017. Inmarsat, 7 June 2018 <https://www.inmarsat.com/press-release/inmarsat-confirms-successful-launch-fourth-global-xpress-satellite/>.

[307] "Inmarsat confirms successful launch of S-band satellite for the European Aviation Network." 29 June 2017. Inmarsat, 7 June 2018 <https://www.inmarsat.com/press-release/inmarsat-confirms-successful-launch-s-band-satellite-european-aviation-network/>.

[308] "Airbus Annual Report 2017 – Connecting the skies." 11 Apr. 2018. Airbus, 4 June 2018: p. 012 <http://www.airbus.com/content/dam/corporate-topics/financial-and-company-information/AIRBUS_Annual_Report_2017.pdf>.

[309] "Company united under single brand following corporate reorganisation." 3 July 2017. Airbus, 4 June 2018 <http://www.airbus.com/newsroom/press-releases/en/2017/06/airbus-flying-as-one.html>.

division earned 10.804 billion € in revenue in the year ending on 31 December 2017, as compared with 11.854 billion € over 2016 (and a further drop from the 13.080 billion € of 2015).[310] The decrease from 2016 to 2017 represents a value of 1050 million € or −8.9% and was attributed to "perimeter changes for defence activities"; however, close to last year's 31%, the Space Systems section contributed 29% of the division's revenue in 2017.[311,312] Further, at the close of 2017, the D&S division's EBIT and EBIT Adjusted were 212 million € and 872 million €, respectively. The EBIT of 212 million € represents an increase from 93 million € in 2016, with the growth attributed to "stable core business performance" as well as inputs from ArianeGroup and MBDA.[313] The division's EBITDA for 2017, furthermore, was 641 million €, as compared with the 390 million € of 2016.[314,315] The Defence and Space division's order backlog fell again from 42.9 billion € in 2015 and 41.5 billion € in 2016 to 37.4 billion € in 2017.[316] Its order intake also dropped from 15.4 billion € in 2016 to 8.9 billion € in 2017.[317] In March of 2017, OneWeb Satellites, a joint venture between Airbus and OneWeb, began construction of its satellite production centre in Florida[318]; as announced in November 2017, Airbus will build two satellites for Türksat that will be based on the Eurostar E3000 platform and will employ Electric Orbit Raising[319]; Airbus Aerial was founded in May 2017 and targets remote sensing, connectivity and cargo drone services.[320]

[310] *Op. cit.* – "Airbus Annual Report 2017 – Connecting the skies." at p. 27 (Registration Document section).

[311] *Op. cit.* – "Airbus Annual Report 2017 – Connecting the skies." at p. 29 (Financial Statements section).

[312] *Op. cit.* – "Airbus Annual Report 2017 – Connecting the skies." at p. 038.

[313] *Op. cit.* – "Airbus Annual Report 2017 – Connecting the skies." at p. 84 (Registration Document section).

[314] *Op. cit.* – "Airbus Annual Report 2017 – Connecting the skies." at p. 27 (Financial Statements section) Figure derived from EBIT + depreciation and amortisation.

[315] "Airbus Financial Statements 2016." 22 Feb. 2017. Airbus, 4 June 2018: p. 27 (Financial Statements section). Figure derived from EBIT + depreciation and amortisation.

[316] *Op. cit.* – "Airbus Annual Report 2017 – Connecting the skies." at p. 81 (Registration Document section).

[317] *Op. cit.* – "Airbus Annual Report 2017 – Connecting the skies." at p. 27 (Registration Document section).

[318] Henry, Caleb. "OneWeb breaks ground on a Florida factory that will build thousands of satellites." 16 Mar. 2017. SpaceNews, 4 June 2018 <http://spacenews.com/oneweb-breaks-ground-on-a-florida-factory-that-will-build-thousands-of-satellites/>.

[319] "Airbus to build Türksat 5A and 5B satellites." 9 Nov. 2017. Airbus, 4 June 2018 <http://www.airbus.com/newsroom/press-releases/en/2017/11/AirbustobuildTurksat5Aand5Bsatellites.html>.

[320] *Op. cit.* – "Airbus Annual Report 2017 – Connecting the skies." at p. 46 (Registration Document section).

Thales Group

The Thales Group's Aerospace operating segment includes the Avionics and Space Global Business Units. The latter is connected to Thales' partnership—known as the Space Alliance—with Rome-based Leonardo (whose name changed in early 2017), built around their mutual ownership in Thales Alenia Space and Telespazio.[321] The Aerospace segment earned 5.985 billion € in revenue for the year 2017, up 3% from the 5.812 billion € of 2016. EBITDA reached 767.9 million € in 2017, compared to 753.8 million € in 2016. Operating profit grew again in 2017 to 601.5 million € from the 571 million € in 2016, representing 5.3% growth. The Thales Aerospace segment however saw a decline in its order backlog, dropping from 9.914 billion € in 2016 to 8.850 billion €. Overall, of Thales' three operating segments, Aerospace represented 38% of the Group's sales in 2017. Thales Alenia Space furthermore signed a number of contracts or took on opportunities in 2017, which can be broken down by application area: in telecommunications, a contract for the build of Inmarsat GX, a very-high-throughput satellite for Inmarsat; in EO, a contract for the MicroCarb satellite's integration in the UK; in science, a study regarding the design of the payload module for the Solar Wind Magnetospheric Ionospheric Link Explorer (SMILE) mission; in exploration, a development contract with ESA "for Space Rider, a new generation European reusable transport system from low orbit"; and in observation and geolocalisation, a "contract for the Argos NEO instrument for the nano-satellite demonstrator Angels for geolocalisation data collection".[322] Furthermore, as announced in May 2017, Thales Alenia Space will build an automated manufacturing facility for photovoltaic assemblies, used on satellites' solar panels.[323]

OHB SE

OHB SE earned 859.7 million € in total revenue for the year 2017, an increase of 18% from the 728.39 million € earned over 2016. OHB's EBITDA rose from 55.08 million € (7.6% of revenue) in 2016 to 58.8 million € in 2017 (6.8% of revenue). Furthermore, operating profit grew from 42.70 million € in 2016 to 44.2 million € in 2017. OHB's Space Systems business unit generated 661.3 million € in non-consolidated total revenues in 2017, up from 559.5 million € in 2016, and the

[321] "Video – Leonardo: Finmeccanica inspired by 'genius'." 2 Jan. 2017. aeronewstv.com, 8 June 2018 <http://www.aeronewstv.com/en/industry/commercial-aviation/3678-leonardo-finmeccanica-inspired-by-genius.html>.

[322] "THALES 2017 Registration Document Including the Annual Financial Report." 3 Apr. 2018. Thales, 8 June 2018 <https://www.thalesgroup.com/sites/default/files/asset/document/2017_registration_document.pdf>.

[323] "Thales Alenia Space To Build A New Automated Facility Dedicated To Photovoltaic Assemblies For Satellite Solar Panels." 5 Feb. 2017. Thales, 8 June 2018 <https://www.thalesgroup.com/en/worldwide/space/press-release/thales-alenia-space-build-new-automated-facility-dedicated>.

unit's order backlog at the end of 2017 reached 2.198 billion €. This represents a considerable increase over the 1.341 billion € backlog at the end of 2016, attributed to several satellite or satellite-related contracts. Notable business events over the past year include, among others, two contracts from ESA and the European Commission for the assembly of a total of 12 satellites for the Galileo programme, valued together at 482 million €; the launch on 12 December 2017 of four OHB-built Galileo programme satellites, for which it is the principal contractor; a contract to MT Mechatronics (subsidiary of MT Aerospace, itself a member of the OHB Group) to build a 40 m radio telescope for the National Astronomical Research Institute of Thailand; a contract from Boeing to MT Aerospace "for further development and assembly work for" NASA's Space Launch System; a contract to OHB System AG from the DLR "for the assembly, testing and launch of the" German national Heinrich Hertz satellite; and a further contract between OHB System AG and Germany "for the installation of a satellite system for global electro-optical reconnaissance".[324]

RUAG

The RUAG Space division of RUAG Group achieved net sales of CHF 365 million for the year 2017 (312 million € as of the 1 January 2018 exchange rate[325]), representing a 6% increase from the CHF 344 million of 2016. The RUAG Space division's EBITDA dropped to CHF 46 million (39.33 million €) in 2017, representing 12.6% of revenue, from CHF 48 million in 2016. Further, operating profit reached CHF 34 million (29 million €) in 2017 from the CHF 32 million of 2016.[326] The Division also saw several major events in 2017, including notably the inauguration of two new facilities in June in Linköping, Sweden, and in July in Titusville, Florida, USA. The more than 4000 m^2 Swedish facility will focus on satellite separation systems, launcher adapters and dispensers[327]; the Florida facility meanwhile will produce satellite structures for OneWeb.[328] And in December 2017,

[324] "European Access to Space – OHB Annual Report 2017." 20 Mar. 2018. OHB, 19 June 2018 <https://www.ohb-system.de/tl_files/system/images/mediathek/downloads/pdf/OHB_GB17_e_s.pdf>.

[325] "xe.com – Current and Historical Rate Tables." data for the Swiss franc from 1 Jan. 2018. xe.com, 19 June 2018 <https://www.xe.com/currencytables/?from=CHF&date=2018-01-01>.

[326] "RUAG Annual Report 2017." 23 Mar. 2018. RUAG, 19 June 2018: pp. 11 & 46 <https://annualreport.ruag.com/sites/ar17/files/media_document/2018-03/RUAG_Annualreport_2017_full_EN.pdf>.

[327] "RUAG Space Sweden opens new facility in Linköping and doubles its onsite capacity." 27 June 2017. RUAG, 19 June 2018 <https://www.ruag.com/en/news/ruag-space-sweden-opens-new-facility-linkoping-and-doubles-its-onsite-capacity>.

[328] "RUAG Space kicks-off production at a new Florida manufacturing facility, celebrating its latest U.S. facility opening." 12 July 2017. RUAG, 19 June 2018 <https://www.ruag.com/en/news/ruag-space-kicks-production-new-florida-manufacturing-facility-celebrating-its-latest-us-facility>.

the prime contractor for the satellites of the Galileo programme—OHB System AG—contracted RUAG Space to provide the Control and Data Units for a further 12 Galileo satellites. These should be delivered over November 2018 to October 2019. RUAG has significant heritage in the Galileo programme and also provided Control and Data Units for the four Galileo satellites launched 12 December 2017.[329]

1.3.3.2 Industrial Developments in the USA

Boeing

Boeing's Defense, Space & Security division earned $21.057 billion for the year ending 31 December 2017, a 7% decrease from the $22.563 billion of 2016. In particular, approximately 79% of the 2017 revenues came from the US Department of Defense, which is a main customer of the division. Its operating earnings for 2017 increased to $2.223 billion (10% of revenue) from $1.966 billion (8.7% of revenue) in 2016. Further, the Defence, Space & Security division's order backlog rose to $49.577 billion at the end of 2017, up from $44.825 billion at the end of 2016.[330] During the year 2017, several new opportunities were secured: under an agreement with the US Air Force announced in February, Boeing will maintain its mission support of the GPS 2A and 2F units on orbit for 5 years[331]; with NASA, a contract announced in February providing up to five further Soyuz crew seats[332]; a public private partnership award announced in May, whereby Boeing will "develop an experimental reusable first stage", its so-called Phantom Express, for the US Defense Advanced Research Projects Agency—the overall programme goal being to reduce costs for launching medium-sized payloads[333]; and a contract with SES to build seven O3b mPower satellites, destined for medium Earth orbit.[334]

---

[329]"Contract renewed: RUAG Space will continue to build the brain of each Galileo satellite." 11 Dec. 2017. RUAG, 19 June 2018 <https://www.ruag.com/en/news/contract-renewed-ruag-space-will-continue-build-brain-each-galileo-satellite>.

[330]"The Boeing Company 2017 Annual Report." 12 Feb. 2018. Boeing, 21 June 2018 <http://s2.q4cdn.com/661678649/files/doc_financials/annual/2017/2017-Annual-Report.pdf>.

[331]Swarts, Phillip. "Boeing, Air Force, pen agreement to oversee GPS 2 for next five years." 2 Feb. 2017. SpaceNews, 21 June 2018 <http://spacenews.com/boeing-air-force-pen-agreement-to-oversee-gps-2-for-next-five-years/>.

[332]Foust, Jeff. "NASA signs agreement with Boeing for Soyuz seats." 28 Feb. 2017. SpaceNews, 21 June 2018 <http://spacenews.com/nasa-signs-agreement-with-boeing-for-soyuz-seats/>.

[333]Foust, Jeff. "DARPA selects Boeing for spaceplane project." 24 May 2017. SpaceNews, 21 June 2018 <http://spacenews.com/darpa-selects-boeing-for-spaceplane-project/>.

[334]"Boeing to Design and Build Seven Medium Earth Orbit Satellites for SES." 11 Sep. 2017. Boeing, 21 June 2018 <http://boeing.mediaroom.com/news-releases-statements?item=130014>.

Lockheed Martin

Lockheed Martin's Space business segment earned $9.473 billion for the year ending 31 December 2017, approximately a 1% increase over the year 2016 earnings of $9.409 billion.[335] The segment's 2017 revenue equals to 19% of Lockheed Martin's total consolidated net sales, and sales for satellite products and services in particular came to 11% of the company's total consolidated net sales in 2017, down from the 13% of 2016 and 15% of 2015. US institutional customers represented 85% of the segment's sales, international customers 14% and US commercial and others 1%.[336] Moreover, the Space segment's operating earnings dropped to $993 million (10.5% of segment revenue) at the end of 2017 from $1.289 billion (13.7% of revenue) at the end of 2016.[337] This decrease of 23% was largely attributed to a 2016 pre-tax gain stemming from the consolidation of AWE Management Limited as well as "declines in space transportation and government satellite work", among other reasons.[338] Further, its order backlog dropped to $17.3 billion at the end of 2017 from $18.8 billion at the end of 2016, attributed to a lower-order volume for government satellite programmes, although offset to a certain degree by orders connected to the Orion programme.[339]

ULA

Following a report in September 2017 that "technical and funding challenges" would necessitate the continued use of Russian RD-180 engines on ULA Atlas V rockets until the mid-2020s,[340] ULA President and CEO Tory Bruno announced the following month that this information was incorrect, stating the company is confident that development on ULA's next-generation Vulcan launcher will meet a 2022 deadline issued by the US Congress to cease orders for Russian engines.[341] As of May 2018,

---

[335] "Lockheed Martin Reports Fourth Quarter And Full Year 2017 Results." 29 Jan. 2018. Lockheed Martin, 22 June 2018 <https://news.lockheedmartin.com/2018-01-29-Lockheed-Martin-Reports-Fourth-Quarter-and-Full-Year-2017-Results>.

[336] "2017 Annual Report Lockheed Martin Corporation." 6 Feb. 2018. Lockheed Martin, 22 June 2018 <https://www.lockheedmartin.com/content/dam/lockheed-martin/eo/documents/annual-reports/2017-annual-report.pdf>.

[337] *Op. cit.* – "Lockheed Martin Reports Fourth Quarter And Full Year 2017 Results."

[338] Foust, Jeff. "Lockheed Martin space sales growth lags." 30 Jan. 2018. SpaceNews, 22 June 2018 <http://spacenews.com/lockheed-martin-space-sales-growth-lags/>.

[339] *Op. cit.* – "2017 Annual Report Lockheed Martin Corporation."

[340] Pasztor, Andy. "Pentagon Faces Delays in Shift Away From Russian Rocket Engines." 4 Sep. 2017. The Wall Street Journal, 20 June 2018 <https://www.wsj.com/articles/pentagon-faces-delays-in-shift-away-from-russian-rocket-engines-1504526402?mod=e2fb>.

[341] Bruno, Tony. "Building on a successful record in space to meet the challenges ahead." (Op-ed) 10. Oct. 2017. SpaceNews, 20 June 2018 <http://spacenews.com/op-ed-building-on-a-successful-record-in-space-to-meet-the-challenges-ahead/>.

ULA was still evaluating the Blue Origin BE-4 engine as a replacement to the RD-180, along with Aerojet Rocketdyne's AR1 engine, although ULA and Blue Origin are in the process of developing a production contract for the BE-4.[342] This would be in line with the National Defense Authorization Act for Fiscal Year 201 (NDAA-18), which limits the use of NDAA-authorised funds used for the EELV programme to development of an American propulsion system "to replace non-allied space launch engines", among other items.[343]

Orbital ATK

As announced 6 June 2017, Orbital ATK was acquired by Northrop Grumman Corporation and became Northrop Grumman Innovation Systems, representing a fourth and new segment of the latter's business activities. Orbital ATK's prior chief operating officer, Blake Larson, becomes corporate vice president and president of the new segment.[344] Financial results under the Orbital ATK name for the full year 2017 were shared via press release, and its structure of three business segments remained unchanged, including the Flight Systems, Defense Systems and Space Systems groups. Space Systems earned $1.284 billion in revenue over 2017, an increase of 3.7% ($46 million) from the $1.238 billion of 2016. Operating earnings also grew, reaching $142.4 million, representing 6.8% growth ($9.1 million) over the $129.5 million of 2016. These gains were attributed "to higher activity on Satellite Systems and Space Components Divisions contracts". The Flight Systems group earned $1.681 billion in revenue, up 12.4% ($185 million) from the $1.497 billion earned in 2016. This was attributed to greater profit margins from 2017 Aerospace Structures contracts.[345]

Notable events over the reporting period include the launch of an Orbital ATK Minotaur-C rocket delivering 10 Planet satellites on 31 October 2017[346] as well as

---

[342]Foust, Jeff. "Blue Origin expects BE-4 qualification tests to be done by year's end." 3 May 2018. SpaceNews, 20 June 2018 <http://spacenews.com/blue-origin-expects-be-4-qualification-tests-to-be-done-by-years-end-2/>.

[343]Messier, Doug. "An Update on the Evolved Expendable Launch Vehicle Program." 6 June 2018. Parabolic Arc, 20 June 2018 <http://www.parabolicarc.com/2018/06/06/eelv-report/>.

[344]"Northrop Grumman Completes Orbital ATK Acquisition, Blake Larson Elected to Lead New Innovation Systems Sector." 6 June 2018. Northrop Grumman, 25 June 2018 <https://news.northropgrumman.com/news/releases/northrop-grumman-completes-orbital-atk-acquisition-blake-larson-elected-to-lead-new-innovation-systems-sector>.

[345]"Orbital ATK Announces Fourth Quarter and Full Year 2017 Financial Results." 22 Feb. 2018. phx.corporate-ir.net, 25 June 2018 <http://phx.corporate-ir.net/phoenix.zhtml?c=81036&p=irol-newsArticle&ID=2333926>.

[346]Foust, Jeff. "Minotaur launches 10 satellites for Planet." 31 Oct. 2017. SpaceNews, 25 June 2018 <http://spacenews.com/minotaur-launches-10-satellites-for-planet/>.

the announcement on 4 January 2018 that Orbital ATK had won its second in-space servicing contract from Intelsat.[347]

DigitalGlobe

Announced 5 October 2017, DigitalGlobe—a commercial high-resolution Earth observation satellite imagery provider—was acquired for $2.4 billion by MacDonald Dettwiler and Associates, which also rebranded itself on this occasion as Maxar Technologies Ltd.[348] The acquisition ultimately resulted in four companies coming together under the new name, as MDA had previously purchased SSL in 2012 and DigitalGlobe had also purchased Radiant Group in 2016.[349] As part of the 2017 acquisition, DigitalGlobe executive vice president for imagery Timothy Hascall becomes the Maxar chief operations officer.[350] Prior to the acquisition, MDA stated that the DigitalGlobe purchase would both "help offset the revenue peaks and valleys of MDA's satellite manufacturing business"[351] as well as improve its product-service offer to US defence and intelligence agencies, including imagery and analysis.[352]

Blue Origin

Blue Origin made progress on multiple fronts in 2017 and 2018. Concerning the New Shepard reusable launch system, the company conducted its seventh and eighth test flights on 12 December 2017 and 29 April 2018.[353] The crew capsule reached 107 km in altitude, which also represents Blue Origin's target for operations.[354] And

---

[347]Erwin, Sandra & Henry, Caleb. "Orbital ATK lands second Intelsat satellite servicing deal." 4 Jan. 2018. SpaceNews, 25 June 2018 <http://spacenews.com/orbital-atk-lands-second-intelsat-satellite-servicing-deal/>.

[348]Henry, Caleb. "MDA closes DigitalGlobe merger, rebrands as Maxar Technologies." 5 Oct. 2017. SpaceNews, 3 July 2018 <http://spacenews.com/mda-closes-digitalglobe-merger-rebrands-as-maxar-technologies/>.

[349]Henry, Caleb. "Meet Maxar, the space industry's newest tech giant." 6 Nov. 2017. SpaceNews, 3 July 2018 <http://spacenews.com/meet-maxar-the-space-industrys-newest-tech-giant/>.

[350]*Op. cit.* – "MDA closes DigitalGlobe merger, rebrands as Maxar Technologies."

[351]Henry, Caleb. "MDA views DigitalGlobe merger as hedge against cyclical satellite market." 23 May 2017. SpaceNews, 3 July 2018 <http://spacenews.com/mda-views-digitalglobe-merger-as-hedge-against-cyclical-satellite-market/>.

[352]Werner, Debra. "MDA seeks to provide extensive support to U.S. intelligence and defense agencies." 8 June 2017. SpaceNews, 3 July 2018 <http://spacenews.com/mda-seeks-to-provide-extensive-support-to-u-s-intelligence-and-defense-agencies/>.

[353]"Crew Capsule 2.0 First Flight." 12 Dec. 2017. Blue Origin, 22 June 2018 <https://www.blueorigin.com/news/news/crew-capsule-2.0-first-flight>.

[354]"Apogee 351,000 Feet." 29 Apr. 2018. Blue Origin, 22 June 2018 <https://www.blueorigin.com/news/news/apogee-351000-feet>.

although a starting date for operations has not been offered, nor the ticket price, Senior Vice President Rob Meyerson announced on 19 June 2018 that Blue Origin plans to sell tickets starting from 2019.[355] In terms of the New Glenn rocket under development—whose first launch is forecasted for quarter 4 of 2020—the company has secured four customers as of March 2018.[356] These include SKY Perfect JSAT, mu Space, Eutelsat and OneWeb, which has reserved five launches.[357] As announced in April 2018, Blue Origin's BE-4 engine—which is to be used in the New Glenn launcher and is also under consideration for ULA's next-generation Vulcan launch vehicle—is expected to finish qualification testing by the end of 2018.[358]

SpaceX

SpaceX's annual financial reporting is not made widely available, but in its latest fundraising activity, the company has earned, according to Crunchbase, $214 million in a series I funding round in April 2018 and approximately $1.9 billion in total funds since 2002.[359] And as according to different sources, its total current valuation ranges from $25 to $28 billion.[360,361] And as of May 2018, SpaceX had relaunched first stages 12 times—demonstrating increasing market acceptance of the technology[362]—but the business case of first stage reusability is influenced by a large range of factors,[363] and as of February 2018, SpaceX does not foresee significant

[355]Foust, Jeff. "Blue Origin plans to start selling suborbital spaceflight tickets next year." 21 June 2018. SpaceNews, 22 June 2018 <http://spacenews.com/blue-origin-plans-to-start-selling-suborbital-spaceflight-tickets-next-year/>.

[356]Henry, Caleb. "Blue Origin switches engines for New Glenn second stage." 29 Mar. 2018. SpaceNews, 22 June 2018 <http://spacenews.com/blue-origin-switches-engines-for-new-glenn-second-stage/>.

[357]Henry, Caleb. "Blue Origin signs Sky Perfect JSAT as fourth New Glenn launch customer." 12 Mar. 2018. SpaceNews, 22 June 2018 <http://spacenews.com/blue-origin-signs-sky-perfect-jsat-as-fourth-new-glenn-launch-customer/>.

[358]*Op. cit.* – "Blue Origin expects BE-4 qualification tests to be done by year's end."

[359]"Overview – SpaceX." Updated regularly. Crunchbase.com, 26 June 2018 <https://www.crunchbase.com/organization/space-exploration-technologies#section-locked-charts>.

[360]Trefis Team. "What's Driving SpaceX's Sky High Valuation?" 30 May 2018. Forbes, 26 June 2018 <https://www.forbes.com/sites/greatspeculations/2018/05/30/whats-driving-spacexs-sky-high-valuation/#1d7a75cbbde9>.

[361]"Equidate – SpaceX." Updated regularly. Equidate, 26 June 2018 <https://equidateinc.com/company/space-exploration-technologies>.

[362]Foust, Jeff. "SpaceX achievements generate growing interest in reusable launchers." 23 May 2018. SpaceNews, 2018. SpaceNews, 26 June 2018 <http://spacenews.com/spacex-achievements-generate-growing-interest-in-reusable-launchers/>.

[363]de Selding, Peter B. "SpaceX's reusable Falcon 9: What are the real cost savings for customers?" 25 Apr. 2016. SpaceNews, 26 June 2018 <http://spacenews.com/spacexs-reusable-falcon-9-what-are-the-real-cost-savings-for-customers/>.

price reductions for launches with used stages in the near future.[364] In terms of overall launch activity, SpaceX achieved a company record of 18 launches over 2017[365] and on 6 February 2018 successfully conducted a demonstration launch of the Falcon Heavy.[366] Looking ahead, the US Air Force has provided a further $40.7 million to SpaceX "for the development of the Raptor rocket propulsion system prototype for the Evolved Expendable Launch Vehicle program", as announced in October of 2017,[367] as well as a contract worth $290 million covering three GPS 3 missions, as announced in March of 2018.[368] The Air Force has also, thirdly, given the company a contract worth $130 million for a satellite launch employing the Falcon Heavy, as announced in June 2018.[369] And on 29 March 2018, the US FCC approved the SpaceX Starlink constellation of over 4400 broadband non-geostationary satellites.[370]

1.3.3.3 Industrial Developments in Russia

Sea Launch

The sale of Sea Launch's assets came to an official close in April 2018, thereby transferring the company's ocean platform, support ship and other equipment and intellectual property rights to S7 Group, a Russian air holding company.[371] This marks the end of a process which initially began in September 2016, when S7 stated its intention to make the purchase, and comes after the December 2017 US

[364]Werner, Debra. "Don't expect deep discounts on preflown SpaceX boosters." 6 Feb. 2018. SpaceNews, 26 June 2018 <http://spacenews.com/dont-expect-deep-discounts-on-preflown-spacex-boosters/>.

[365]Henry, Caleb. "SpaceX concludes 2017 with fourth Iridium Next launch." 22 Dec. 2017. SpaceNews, 26 June 2018 <http://spacenews.com/spacex-concludes-2017-with-fourth-iridium-next-launch/>.

[366]Foust, Jeff. "Updated | SpaceX successfully launches Falcon Heavy." 6 Feb. 2018. SpaceNews, 26 June 2018 <http://spacenews.com/spacex-successfully-launches-falcon-heavy/>.

[367]Foust, Jeff. "Air Force adds more than $40 million to SpaceX engine contract." 21 Oct. 2017. SpaceNews, 26 June 2018 <http://spacenews.com/air-force-adds-more-than-40-million-to-spacex-engine-contract/>.

[368]Erwin, Sandra. "Air Force awards big launch contracts to SpaceX and ULA." 14 Mar. 2018. SpaceNews, 26 June 2018 <http://spacenews.com/air-force-awards-big-launch-contracts-to-spacex-and-ula/>.

[369]Erwin, Sandra. "SpaceX wins $130 million military launch contract for Falcon Heavy." 21 June 2018. SpaceNews, 26 June 2018 <http://spacenews.com/spacex-wins-130-million-military-launch-contract-for-falcon-heavy/>.

[370]Henry, Caleb. "FCC approves SpaceX constellation, denies waiver for easier deployment deadline." 29 Mar. 2018. SpaceNews, 26 June 2018 <http://spacenews.com/us-regulators-approve-spacex-constellation-but-deny-waiver-for-easier-deployment-deadline/>.

[371]Henry, Caleb. "S7 closes Sea Launch purchase, future rocket TBD." 17 Apr. 2018. SpaceNews, 28 June 2018 <http://spacenews.com/s7-closes-sea-launch-purchase-future-rocket-tbd/>.

Department of State approval of the transfer.[372] New owner S7 also announced in December 2017 that launches would resume as of 2019[373]; however, complicating the matter is that Sea Launch had used the Zenit rocket, whose availability was disrupted following the 2014 Russian annexation of Crimea from Ukraine.[374] The Soyuz-5 is another possibility, considered an "analog" to the Zenit, but is only expected to be ready in 4–5 years.[375]

1.3.3.4 Industrial Developments in Japan

Mitsubishi Electric Co. (Information and Communication Systems)

Mitsubishi Electric Co. (Melco) of Japan develops satellites within its Information and Communication Systems (ICS) business segment; however, as this segment does not separate satellite-related revenue from its telecommunication, information systems and electronic system business, it should only be seen as generating a portion of the total revenue earned by this segment. For the year ending 31 March 2017, Melco earned ¥4.239 trillion (35.6 billion € as of 31 March 2017[376]), down from the ¥4.394 trillion of the year ending 31 March 2016. Its ICS segment in 2017 represented 9.2% of total company sales; it earned ¥447.7 billion in net sales, which was down 20% from the ¥561.119 billion the previous year. The ICS operating income dropped to ¥12.7 billion, dropping a second year in a row from the ¥14.9 billion of 2016. These declines were attributed to drops in sales in communications infrastructure equipment as well as in system integrations.[377]

Mitsubishi Heavy Industries

Chosen by JAXA in 2014, Mitsubishi Heavy Industries (MHI) is the prime contractor of the next-generation H3 launcher, a next-generation rocket which is supposed to halve the price per kilogramme.[378] According to MHI, the H3 will be capable of

---

[372]Henry, Caleb. "Sea Launch CEO Sergey Gugkaev to leave company when S7 purchase closes." 13 Mar. 2018. SpaceNews, 28 June 2018 <http://spacenews.com/sea-launch-ceo-sergey-gugkaev-to-leave-company-when-s7-purchase-closes/>.

[373]"Sea Launch floating spaceport's new owner to resume launches." 12 Dec. 2017. TASS, 28 June 2018 <http://tass.com/science/980483>.

[374]*Op. cit.* – "Sea Launch CEO Sergey Gugkaev to leave company when S7 purchase closes."

[375]Op. cit. – "S7 closes Sea Launch purchase, future rocket TBD."

[376]"xe.com – Current and Historical Rate Tables." data for the Japanese Yen from 31 Mar. 2017. xe.com, 29 June 2018 <https://www.xe.com/currencytables/?from=JPY&date=2017-03-31>.

[377]"Mitsubishi Electric 2017 Annual Report." 29 June 2017. Mitsubishi Electric, 29 June 2018 <http://www.mitsubishielectric.com/en/investors/library/annual_report/pdf/ar2017.pdf>.

[378]Henry, Caleb. "MHI says H3 rocket development on track for 2020." 26 June 2017. SpaceNews, 29 June 2018 <http://spacenews.com/mhi-says-h3-rocket-development-on-track-for-2020/>.

delivering 10–20 satellites per launch and lift a maximum of 7 metric tons to a geostationary transfer orbit. It is to begin operations in 2020, and MHI aims to reach a launch rate of two launches per 3-week period. With current infrastructure limiting MHI to about four launches per year, the company announced in March of 2018 it will convert both the H2A and H2B launch pads at the Tanegashima spaceport to enable the desired launch frequency.[379]

NEC Corporation

NEC Corporation's satellite-related revenue is based within its Public Business segment under its Public Infrastructure business line, along with several other sectors including air traffic control and postal tracking; as such, its satellite-related business should only be seen as generating a portion of the total revenue earned by this segment and the company as a whole. For the year ended 31 March 2017, NEC earned ¥2.655 trillion (22.3 billion € as of 31 March 2017[380]), dropping from the ¥2.825 trillion of the previous reporting year. The Public Business segment earned ¥766.2 billion in revenue, representing 29% of total NEC revenue as well as a 2.6% drop from the previous year's ¥786.5 billion. The segment earned moreover ¥33.2 billion in operating profit for the year, a negative 37.8% drop from the ¥53.4 of the previous year. This drop in operating profit was attributed to weaker profitability of the space business.[381]

Announced in May 2017, Sky Perfect JSAT has agreed to invest in LeoSat, which aims to start launching in 2019 a constellation of 108 high-throughput communications satellites at LEO.[382,383]

Sky Perfect JSAT

Tokyo-based satellite operator Sky Perfect JSAT's activities are based on a Media Business as well as Space & Satellite Business, and as of July 2018, Sky Perfect

---

[379]Henry, Caleb. "Japan to modernize spaceport launch pads to support H3 rocket." 23 Mar. 2018. SpaceNews, 29 June 2018 <http://spacenews.com/japan-to-modernize-spaceport-launch-pads-to-support-h3-rocket/>.

[380]*Op. cit.* – "xe.com – Current and Historical Rate Tables." data for the Japanese Yen from 31 Mar. 2017.

[381]"NEC Annual Report 2017." 22 June 2017. NEC, 29 June 2018 <https://www.nec.com/en/global/ir/pdf/annual/2017/ar2017-e_two.pdf>.

[382]Foust, Jeff. "Japanese satellite operator Sky Perfect JSAT will invest in broadband company LeoSat." 11 May 2017. SpaceNews, 29 June 2018 <http://spacenews.com/japanese-satellite-operator-sky-perfect-jsat-will-invest-in-broadband-company-leosat/>.

[383]"SKY Perfect JSAT and LeoSat Sign Strategic Partnership & Investment Agreement." 11 May 2017. BusinessWire, 29 June 2018 <https://www.businesswire.com/news/home/20170510006791/en/SKY-Perfect-JSAT-LeoSat-Sign-Strategic-Partnership>.

JSAT has a fleet of 18 satellites.[384] For the year ending 31 March 2017, the company earned ¥192.9 billion in overall revenues ($1.7 billion), an increase of 18.4% over the ¥162.9 billion of the previous reporting year. EBITDA reached ¥47.7 billion for the year ended 31.3.2017 ($425.5 million) (24.7% of revenue), up from the ¥46.7 billion (28.6% of revenue) of the previous reporting year, and operating profit grew to ¥24.4 billion ($217.8 million) in 2017, up from ¥24.2 billion in 2016. In terms of segment, the Space & Satellite Business represented 81.8% of operating profit earned and 42.9% of total revenue earned.[385] Sky Perfect JSAT has made a series of investments in recent years, including in LeoSat Enterprises in May 2017—to support its Ka-band constellation[386]—and in February 2018 in KVH Industries, purchasing $4.5 million in stock.[387]

1.3.3.5 Industrial Developments in China

AsiaSat

AsiaSat of Hong Kong earned HK$1.354 billion (144.4 million € as of 1 January 2018[388]) for the year 2017, a 6% increase over the HK$1.272 billion of 2016. HK $1.309 billion or 96.7% of the 2017 revenue represents recurring "income from provision of satellite transponder capacity". AsiaSat's gross profit likewise grew about 6.5% to HK$714 million in 2017 (52.7% of revenue) from HK$645.0 million in 2016 (50.7% of 2016 revenue). Moreover, operating profit also grew to HK$642 million in 2017 from HK$511.3 million in 2016. Revenue gains in 2017 were attributed in financial reporting to the February 2017 full lease of AsiaSat 8's Ku-band payload as well as greater data services demand and the migration from Standard to High Definition broadcasting.[389] In fleet activity, the AsiaSat 9 satellite—built by Space Systems Loral—was launched 29 September 2017 and replaces

384Henry, Caleb. "Sky Perfect JSAT mulls future technology investments." 1 July 2018. SpaceNews, 4 July 2018 <http://spacenews.com/sky-perfect-jsat-mulls-future-technology-investments/>.

385"SKY Perfect JSAT Annual Report 2017 – Active on a Universal Scale." 23 June 2017. SKY Perfect JSAT, 4 July 2018 <https://www.skyperfectjsat.co.jp/load_pdf.php?pTb=t_annual_&pRi=78&pJe=1>.

386Henry, Caleb. "LeoSat confident Jsat investment will spur other investors." 14 May 2017. SpaceNews, 4 July 2018 <http://spacenews.com/leosat-confident-jsat-investment-will-spur-other-investors/>.

387Henry, Caleb. "Sky Perfect Jsat investing $4.5 million in maritime satcom company KVH." 28 Feb. 2018. SpaceNews, 4 July 2018 <http://spacenews.com/sky-perfect-jsat-investing-4-5-million-in-maritime-satcom-company-kvh/>.

388"xe.com – Current and Historical Rate Tables." data for the Hong Kong Dollar from 1 Jan. 2018. xe.com, 2 July 2018 <https://www.xe.com/currencytables/?from=HKD&date=2018-01-01>.

389"AsiaSat Annual Report 2017." 23 Mar. 2018. AsiaSat, 2 July 2018 <https://www.asiasat.com/sites/default/files/2017-annual-report-en.pdf>.

AsiaSat 4. The new unit offers high-throughput Ka-band service and provides new coverage for Indonesia, Mongolia and Myanmar.[390,391]

DFH-4-Derived Satellite Launches

The year 2017 saw the launch of three communications satellites based on the DFH-4 platform, developed by CAST, including:

- Shijian 13 (also known as SJ-13 & ChinaSat 16) launched 12 April 2017[392]
- ChinaSat 9A, launched 18 June 2017[393]
- Alcomsat 1, 10 December 2017[394]

Shijian 13 is intended to provide Internet connectivity to rural areas in China as well as to rail and air travellers; ChinaSat 9A is owned and operated by China sitcom and will provide direct-to-home TV service; Alcomsat 1 is Algeria's first communications satellite and employs over 30 operational transponders.

1.3.3.6 Industrial Developments in India

Antrix

ISRO's Antrix commercial arm earned ₹19.911 billion in revenue for the year ending 31 March 2017 (287 million € as of 31 March 2017[395]), representing a 3.5% increase over the previous reporting period. Gross profit grew from ₹3104 million for the year ending 31.3.2016 to ₹3352 million for the year ending 31.3.2017, and net profit grew from ₹2054 million for the year ending 31.3.2016

[390] Henry, Caleb. "ILS Proton to launch AsiaSat-9 on Sept. 28." 31 July 2017. SpaceNews, 2 July 2018 http://spacenews.com/ils-proton-to-launch-asiasat-9-on-sept-28/

[391] Henry, Caleb. "ILS Proton launches AsiaSat-9, completes 2017 commercial manifest." 29 Sep. 2017. SpaceNews, 2 July 2018 <http://spacenews.com/ils-proton-launches-asiasat-9-completes-2017-commercial-manifest/>.

[392] Clark, Stephen. "China's highest-capacity communications satellite launched into orbit." 12 Apr. 2017. Spaceflight Now, 2 July 2018 <https://spaceflightnow.com/2017/04/12/chinas-highest-capacity-communications-satellite-launched-into-orbit/>.

[393] Clark, Stephen. "Chinese TV broadcasting satellite reaches operational orbit after off-target launch." 13 July 2017. Spaceflight Now, 2 July 2018 <https://spaceflightnow.com/2017/07/13/chinese-tv-broadcasting-satellite-reaches-operational-orbit-after-off-target-launch/>.

[394] Clark, Stephen. "Chinese rocket launches with first Algerian communications satellite." 10 Dec. 2017. Spaceflight Now, 2 July 2018 <https://spaceflightnow.com/2017/12/10/chinese-rocket-launches-with-first-algerian-communications-satellite/>.

[395] "xe.com – Current and Historical Rate Tables." data for the Indian Rupee from 31 Mar. 2017. xe.com, 3 July 2018 <https://www.xe.com/currencytables/?from=INR&date=2017-03-31>.

to ₹2180 million for the year ending 31.3.2017.[396] Moreover, the Indian government's Department of Space reports that over the period 2016–2017, Antrix earned 34 million € in revenue associated with commercial launches for foreign country clients, including Algeria, Canada, Germany, Indonesia, Israel, the Netherlands, Kazakhstan, the UAE and the USA.[397]

1.3.3.7 Other Global Industrial Developments

Maxar Technologies, Formerly MacDonald, Dettwiler and Associates

On 5 October 2017, Canada's MacDonald, Dettwiler and Associates announced its acquisition of DigitalGlobe. The purchase was reportedly worth $2.4 billion, and on the same occasion, the firm rebranded itself as Maxar Technologies Ltd.[398] This move results in four companies coming together under the new Maxar Technologies name, as MDA had previously purchased SSL in 2012 and DigitalGlobe had also purchased Radiant Group in 2016.[399] As part of the 2017 acquisition, DigitalGlobe executive vice president for imagery Timothy Hascall becomes the Maxar chief operations officer, and William McCombe of SSL becomes the chief financial officer of the new Maxar.[400] MDA stated that the purchase would both "help offset the revenue peaks and valleys of MDA's satellite manufacturing business"[401] as well as improve its product-service offer to US defence and intelligence agencies, including imagery and analysis.[402]

The company earned $1.631 billion (in US currency, no longer reported in Canadian dollars) for the year 2017, representing an increase of 4.7% over the $1.558 billion of 2016. EBITDA grew 41.5% to $378.7 million in 2017 from the $267.6 million of 2016; net profit, however, dropped from $105.6 million to $100.4 million over 2016–2017. Moreover, the order backlog rose from $1.8 billion in 2016 to $3.3 billion for the year 2017, incorporating the $1.7 billion DigitalGlobe backlog. And in line with the DigitalGlobe merger, Maxar adapted its reporting to include the following segments: Space Systems, Imagery and Services. Space Systems earned total revenues of $1.27 billion, up from $1.421 billion in 2016. The Imagery segment earned $230.1 million in total revenues in 2017, up from $41.8

396"ANTRIX Annual Report 2016–17." 24 July 2017. ANTRIX, 3 July 2018 <http://www.antrix.co.in/sites/default/files/article-attachments/ANNUALREPORT-2016-17-English.pdf>.

397"Commercial Launching of Satellite." 4 Jan. 2018. Press Information Bureau, Government of India, Department of Space, 3 July 2018 <http://pib.nic.in/newsite/PrintRelease.aspx?relid=175312>.

398*Op. cit.* – "MDA closes DigitalGlobe merger, rebrands as Maxar Technologies."

399*Op. cit.* – "Meet Maxar, the space industry's newest tech giant."

400*Op. cit.* – "MDA closes DigitalGlobe merger, rebrands as Maxar Technologies."

401*Op. cit.* – "MDA views DigitalGlobe merger as hedge against cyclical satellite market."

402*Op. cit.* – "MDA seeks to provide extensive support to U.S. intelligence and defense agencies."

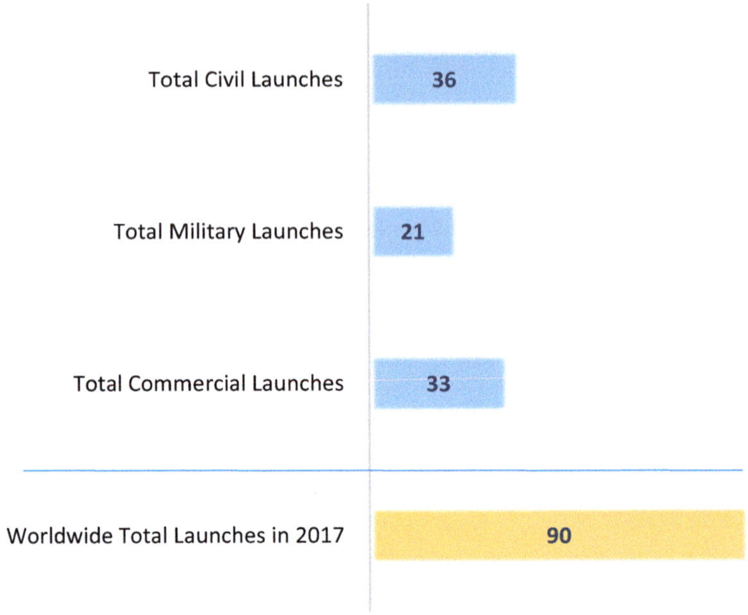

Fig. 1.9 Total worldwide orbital launches by status (Source: ESPI Database)

million in 2016. And the Services segment earned $144.6 million in 2017, likewise up from the $99.7 million of 2016.[403]

1.3.4 Industrial Overview

1.3.4.1 Launch Sector

The Launch sector is an enabler rather than a primary economic activity. With the growth in low-cost launch services, the marginal revenue the launch sector generates is becoming a more important factor to watch.

In 2017, there were 90 launch attempts (Fig. 1.9), and the launch activity increased from 2016, which counted 85 launches. Five of the 90 launches failed, including four government launches: the Meteor-M 2-1 satellite with 18 secondary payloads carried by a Soyuz 2.1b, a PSLV-XL carrying the IRNSS-1H satellite, a Long March 5A carrying the Shijan 18-01 satellite, an Electron carrying a telemetry

[403]"Maxar Technologies – 2017 Annual Report." 22 Feb. 2018. Maxar Technologies, 4 July 2018 <http://s22.q4cdn.com/683266634/files/doc_financials/annual/Maxar-Annual-Report-2017-Final-R1[4].pdf>.

Table 1.2 Worldwide payloads launched in 2017 per country and commercial status

Launch country	Commercial	Governmental civil	Military	Education	Dual	Grand total
USA	72	22	18	42		154
India	113	10		14	1	138
Russia	81	23	6	5		115
China	10	11	9	2	2	34
Europe	11	8	1			20
Japan		6	1		1	8
Others	2					2
Grand total	**289**	**80**	**35**	**63**	**4**	**471**

Source: ESPI database

package called Humanity Star and a SS-520 Upgrade carrying the TRICOM-1 satellite.[404]

The worldwide launch activity (Fig. 1.9) accounted for 33 commercial launches in 2017 (37% of the total) and 57 non-commercial (63% of the total), conducted by the USA (representing 32% of all launches), Russia (21%), China (20%), Europe (12.2%), Japan (7.7%), India (5.5%) and New Zealand (1.1%). The total amount of revenues estimated from the 33 commercial orbital launches are at $3 billion for 2017, compared with $2.5 billion in 2016.[405]

The nations with the highest launch rates remained the same as in 2016 with a shifted ranking.

Standing on the number of commercial launches provided by the FAA's Annual Compendium of Commercial Space Transportation, it should be noted that its definition of a commercial launch includes either one where the primary payload's launch contract was awarded according to a fair and open process or where the launch is privately financed by a private actor without government support.[406]

When looking into the number of payloads launched by specific countries (Table 1.2 and Fig. 1.10), India held the first position with the number of payloads designed to generate revenue, while the situation of 2016 showed the USA in the leading position. Over a total of 471 payloads, 289 were commercial, 80 were governmental, 35 were military, and an increased number of 63 related to education and 4 considered dual use.[407] As was the case in 2016, Europe was in second position in terms of number of commercial launches in 2017 (24% of the total), followed by Russia (9%) and New Zealand (1%). The USA was at the first position with 21 commercial launches (64%), of which 17 were conducted by SpaceX, launching 7 payloads to GEO, 9 to LEO and 1 to SSO. Three out of nine LEO

[404]Federal Aviation Administration. The Annual Compendium of Commercial Space Transportation: 2018. Washington DC: FAA, Jan. 2018: 39–40.

[405]Ibid.

[406]Ibid. at p. 93.

[407]ESPI database.

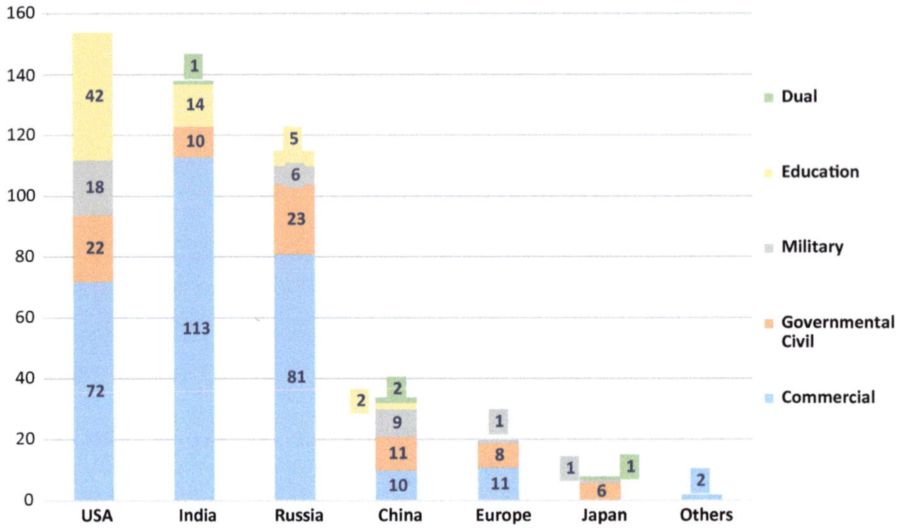

Fig. 1.10 Total payloads launched in 2017 by country and market (Source: ESPI Database)

launches were for the ISS Commercial Resupply Services programme (NASA CRS), and it was the first provider to reuse the first stage of the launcher. India tripled the Russian record of the largest number of satellites launched in a single mission by launching 104 satellites in February, carried by a PSLV-XL.[408] The new launch vehicle Electron was also tested, unsuccessfully, but Rocket Lab announced on 7 August 2018 an agreement with a Dubai-based smallsat company for ten launches. The agreement foresees launches starting at the end of 2019 from Rocket Lab's New Zealand launch site.[409]

In terms of the global share of payloads launched in 2017 (Table 1.2 and Fig. 1.10), the total number of spacecraft doubled from 2016, including the increasing number of small satellites that are changing this market sector. In relation to this increase, in February 2018, ESA announced that five companies will study potential small launch vehicles for the agency's Future Launchers Preparatory Programme (FLPP); a European microlauncher for commercial purposes could meet the growing demand for services related to small satellites.[410]

---

[408]Federal Aviation Administration. The Annual Compendium of Commercial Space Transportation: 2018. Washington DC: FAA, Jan. 2018: 40.

[409]Jeff Foust "Rocket Lab announces order for 10 Electron launches from new Dubai company", Space News 8th August 2018. <https://spacenews.com/rocket-lab-announces-order-for-10-electron-launches-from-new-dubai-company/> 30th August 2018.

[410]ArianeGroup, MT Aerospace, European Launch Vehicle, Deimos and PLD Space. Caleb Henry "ESA awards five smallsat launcher study contract", Space News 8th February 2018. <https://spacenews.com/esa-awards-five-smallsat-launcher-study-contracts/> 30th August 2018.

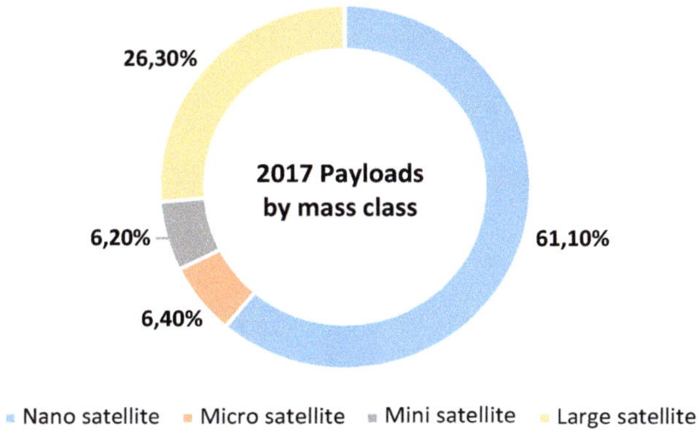

Fig. 1.11 Distribution of the payloads launched in 2017 by mass class (Source: ESPI database)

Table 1.3 Distribution of the payloads launched in 2017 by mass class

Payloads by mass class	Number of spacecraft in 2017	Percentage (%)	Average mass (kg)	Total mass (kg)
Nano satellite	288	61.1	4.3	1.238
Micro satellite	30	6.4	43.3	1.300
Mini satellite	29	6.2	266.6	7.732
Large satellite	124	26.3	3149.1	390.483
	471	**100.0**	**850.9**	**400.754**

Source: ESPI database

In 2017, there were also come changes in the distribution of payload sizes (Fig. 1.11 and Table 1.3). The number of payloads increased from 222 (2016) to 471, and the total mass is approximately valued at the same weight, i.e. 390,040 kg in 2016 and 400,754 in 2017. The average mass of spacecraft launched into orbit is reduced to 850.9 kg in 2017 compared with 1756.9 of 2016. Figure 1.11 shows the large share of Nano satellites launched, representing 61.1% of the total in 2017; moreover, the number of Nano satellites launched in 2017 is higher than the total number of spacecraft launched in 2016.

Once again, Arianespace conducted the most launches to GEO in 2017, with a 29.73% share, followed by SpaceX with an 18.92% share (Fig. 1.12 and Table 1.4). China Aerospace Science Corporation (CASC), the US United Launch Alliance, Russia's International Launch Service and Mitsubishi Heavy Industries shared the same number of GEO satellites placed in orbit (i.e. 4), representing a 10.81% share. Arianespace placed 11 commercial telecommunication satellites into GEO orbit using 9 Ariane 5 ECA launchers and 2 Soyuz-ST-B Fregat-MT launchers, carrying a total of 50,195.00 kg. SpaceX had seven launches with Falcon 9 v1.2, lifting telecommunication satellites (EchoStar 23, SES 10, Inmarsat 5 F4, BulgariaSat 1, Intelsat 35e, EchoStar 105/SES 11, Koreasat 05A) for a total of 36,100.00 kg. China's CASC had four launches, three with Long March 3B E (CZ-3B/G2) and one with a Long March 5 (CZ-5) for a total mass of 20,300.00 kg. Considering the partial

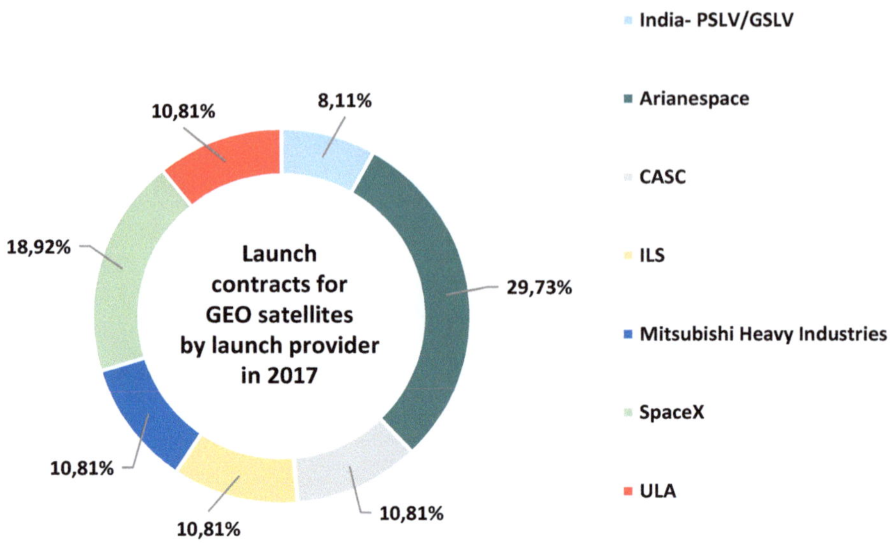

Fig. 1.12 Share of launch contracts for GEO satellites in 2017 by launch service provider. (Source: ESPI database)

Table 1.4 Launch contracts for GEO satellites in 2017 by launch service provider

Service provider	GEO
India-PSLV and GSLV	3
Arianespace	11
CASC	4
ILS	4
Mitsubishi Heavy Industries	4
SpaceX	7
ULA	4
Grand total	**37**

Source: ESPI database

failure of ZhongXing 09A/ChinaSat 9A and the failure of the ShiJian 18 (both commercial telecommunication satellites), the two successful launches of governmental civil technology demonstration satellites were of Tongxin Jishu Shiyan 02 and the ShiJian 13. The ULA had three launches with Atlas-5 (two launches, including one for the military SBIRS-GEO 03 satellite and the governmental telecommunication TDRS 13 satellite with the 401 Atlas version and another for the military telecommunication SDS 402/Quasar 21 satellite with the 421 version) and one for a military telecommunication satellite (WGS 09) with a Delta-4M+ (5,4 upgrade), representing four in total and carrying a total mass of 19,841.00 kg to GEO. Russia's International Launch Services conducted three launches with a Proton-M Briz-M (Ph.3) for three telecommunication satellites (EchoStar 21, Blagovest 11L, Amazonas 05) and one launch with a Proton-M Briz-M (Ph.4) carrying a commercial telecommunication satellite AsiaSat 9. The total amount of transported satellite mass was 23,912.00 kg. Mitsubishi Heavy Industries carried 16,700.00 kg of total mass for three navigation satellites (QZS 02/03/04) and one telecommunication satellite (Kirameki 02). And lastly, India's PSLV-XL launcher lifted the navigation satellite IRNSS 1H, while its GSLV Mk.2 and Mk.3 launched the two telecommunication satellites GSat 09 and GSat 19. With a share of 8.11%, India carried a total mass of 6756.00 kg into GEO orbit.[411]

1.3.4.2 Satellite Manufacturing Sector

Looking at the market share of satellites launched and ordered in a given year provides a good indication of the vitality of domestic space industries, while also providing clues of global trends in the space industry.

In 2017, 471 payloads were launched (including 445 satellites and CubeSats, 26 transfer vehicles including crewed and cargo, or hardware missions to the ISS, ISS infrastructure and technology demonstration). Europe, with its 88 satellites mostly built for communications and 4 for Galileo, accounted for 18.68% of the payloads launched. As Fig. 1.13 shows, commercial satellites are predominant in the 2017 market.

India held the primacy for the largest amount of total satellite mass launched at 11,678.8 kg, although it represented a smaller number of satellites produced, compared with Europe (which represented 85,989 kg in total) and the USA (143,393 kg in total). The same evaluation can be made in comparison with human spaceflight missions, where the total mass carried up was 45,972 kg. The overall mass manufactured by the selected industries stands at 293,813.8 kg (Fig. 1.14).

The main manufacturers in Europe were Airbus, OHB System and Thales Alenia Space, which produced 57 satellites total, 50 of which being commercial. Out of 109 payloads only 4 were dual use (2 from China, 1 from Japan and 1 from India) and 7 related to the military market (Fig. 1.13).

In 2017, the rise of the number of CubeSats is impressive, representing a total of 194 out of 445 satellites launched, all of which being commercial (Fig. 1.15) (for comparison, in 2016 the total amount of CubeSats was 84). Planet was the main actor with 140 commercial CubeSats launched; from a European perspective, the principal manufacturer was GOMSpace with 3 CubeSats out of a total of 5 (the others including 1 from D-Orbit and 1 from German Orbital System).

In 2017, the largest number of satellites had LEO as their destination, followed by GEO (Fig. 1.16). In terms of mass launched to GEO (Fig. 1.17), Boeing (representing 43,330 kg launched), Space System Loral (31,701 kg) and Thales Alenia Space (22,245 kg) launched the largest amount of satellite mass. Despite the 37 GEO satellites launched, 2017 was less fruitful for the leading satellite manufacturers in Europe, with the number of contracts awarded in the region for geostationary communication satellites dropping to only four. However, from a satellite

[411]ESPI database.

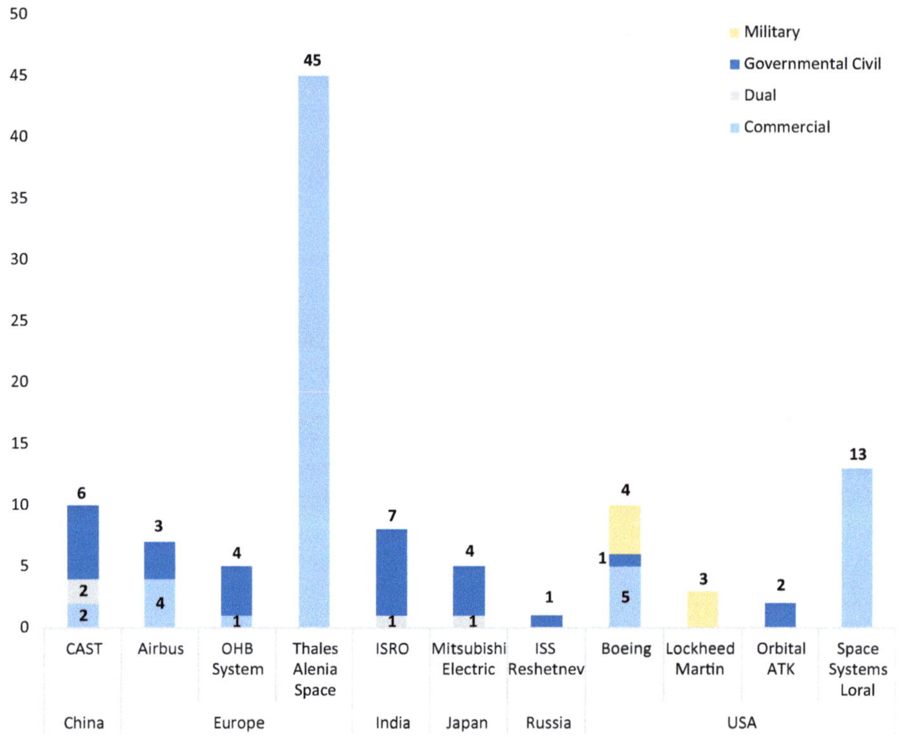

Fig. 1.13 Satellites launched in 2017 by selected manufacturer and commercial status (Source: ESPI database)

operator perspective, they can now improve profit margins through serving the same clients without reintegrating their fleet of GEO satellites. Moreover, the telecommunication market is going through a general diversification, aiming to link constellations in a mix of LEO and MEO orbits.[412] The joint venture between Airbus and the broadband startup OneWeb is a key figure from the market revolution perspective, aiming to launch a small LEO telecommunication constellation composed of 900 satellites (with a maximum weight of 145 kg per satellite[413]) to begin launching at the end of 2018. Arianespace will launch the first satellite generation—with 21 Soyuz launches already contracted—and is willing to carry the second generation with the

[412]Bryan Berger "Are Geo satellites orders still a good measure of industry health?", Space News 15th September 2017. <https://spacenews.com/are-geo-satellite-orders-still-a-good-measure-of-industry-health/> 3rd of September 2018.

[413]Caleb Henry "OneWeb says regulatory concerns main reasons it's foregoing inter-satellite links" 2nd of July 2018. <https://spacenews.com/oneweb-says-regulatory-concerns-main-reason-its-for going-inter-satellite-links/> 3rd of September 2018.

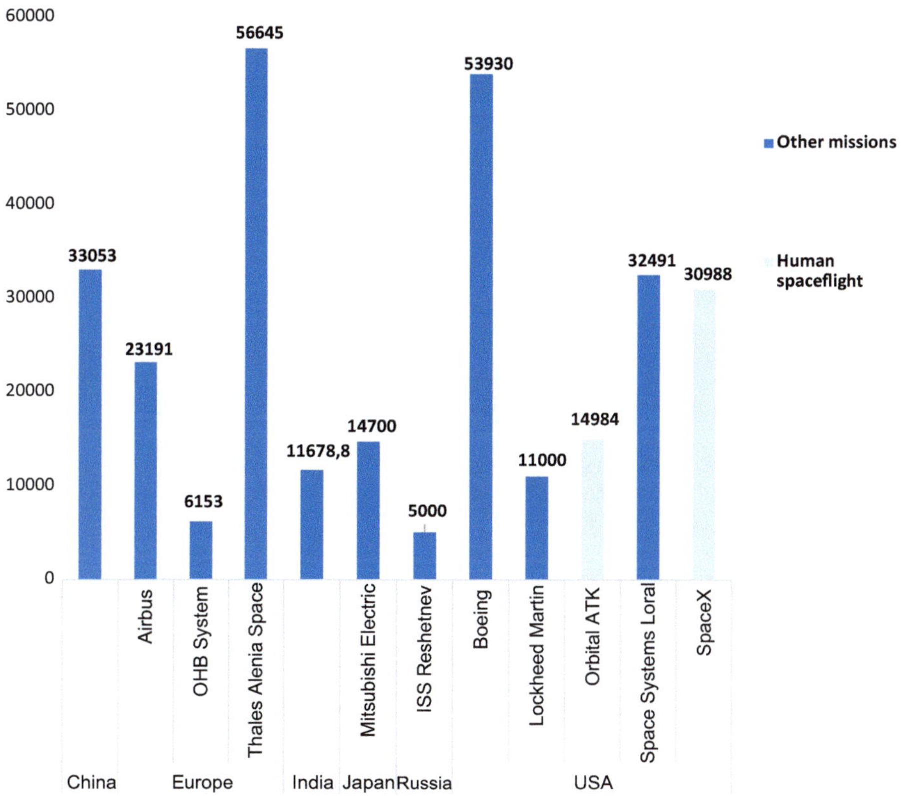

Fig. 1.14 Comparison of human spaceflight mission total mass and other missions' total mass in 2017 (Source: ESPI database)

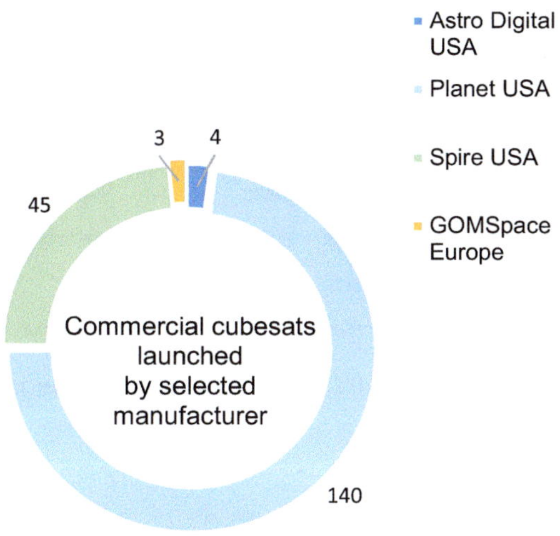

Fig. 1.15 Commercial CubeSats launched in 2017 by selected manufacturer (Source: ESPI database)

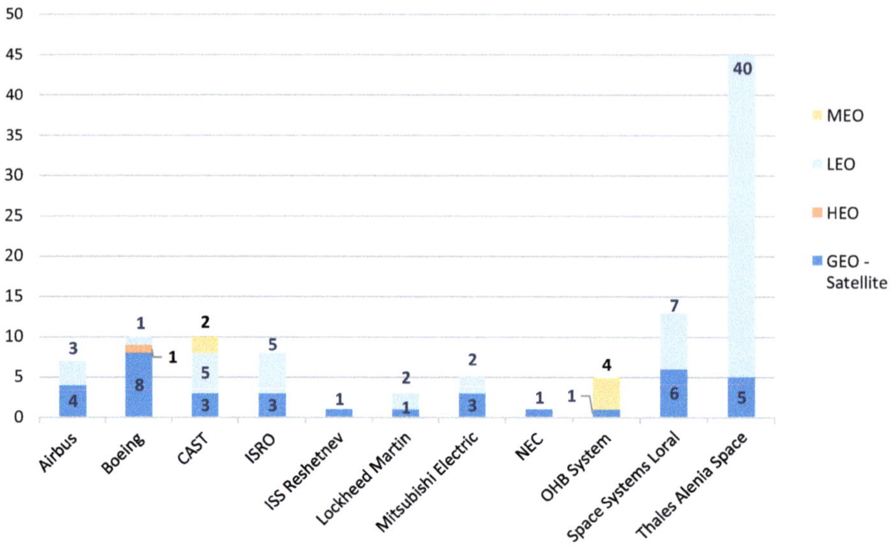

Fig. 1.16 Satellites launched in 2017 by manufacturer and orbit types (Source: ESPI database)

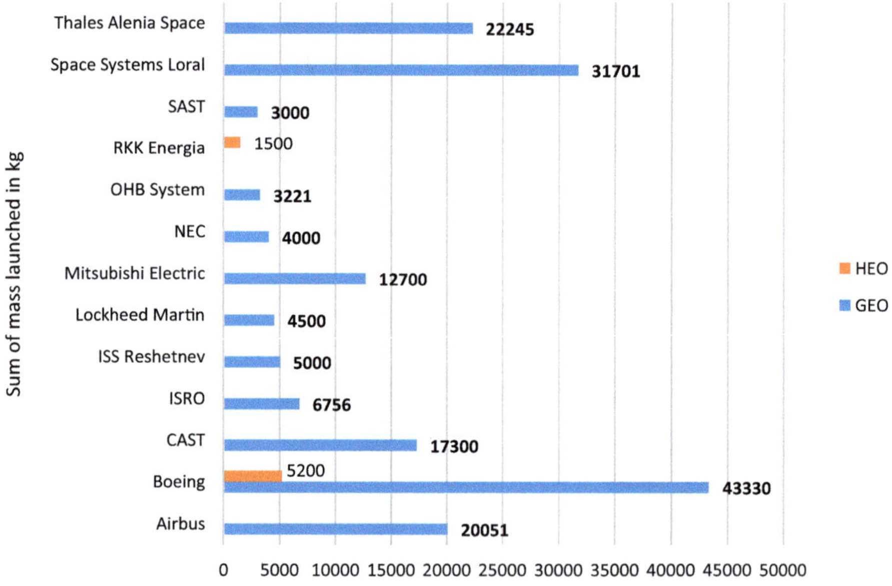

Fig. 1.17 Satellite mass launched in GEO and HEO by manufacturer in 2017 (Source: ESPI database)

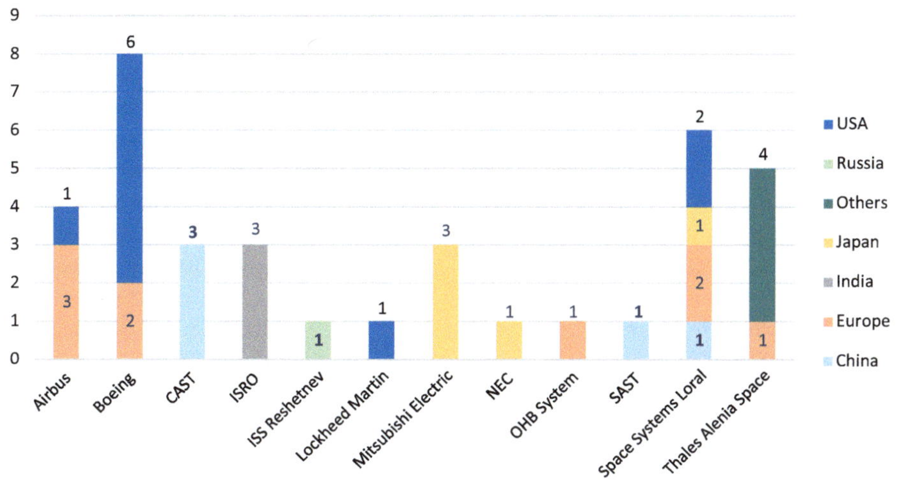

Fig. 1.18 GEO satellite orders in 2017 by manufacturer and customer region (Source: ESPI database)

Ariane 6.[414] In this regard, the analysis concerning exclusively GEO orders won't reflect the prosperity of the satellite sector in the future (Fig. 1.18).

In 2017, ten new geostationary and MEO commercial satellite contracts were signed, despite the lower number in comparison with 2016. In Europe, Airbus won two GEO communication satellite orders for Türksat 5A and 5B, Thales Alenia Space won a contract for an Inmarsat communication satellite Inmarsat-5 F5 (GX 5), and OHB received an order for Heinrich Hertz (H2Sat), to be operated by DLR and launched in 2021 with an Ariane 5 ECA.[415] The US Space System Loral won two contracts for Star One D2 and Jupiter 3, while Boeing was contracted for JCSat 18/Kacific 1. In 2018, the contracts for commercial satellites expanded to 12, with Airbus winning 2 contracts to launch Hotbird 13F and 13G, while Thales Alenia is confirmed with 1 contract for Eutelsat's Konnect VHTS. They are to be launched in 2021.[416]

1.3.4.3 Satellite Operators Sector

The ranking of the five largest satellite fleet operators—being SES (Luxemburg), Intelsat S.A. (Luxemburg), Eutelsat (France), Telesat (Canada) and Echostar

[414]Caleb Henry "OneWeb shifts the first launch to year's end", Space News, 1st of May 2018. <https://spacenews.com/oneweb-shifts-first-launch-to-years-end/>, 3rd September 2018.

[415]Gunter's space page "Recently awarded GEO-Sat contract".<https://space.skyrocket.de/doc_sat/sat-contracts.htm> 4th September 2018.

[416]Ibid.

Satellite Services (USA)—remained relatively the same in 2017 as in 2016.[417] SES earned 2.03 billion € in revenue, with 45 satellites in orbit and saw 3 satellites launched in 2017, while Intelsat earned $2.14 billion (1.87 billion €) in revenue, with 54 satellites in orbit and saw 3 satellites launched in 2017. Eutelsat, whose revenue on 30 June 2017 was estimated to be 1.477 billion €, has 38 satellites in orbit, with 1 satellite launched in 2017, and has plans to launch 7 further satellites over the next 3 years. Eutelsat also commissioned the first LEO satellites for Internet of Things, and Eutelsat LEO for Objects (ELO) is scheduled for launch in 2019; it will provide narrowband connectivity for the objects.[418] Echostar (through its Hughes subsidiary) earned $1.89 billion (1.7 billion €), with 26 satellites in orbit, the most recent of which launched in October 2017. Telesat, with $927 million (835 million €) in revenue, launched one satellite in 2017 and owns 16 satellites in orbit.

1.4 The Security Dimension

1.4.1 Overview, Military Space Spending and Launch

The Space Report 2018 reports that overall government space spending grew 4.8% from the $72.7 billion of 2016 to $76.2 billion over 2017.[419] Of this, it finds military space spending in 2017 to be $27.652 billion, a 3.5% decrease of about $1 billion from the $28.655 billion reported for 2016. These figures show that military space spending in 2017 represented 36.3% of the 2017 overall government space spending, which is a drop from the 39.4% it represented in 2016. However, as is typical with the nature of dual-use technology in space activity, there is a risk that certain military activities have been already included in larger budgets, which can result in double counting.

Moreover, according to the Space Report figures, US military space spending (via the Department of Defense) represented 74.8% ($20.688 billion) of global military space spending in 2017, a slight drop from the 76.8% ($22 billion) in 2016. Non-US global military spending increased from 23.2% ($6.655 billion) in 2016 to 25.2% ($6.964 billion) in 2017.[420] However, as in previous years, a direct comparison of these figures in fixed dollar values does not present a clear picture of the differing countries' relative space defence efforts, since fluctuating exchange rates, variations

[417]"World Teleport Association Publishes Top Operator Rankings For 2017." 9 Jan. 2018. WTA, 12 Oct. 2018 <https://www.worldteleport.org/news/381555/WORLD-TELEPORT-ASSOCIATION-PUBLISHES-TOP-OPERATOR-RANKINGS-FOR-2017.htm>.

[418]"Eutelsat Commissions ELO, Its First Low Earth Orbit Satellite Designed For The Internet Of Things" 8th March 2018.Eutelsat12th October <http://news.eutelsat.com/pressreleases/eutelsat-commissions-elo-its-first-low-earth-orbit-satellite-designed-for-the-internet-of-things-2440770>.

[419]The Space Report 2018. Colorado Springs: The Space Foundation, 2018: p. 8.

[420]The Space Report 2018. Colorado Springs: The Space Foundation, 2018: p. 15.

in purchasing power and different employment costs distort the impact of these investment amounts.

In terms of launch activity, the Space Report finds that military activity across the USA, Russia, China, Europe, Japan and India accounted for a total of 22 launch attempts in 2017. Further, the military spacecraft launched over the year account for a total value of approximately $13 billion, and the launch services in 2017 for military activity were valued at $1.6 billion.[421]

1.4.2 Europe

The European Commission released its Space Strategy in late 2016, comprising four strategic goals, with the overarching aim of building a sustainable space economy. In addition to this, it aims to advance Europe's Space Situational Awareness capability and Governmental Satellite Communications (GOVSATCOM) programme.[422] Moreover, on 30 November 2016, the European Commission released its European Defence Action Plan, comprising three main pillars: (1) launching a European defence fund, (2) fostering investments in defence supply chains and (3) reinforcing the single market for defence. Moreover wherever appropriate, the Commission will promote civil and military synergies within EU policies.[423,424]

With regard to Governmental Satellite Communications (GOVSATCOM), the initiative was defined in 2013 by the European Council as one of the four capability development programmes and is planned to "Provide by 2018 EDA Member States and European CSDP actors with access to a GOVSATCOM capability based on existing, pooled, governmental SATCOM resources". It is highlighted in both the Space Strategy and the Defence Action Plan, and moreover, "GOVSATCOM's objective is to ensure in both the civil and military environment reliable, secure and cost-effective satellite communication services for EU and national public authorities managing security critical missions and operations."[425] Ultimately, it is

[421]The Space Report 2018. Colorado Springs: The Space Foundation, 2018: p. 5.

[422]"Space." Regularly updated. European Commission, Growth, Sectors, Space, 18 July 2018 (retrieved) <http://ec.europa.eu/growth/sectors/space_en>.

[423]"European Defence Action Plan: Towards a European Defence Fund." 30 Nov. 2016. European Commission Press Release Database, 18 July 2018 <http://europa.eu/rapid/press-release_IP-16-4088_en.htm>.

[424]"Communication From The Commission To The European Parliament, The European Council, The Council, The European Economic And Social Committee And The Committee Of The Regions – European Defence Action Plan." 30 Nov. 2016. European Commission, 18 July 2018 <https://eeas.europa.eu/sites/eeas/files/com_2016_950_f1_communication_from_commission_to_inst_en_v5_p1_869631.pdf>.

[425]"Governmental Satellite Communications (GovSatcom)." 15 June 2017. European Defence Agency, 18 July 2018 <https://www.eda.europa.eu/what-we-do/activities/activities-search/governmental-satellite-communications-(govsatcom)>.

envisaged as a "capability that is placed in between the commercial satellite communication market and the highly protected military satellite communication capability".[426] Its inclusion within the European Space Policy underlines its importance for the European Commission, and important steps were taken in 2017: In March, both a GOVSATCOM Common Staff Requirements and a connected Business Case were approved by the European Defence Agency (EDA) Steering Board, and in June the Outline Description for the Governmental Satellite Communications Pooling and Sharing demonstration project was additionally accepted, with the project led by Spain and comprising 14 EDA member states as well as Norway. This demonstration project aims to begin offering services to governmental users by the summer of 2018. It will provide the platform of the project ultimately to be led by the European Commission, which is anticipated to be operational in 2020.[427] Lastly, near the end of 2017, an implementing arrangement was signed between the EDA and ESA toward supporting the European Commission in the GOVSATCOM initiative.[428]

A further development in 2017 included "the elaboration of a Military Satellite Navigation Policy to scope the potential European secure use of positioning, navigation & timing information"—which is the first of its kind and was adopted by the EDA Steering Board on 23 March 2017. A "military user needs document" was requested by member states to follow.[429,430]

In terms of military space spending in 2017, France's military space programme held a 300 million € budget for the year, while Germany's Ministry of Defense provided 32.64 million € to the DLR for defence-related activity.[431]

1.4.3 The USA

The US Department of Defense (DoD) *unclassified* space budget reached an estimated $10.1 billion for the fiscal year 2017, up 5% from the recorded $9.7 billion for

[426]"Future European GOVSATCOM programme takes next step." 12 Sep. 2017. European Defence Agency, 18 July 2018 <https://eda.europa.eu/info-hub/press-centre/latest-news/2017/09/12/future-european-govsatcom-programme-takes-next-step>.

[427]Pultarova, Tereza. "Govsatcom demonstration aims to start service next summer." 8 Nov. 2017. SpaceNews, 18 July 2018 <http://spacenews.com/govsatcom-demonstration-aims-at-start-service-next-summer/>.

[428]"Chief Executive Domecq at ESA." 5 Feb. 2018. European Defence Agency, 18 July 2018 <https://eda.europa.eu/info-hub/press-centre/latest-news/2018/02/05/chief-executive-domecq-at-esa>.

[429]"European Defence Agency Annual Report 2017." 12 Mar. 2018. European Defence Agency, 18 July 2018 <https://eda.europa.eu/docs/default-source/eda-annual-reports/eda-2017-annual-report-final>.

[430]Domecq, Jorge & Wörner, Johann-Dietrich. "Space and Security: Crucial synergies for European citizens." 13 Dec. 2017. *European Defence Matters*, Issue 13, 18 July 2018: p. 20 <https://www.eda.europa.eu/docs/default-source/eda-magazine/edmissue13lowresweb>.

[431]The Space Report 2018. Colorado Springs: The Space Foundation, 2018: p. 12.

2016.[432] Further, the Missile Defense Agency budget for fiscal year 2017 was $7.9 billion,[433] up $421.8 million from the 2017 DoD President's Budget Submission released in February 2016.[434]

As indicated in the US Air Force's 2019 budget request[435] released February 2018 and in media reporting, the Air Force is aiming to end and transition from the Space-Based Infrared System (SBIRS) following the launch of its 6th satellite in 2022, with funding for the originally planned 7th and 8th units being redirected to a new programme known as "Next-Generation Overhead Persistent Infrared" (OPIR).[436] According to the Air Force, it "is implementing rapid procurement authorities and is targeting the first Next-Gen OPIR launch in 2023".[437] Further, the follow-on OPIR programme is seen as a "pacesetter" in a broader initiative toward reforms for acquisitions in the US Air Force.[438] On 23 March 2018, a National Space Strategy was released by the White House, focusing on "four pillars" concerning "more resilient space architectures"; "deterrence and warfighting options"; "foundational capabilities, structures and processes"; and "conducive domestic and international environments".[439]

Further, the National Defense Authorization Act for Fiscal Year 2018 (NDAA-18) was signed into law on 12 December 2017 by US President Donald Trump and

---

[432]"Aeronautics and Space Report of the President Fiscal Year 2016 Activities." NASA, 10 July 2018 (retrieved) <https://history.nasa.gov/presrep2016.pdf>.

[433]Williams, Ian. "FY 2017 Missile Defense Agency Budget Tracker." 29 Mar. 2017 (modified 15 June 2018). CSIS Missile Defense Project – Missile Threat, 10 July 2018 <https://missilethreat.csis.org/fy-2017-missile-defense-agency-budget-tracker/>.

[434]"Department of Defense – Fiscal Year (FY) 2017 President's Budget Submission – Missile Defense Agency." 15 Feb. 2016. US Department of Defense, 10 July 2018: pg. 36 (PDF count, not number on page) <https://comptroller.defense.gov/Portals/45/Documents/defbudget/FY2017/budget_justification/pdfs/03_RDT_and_E/MDA_RDTE_MasterJustificationBook_Missile_Defense_Agency_PB_2017_1.pdf>.

[435]"Department of Defense – Fiscal Year (FY) 2019 Budget Estimates – Air Force." 1 Feb. 2018. US Department of the Air Force, 10 July 2018 <http://www.saffm.hq.af.mil/Portals/84/documents/FY19/Proc/Air%20Force%20Space%20Procurement%20FY19.pdf?ver=2018-02-12-190223-850>.

[436]Erwin, Sandra. "The end of SBIRS: Air Force says it's time to move on." 19 Feb. 2018. SpaceNews, 10 July 2018 <http://spacenews.com/the-end-of-sbirs-air-force-says-its-time-to-move-on/>.

[437]"AF plans to accelerate defendable space with Next-Gen OPIR." 4 May 2018. Secretary of the Air Force Public Affairs, 10 July 2018 <http://www.af.mil/News/Article-Display/Article/1512949/af-plans-to-accelerate-defendable-space-with-next-gen-opir/>.

[438]Erwin, Sandra. "Air Force confident it can build satellites faster. A bigger concern: software." 28 Apr. 2018. SpaceNews, 10 July 2018 <http://spacenews.com/air-force-confident-it-can-build-satellites-faster-a-bigger-concern-software/>.

[439]"President Donald J. Trump is Unveiling an America First National Space Strategy." 23 Mar. 2018. whitehouse.gov – Infrastructure & Technology, 10 July 2018 <https://www.whitehouse.gov/briefings-statements/president-donald-j-trump-unveiling-america-first-national-space-strategy/>.

marks a period of evolution in the US military's space organisation.[440] Having been approved by the US Congress' House of Representatives, the controversial proposal for a so-called Space Corps ultimately did not get through the Senate, but significant reform was nevertheless introduced. Firstly, the position of Principal Department of Defense Space Advisor was removed from the Secretary of the US Air Force.[441] In connection to this action, the 2018 NDAA also directs the Deputy Secretary of Defense to establish a plan to create a separate military department for national security space activities, by the end of 2018.[442] Additionally, Deputy Defense Secretary Patrick Shanahan has disbanded the Defense Space Council, and the position of commander of the Air Force Space Command was established as a Joint Force Space Component Commander under the US Strategic Command. The service term of the commander of the Air Force Space Command was also extended to 6 years.[443] And included in this position's scope of responsibility is the oversight of acquisitions for defence in space as well as the procurement for the Department of Defense of all commercial satellite communications.[444]

1.4.4 Russia

Russian defence spending had followed a general upward curve for the past decade, reaching a peak $60.825 billion in 2016 (4.5% of GDP), but shrinking 30.5% to $42.278 billion in 2017 (3.1% of GDP).[445] And in late December 2017, the 2018 budget was announced to be $46 billion (2.8% of GDP),[446] representing growth again, but not equalling the 2016 values.

The recent pattern of overall growth is itself attributed to a strong general rearmament initiative. The latest State Armament Programme, for 2018–2027, was

[440]"Remarks by President Trump at Signing of H.R. 2810, National Defense Authorization Act for FY2018." 12 Dec. 2017. whitehouse.gov – National Security & Defense, 10 July 2018 <https://www.whitehouse.gov/briefings-statements/remarks-president-trump-signing-h-r-2810-national-defense-authorization-act-fy2018/>.

[441]Erwin, Sandra. "2018 a big year of transition for military space." 28 Dec. 2017. SpaceNews, 10 July 2018 <http://spacenews.com/2018-a-big-year-of-transition-for-military-space/>.

[442]Cowen-Hirsch, Rebecca. "NDAA: A Step Forward to Stronger, Unified Satcom." 22 Mar. 2018. Via Satellite, 10 July 2018 <https://www.satellitetoday.com/government-military/2018/03/22/ndaa-a-step-forward-to-stronger-unified-satcom/>.

[443]Erwin, Sandra. "DoD delivers report to Congress on space reforms: Air Force acquisition system a big problem." 6 Mar. 2018. SpaceNews, 10 July 2018 <http://spacenews.com/dod-delivers-report-to-congress-on-space-reforms-air-force-acquisition-system-a-big-problem/>.

[444]*Op. cit.* – "NDAA: A Step Forward to Stronger, Unified Satcom."

[445]"Russia Military Power." 28 June 2017. US Defense Intelligence Agency, 11 July 2018 <http://www.dia.mil/Portals/27/Documents/News/Military%20Power%20Publications/Russia%20Military%20Power%20Report%202017.pdf>.

[446]"Russia to shell out $46 bln on defense spending in 2018." 22 Dec. 2017. TASS, 11 July 2018 <http://tass.com/defense/982575>.

approved in December 2017 and gives considerable attention to military satellite capabilities, including a new reconnaissance system forecasted to begin operations in 2018 and a possible antisatellite missile system.[447] According to media sources as of October 2017, Russia had 81 military satellites, second only to the USA[448] and 55 of which were launched over 2012–2017. Furthermore, it had put in orbit two of its new Tundra units as of the end of 2017, with the goal being to place six in operation by 2020.[449] Moreover, Russian President Vladimir Putin has stated in May of 2017 the goal of having a minimum of 15 satellites operational by 2020 capable of monitoring ballistic missile launch.

In terms of larger space objectives, the US Director of National Intelligence Mr. Daniel Coats has stated, also in May 2017, "Russia aims to improve intelligence collection, missile warning, and military communications systems to better support situational awareness and tactical weapons targeting. Russian plans to expand its imagery constellation and double or possibly triple the number of satellites by 2025".[450]

1.4.5 *China*

China's military space capability attracts significant commentary and its intentions in space draw much speculation, as the country's space sector has long been intimately connected to the People's Liberation Army (PLA). Currently, two high-level directing documents with space implications for national defence activity are in place, including "China's Military Strategy"—the country's first military strategy white paper, released in May 2015 by the Ministry of National Defence (MOD)[451]—as well as the fourth version of its White Paper on space activities, released in

---

447 "New Research Division Publication: "The Russian State Armament Programme, 2018–2027." 3 May 2018. NATO Defense College, 11 July 2018 <http://www.ndc.nato.int/news/news.php?icode=1167#>.

448 Luzin, Pavel. "Space Power: What is Russia's Military Strategy in Outer Space?" 25 Oct. 2017. intersectionproject.eu, 11 July 2018 <http://intersectionproject.eu/article/security/space-power-what-russias-military-strategy-outer-space>.

449 *Op. cit.* – "New Research Division Publication: "The Russian State Armament Programme, 2018–2027."

450 Blank, Stephen. "Space and the Russian Military: New Trends." 14 June 2017. The Jamestown Foundation, 11 July 2018 <https://jamestown.org/program/space-russian-military-new-trends/>.

451 Minnick, Wendell. "White Paper Outlines China's Ambitions." 27 May 2015. Defense News, 16 July 2018 <http://www.defensenews.com/story/defense/policy-budget/warfare/2015/05/26/china-us-pentagon-taiwan-report-south-east-sea-islands-reefs-s400-su35-missiles-satellite-space-deterrence/27957131/>; *See also* "China's Military Strategy." 26 May 2015. Ministry of National Defense – The People's Republic of China, 16 July 2018 <http://eng.mod.gov.cn/Database/WhitePapers/index.htm>.

December 2016.[452] "China's Military Strategy" of 2015 observes that "the first signs of weaponization of outer space have appeared" and states that "Outer space and cyber space have become new commanding heights in strategic competition among all parties". Further, as per the 2016 White Paper on space activities, China is "to meet the demands of economic, scientific and technological development, national security and social progress". The White Paper's Vision additionally aims to "To build China into a space power in all respects" as well as "to effectively and reliably guarantee national security".[453]

Furthermore, all Chinese government-issued statements are consistent with a public stance against the militarisation of space, as highlighted in the Military Strategy, which states "China has all along advocated the peaceful use of outer space, opposed the weaponization of and arms race in outer space, and taken an active part in international space cooperation".[454]

However, it is in the above context over recent years that an apparently large number of technology development and testing programmes have occurred, comprising systems of a clearly ASAT purpose (such as direct-ascent ASAT missiles) as well as technologies ostensibly conceived for debris removal, inspection, repair and refuelling, but which possess clear potential counterspace applications. China's counterspace weapons development activity is outlined in the CSIS "Space Threat Assessment 2018",[455] while the US Director of National Intelligence's "Worldwide Threat Assessment of the US Intelligence Community", published February 2018, notes that "Chinese destructive ASAT weapons probably will reach initial operational capability in the next few years".[456]

Moreover, western analysts' commentaries identify the following major components in China's approach, including its projection as committed to the peaceful use of space, its development of counterspace capabilities and its aim—quoting the US-China Economic and Security Review Commission in the "Space Threat

---

[452]"China's Space Activities in 2016 – the 4th version of white paper." 30 Jan. 2017. China National Space Administration (powerpoint hosted online by unoosa.org), 16 July 2018 <http://www.unoosa.org/documents/pdf/copuos/stsc/2017/tech-01E.pdf>.

[453]"Full text of white paper on China's space activities in 2016." 28 Dec. 2016. The State Council – english.gov.cn, 16 July 2018 <http://english.gov.cn/archive/white_paper/2016/12/28/content_281475527159496.htm>.

[454]"IV. Building and Development of China's Armed Forces." (Excerpt from China's Military Strategy of May 2015) 26 May 2015. Chinese Ministry of Defense, 16 July 2018 <http://eng.mod.gov.cn/Database/WhitePapers/2015-05/26/content_4586713.htm>.

See also: "China's Military Strategy." 26 May 2015. Chinese Ministry of Defense, 16 July 2018 <http://eng.mod.gov.cn/Database/WhitePapers/2014.htm>.

[455]Harrison, Todd; Johnson, Kaitlyn; & Roberts, Thomas G. "Space Threat Assessment 2018." 12 Apr. 2018. Center for Strategic and International Studies (CSIS), 16 July 2018 <https://aerospace.csis.org/wp-content/uploads/2018/04/Harrison_SpaceThreatAssessment_FULL_WEB.pdf>.

[456]Coats, Daniel R. "Worldwide Threat Assessment of the US Intelligence Community." 13 Feb. 2018. US Office of the Director of National Intelligence, 16 July 2018 <https://www.dni.gov/files/documents/Newsroom/Testimonies/2018-ATA%2D%2D-Unclassified-SSCI.pdf>.

Assessment 2018"—"to deter U.S. strikes against China's space assets, deny space superiority to the United States, and attack U.S. satellites". The latter is likewise informed by a perceived vulnerability of US space systems.[457]

1.4.6 *Japan*

Japan's defence budget has increased gradually from 2012 to 2018; moreover, from 2016 to 2017 and from 2017 to 2018, it has seen 0.8% growth for each year. In 2018, the budget stood at ¥4.94 trillion (37.2 billion € as of mid-December 2017, the same month the figures were published[458]), up from the ¥4.90 trillion of 2017.[459] In 2017, Ministry of Defense general space-related activities had a total budget of ¥42.7 billion, and space-related programmes within Ballistic Missile Defense had a budget of ¥40 billion. Within general space activities, budgetary items included Space Situational Awareness, satellite communications, "Use of commercial imagery satellites and meteorological satellite information", "Research for the enhancement of C4ISR* functions through the use of outer space" and lastly "Dispatch of personnel to the U.S. Air Force Space Operations Course".[460] Among these, the largest recipient was satellite communications, encompassing among other activities the partial procurement of a defence communications satellite (to follow Superbird C2).

As stated, the Ballistic Missile Defense budget for 2017 reached ¥40 billion, reflecting its priority within Japanese strategic military thinking. Furthermore, this priority is additionally seen in the Ministry of Defense Annual White Paper for 2017 which raises directly the threat of space debris and antisatellite systems.[461] The White Paper furthermore states that Japan aims to improve "information collection capability using satellites" as well as "improve the survivability of satellites though such initiatives as space situational awareness".[462] And as announced in January

---

[457] *Op. cit.* – "Space Threat Assessment 2018."

[458] "xe.com – Current and Historical Rate Tables." data for the Japanese Yen from 15 Dec. 2017. xe.com, 12 July 2018 <https://www.xe.com/currencytables/?from=JPY&date=2017-12-15>.

[459] "Defense Programs and Budget of Japan – Overview of FY2018 Budget." Dec. 2017. Japanese Ministry of Defense, 12 July 2018 <http://www.mod.go.jp/e/d_budget/pdf/300329.pdf>.

[460] "Defense Programs and Budget of Japan – Overview of FY2017 Budget Bill." Aug. 2016. Japanese Ministry of Defense, 12 July 2018 <http://www.mod.go.jp/e/d_budget/pdf/290328.pdf>.

[461] "Section 4 – Outer Space and Security." (Excerpt from Defense of Japan 2017 (Annual White Paper)) Japanese Ministry of Defense, 12 July 2018 (retrieved) <http://www.mod.go.jp/e/publ/w_paper/pdf/2017/DOJ2017_1-3-4_web.pdf>.

See also: "Defense of Japan 2017." (Annual White Paper) Japanese Ministry of Defense, 12 July 2018 (retrieved) <http://www.mod.go.jp/e/publ/w_paper/2017.html>.

[462] "Chapter 2 – Building a Dynamic Joint Defense Force." (Excerpt from Defense of Japan 2017 (Annual White Paper)) Japanese Ministry of Defense, 12 July 2018 (retrieved) <http://www.mod.go.jp/e/publ/w_paper/pdf/2017/DOJ2017_2-2-1_web.pdf>.

2018, Japan has requested to purchase four Standard Missile-3 (SM-3) Block IIA missiles from the US government—a system used as defence against ballistic missiles.[463] According to the US Missile Defense Agency, the system is "developed jointly by a Japanese and U.S. government and industry team, is vitally important to both our nations and will ultimately improve our ability to defend against increasing ballistic missile threats around the world".[464]

1.4.7 India

Separate from the civil space activities of ISRO, India is developing a Ballistic Missile Defence (BMD) Programme through its Defence Research Development Organization (DRDO) to counter threats raised by Pakistan's Strategic Missile Group which is developing its own medium-range ballistic missiles. The BMD Programme is a two-tiered system to provide high- and low-altitude cover against incoming ballistic missiles. While the DRDO's Advanced Air Defence (AAD) system is optimised for surface-to-air strikes against aircraft and UAVs at endoatmospheric altitudes between 20 and 40 km, the Prithvi Air Defense (PAD) missile provides exoatmospheric defence at altitudes of 50-80 km.[465] On 28 December 2017, the DRDO ran its third live test of the AAD and successfully intercepted "an incoming Prithvi ballistic missile within 30 km of the earth atmosphere".[466] The second test had occurred 1 March 2017 and was also successful.[467]

It should be noted that the PAD missile can also be seen as a further step for India in developing its own antisatellite capabilities. In this pursuit, the DRDO is looking at the feasibility of developing such an antisatellite vehicle by integrating its Angi-3 missile with its PAD. If it succeeds, the antisatellite missile would have an effective range of about 1400–1500 km and would advance India's missile capabilities to be

---

[463]"Japan – Standard Missile-3 (SM-3) Block IIA Missiles." 9 Jan. 2018. Defense Security Cooperation Agency, 12 July 2018 <http://www.dsca.mil/major-arms-sales/japan-standard-missile-3-sm-3-block-iia-missiles>.

[464]Swarts, Phillip. "U.S., Japan hope next-generation interceptor will be future of missile defense." 7 Feb. 2017. SpaceNews, 12 July 2018 <https://spacenews.com/u-s-japan-hope-next-generation-interceptor-will-be-future-of-missile-defense/>.

[465]Gady, Franz-Stefan. "Did India Hide a Failed Supersonic Missile Test?" 26 May 2016. The Diplomat, 13 July 2018 <http://thediplomat.com/2016/05/did-india-hide-a-failed-supersonic-missile-test/>.

[466]Gady, Franz-Stefan. "India's Advanced Air Defense Interceptor Destroys Incoming Ballistic Missile in Test." 28 Dec. 2017. The Diplomat, 13 July 2018 <https://thediplomat.com/2017/12/indias-advanced-air-defense-interceptor-destroys-incoming-ballistic-missile-in-test/>.

[467]"India Tests Ballistic Missile Defense System." 6 Mar. 2017. Missile Threat – Center for Strategic and International Studies, 13 July 2018 <https://missilethreat.csis.org/india-tests-ballistic-missile-defense-system/>.

on a par with the USA and China.[468] Moreover, India appears to have been pursuing antisatellite technology for several years already—The DRDO is quoted announcing already in 2012 that "we have developed all the building blocks for an anti-satellite (ASAT) capability".[469]

India's pursuit of this technology as well as its BMD Programme is largely influenced by the general global increase in militarisation of the space environment, as well as by the missile programmes of China and its neighbour Pakistan.[470] Further, Lt. Gen. PM Bali, Director General, Perspective Planning, of the Indian Army was quoted stating in February 2017 that (*as paraphrased in the source text*) the developing global and regional conditions require the country to possess space-based military assets as well as develop a dedicated military space programme.[471]

1.4.8 North Korea

Following its last launch in February 2016 of the Kwangmyongsong 4 Earth observation satellite into Sun-synchronous LEO on its Unha-3 rocket,[472,473] North Korea has launched no new satellites in 2017 or the first half of 2018.[474] However, its missile and nuclear testing activities continued into 2017 with a series of missile tests including three intercontinental ballistic missile (ICBM) tests and its 6th nuclear test.[475] The ICBM flight tests were the first for the state led by Kim Jung-un, with the first test announced by North Korea on 4 July 2017; North Korea state television highlighted that the country is now "a full-fledged nuclear power that possesses [a] powerful inter-continental ballistic rocket capable of hitting any part of the

[468]"India Contemplates Anti-Satellite Vehicle Integration with Agni-III Ballistic Missile." 15 Oct. 2013. Missile Threat – Center for Strategic and International Studies, 13 July 2018 <http://missilethreat.com/india-contemplates-anti-satellite-vehicle-integration-with-agni-iii-ballistic-missile/>.

[469]Unnithan, Sandeep. "'India attains the capability to target, destroy space satellites in orbit'." 28 Apr. 2012. India Today, 13 July 2018 <https://www.indiatoday.in/magazine/nation/story/20120507-agni-v-launch-india-takes-on-china-drdo-vijay-saraswat-758208-2012-04-28>.

[470]*Op. cit.* – "India's Advanced Air Defense Interceptor Destroys Incoming Ballistic Missile in Test."

[471]"India to get all its energy from the moon by 2030? This ISRO prof thinks so." 19 Feb. 2017. Business Standard, 13 July 2018 <https://www.business-standard.com/article/current-affairs/india-to-get-all-its-energy-from-the-moon-by-2030-this-isro-prof-thinks-so-117021800715_1.html>.

[472]Wall, Mike. "North Korea Launches Satellite to Space." 8 Feb. 2016. Space.com, 13 July 2018 <https://www.space.com/31860-north-korea-satellite-launch.html>.

[473]Krebs, Gunter Dirk. "Kwangmyŏngsŏng 4 (KMS 4)." 29 Dec. 2017. Gunter's Space Page, 13 July 2018 <http://space.skyrocket.de/doc_sdat/kwangmyongsong-4.htm>.

[474]*Op. cit.* – "Space Threat Assessment 2018."

[475]"North Korea: A Timeline of Missile and Warhead Tests in 2017." Time.com, 26 Feb. 2018 (retrieved) <http://time.com/5040375/north-korea-nuclear-missile-tests-2017/>.

world".[476] The third test flight in November flew 960 km and reached an altitude of 4500 km and is considered capable of reaching any point in the continental USA.[477] Moreover, the latest launch in 2016 as well as North Korea's satellite deployment campaign in general are seen largely by many nations as well as the UN as a cover for missile systems testing.[478,479] Furthermore, deployment of the American-built Terminal High Altitude Area Defense (THAAD) system began in South Korea in early 2017, but sparked significant diplomatic tension with China.[480]

---

[476]Campbell, Charlie. "How North Korea's ICBM Test Could Bring Negotiators Back to the Table." 5 July 2017. Time.com, 26 Feb. 2018 <http://time.com/4844829/north-korea-missile-icbm-negotiation/>.

[477]*Op. cit.* – "North Korea: A Timeline of Missile and Warhead Tests in 2017."

[478]*Op. cit.* – "North Korea Launches Satellite to Space."

[479]Onyanga-Omara, Jane. "North Korea vows to launch more satellites into orbit." 30 Oct. 2017. USA Today, 13 July 2018 <https://eu.usatoday.com/story/news/world/2017/10/30/north-korea-vows-launch-more-satellites-into-orbit/812313001/>.

[480]Westcott, Ben & Suk, Lauren. "China, South Korea end year-long diplomatic feud over missile system." 31 Oct. 2017. CNN, 13 July 2018 <https://edition.cnn.com/2017/10/31/asia/china-south-korea-thaad/index.html>.

Chapter 2
Global Space Activities: Missions and Technological Developments

2.1 Space Transportation

2.1.1 Europe

As Europe's launch sector continues substantial development to grow more competitive in the global market, Arianespace's Vega launcher—manufactured by ELV (a joint venture[1] between the Italian Space Agency and Avio)—has launched a total of 12 times since 2012, all successfully.[2] Over the reporting period in particular (including up to August 2018), Vega launched three times in 2017 and once in 2018.[3] Looking to the future, and as part of overall efforts to increase the Vega's competitiveness on the launch market, ESA and Avio have been developing since 2016 a new Small Spacecraft Mission System adaptor (SMSS), which can deliver cubesats and microsats up to 400 kg to orbit. In 2018, Arianespace secured four customers on the first SMSS, scheduled for 2019, including Spaceflight Inc. (American), ISIS (Dutch), and the Italian Sitael and D-Orbit. The SMSS is intended for one to two flights per year. Moreover, the P120 C engine under development by ArianeGroup (as of July 2017, the corporate name of Airbus Safran Launchers was changed to ArianeGroup[4]) and Avio, intended for both the next-generation Vega C and Ariane 6, is to be produced in up to 35 units per year and "will provide scale meaningful enough to drive prices lower, according to Arianespace and Avio

[1] "Frequently Asked Questions on Vega." 10 May 2017. ESA, 7 Sep. 2018 <https://www.esa.int/Our_Activities/Space_Transportation/Launch_vehicles/Frequently_asked_questions_on_Vega>.

[2] Graham, William. "Arianespace's Vega rocket launches ESA's Aeolus." 22 Aug. 2018. NASA Spaceflight, 7 Sep. 2018 <https://www.nasaspaceflight.com/2018/08/arianespaces-vega-rocket-esa-aeolus-launch/>.

[3] "Gunter's Space Page—Vega (P80 based)." Data from 2017 and 2018. Gunter's Space Page, 7 Sep. 2018 <https://space.skyrocket.de/doc_lau/vega_p80.htm>.

[4] "Airbus Safran Launchers to Become ArianeGroup." 17 May 2017. Safran, 5 June 2018 <https://www.safran-group.com/media/airbus-safran-launchers-become-arianegroup-20170517>.

E. Burger, G. Bordacchini, *Yearbook on Space Policy 2017*, Yearbook on Space Policy, https://doi.org/10.1007/978-3-030-05417-5_2

officials".[5] Lastly, with the Vega C still scheduled for its first flight in 2019, Avio is also developing a further launcher, known as Vega E (Evolution), expected for launch in 2024; ESA has additionally committed to the latter in a contract with Avio valued at 53 million €. According to SpaceNews.com, this reportedly represents only a "first tranche".[6]

Regarding activities in French Guiana, CNES signed two contracts on Ariane 6 development in June 2017: the first concerning Ariane 6 mechanical ground transport systems, signed with APCO Technologies of Switzerland, and the second for the development of cryogenic interfaces between the launcher and the ground systems, signed with a consortium including Latecoere Services (as lead), Air Liquide and Cegelec Projets Espace et Caraïbes.[7] Further, in reaffirming and strengthening their cooperation, CNES and ArianeGroup signed a cooperation agreement on 15 September 2017, whereby the two actors will "pursue close cooperation under the roadmap designed to shape the long-term evolution of the Ariane launcher family".[8] In terms of technical activities at Guiana Space Centre, the first test of the Ariane 6 and Vega C common P120 C engine (under development by ArianeGroup and Avio) took place 16 July 2018 on CNES' BEAP solid booster test stand; "The test was designed to validate the booster's design, fabrication and performance, as well as its overall behaviour and components".[9] Moreover, the engine can be used on Vega C's first stage and for the strap-on boosters for Ariane 6; this common production model is aimed at "aggressively reducing production costs" for the launchers.[10]

The Ariane 5 remained a reliable launch option in 2017, with the Ariane 5 ECA conducting five launches, lifting nine commercial payloads to orbit (including Intelsat 32e, Telkom 03s, Koreasat 07, SDGC 1, Eutelsat 172B, ViaSat 02, Hellas-Sat 3/Inmarsat-S-EAS, BSat 4A and Intelsat 37e) and a single civil governmental telecommunication satellite to LEO (GSat 17). In addition, the Ariane 5 ES launcher saw one launch, placing four Galileo navigation satellites (Galileo FOC-15, FOC-16, FOC-17 and FOC-18) to medium Earth orbit (MEO) for the European

[5]Henry, Caleb. "Vega's long-awaited (small) successes." 8 Aug. 2018. SpaceNews, 7 Sep. 2018 <https://spacenews.com/vegas-long-awaited-small-successes/>.

[6]Henry, Caleb. "ESA pours $107 million into Vega E and a reusable spaceplane." 30 Nov. 2017. SpaceNews, 7 Sep. 2018 <https://spacenews.com/esa-pours-107-million-into-vega-e-and-a-reusable-spaceplane/>.

[7]"CNES signs two Ariane 6 development contracts for French Guiana." 21 June 2017. CNES, 6 Sep. 2018 <https://presse.cnes.fr/sites/default/files/drupal/201706/default/cp101-2017_-_siae_guyane_contrats_a6_va.pdf>.

[8]"CNES and ArianeGroup step up Cooperation in the Launch Vehicles Sector." 21 Sep. 2017. CNES, 6 Sep. 2018 <https://presse.cnes.fr/en/cnes-and-arianegroup-step-cooperation-launch-vehicles-sector>.

[9]"[Ariane 6] New P120c Engine Successfully Tested." 16 July 2018. CNES, 6 Sep. 2018 <https://cnes.fr/en/ariane-6-new-p120c-engine-successfully-tested>.

[10]Henry, Caleb. "Ariane 6 is nearing completion, but Europe's work is far from over." 15 Aug. 2018. SpaceNews, 6 Sep. 2018 <https://spacenews.com/ariane-6-is-nearing-completion-but-europes-work-is-far-from-over/>.

Commission.[11] Moreover, ArianeGroup and Arianespace have an order for ten further Ariane 5 launch vehicles, all to be launched from the Guiana Space Centre. The order comes to a value of over one billion euros for the European space industry, and it leads to a total of 23 Ariane 5 launchers to be built or under production.[12] These will be launched over 2020 to 2022, creating a 3-year overlap with Ariane 6 operations.[13]

Concerning the Ariane 6, ArianeGroup continued development in 2017, which it is building as the lead contractor and under the oversight of ESA.[14,15] As reported, the new launcher is envisioned to begin operations in 2020 and cost about half the price of its predecessor, the Ariane 5.[16] Furthermore, it is built for approximately five institutional missions annually, and according to Arianespace these missions are there up to and after 2022, coming in particular from ESA, EUMETSAT, governments and the European Commission.[17] Moreover, ArianeGroup also announced in December 2017 that it will start production of the first Ariane 6 vehicle; the first unit will be an Ariane 62, which is one of the two configurations of the new launch vehicle.[18] And in September of 2017, ESA signed the first launch contract for the Ariane 6, to launch a total of four Galileo satellites on two Ariane 62 launchers near the end of 2020 and in the middle of 2021.[19] Eutelsat had also expressed interest in June of 2017 in launching with the new launch vehicle.[20]

In other launcher activity, development continues on the Callisto rocket, whose first version will measure just over 13 m and as a project aims overall at studying "different aspects of recovery and reuse". The project is a collaborative effort

---

[11]ESPI database.

[12]"Arianespace and ArianeGroup Kick Off Production for the Final 10 Ariane 5 Launchers Across Europe's Space Industry in Parallel to the Operational Debut of Ariane 6." 9 Jan. 2018. ArianeGroup, 5 June 2018 <https://www.ariane.group/en/news/arianespace-and-arianegroup-kick-off-production-for-the-final-10-ariane-5-launchers-across-europes-space-industry-in-parallel-to-the-operational-debut-of-ariane-6/>.

[13]Henry, Caleb. "Ariane 5 down to two dozen launches before Ariane 6 takes over." 16 Jan. 2018. SpaceNews, 5 June 2018 <http://spacenews.com/ariane-5-down-to-two-dozen-launches-before-ariane-6-takes-over/>.

[14]"CNES and ArianeGroup Step Up Cooperation in the Launch Vehicles Sector." 20 Sep. 2017. ArianeGroup, 5 June 2018 <https://www.ariane.group/en/news/cnes-and-arianegroup-step-up-cooperation-in-the-launch-vehicles-sector/> .

[15]Henry, Caleb. "ESA signs up as Ariane 6 inaugural customer with two Galileo missions." 14 Sep. 2017. SpaceNews, 5 June 2018 <http://spacenews.com/esa-signs-up-as-ariane-6-inaugural-customer-with-two-galileo-missions/>.

[16]Henry, Caleb. "Eutelsat awards three launches to Arianespace, hints at Ariane 6 interest." 2 June 2017. SpaceNews, 5 June 2018 <http://spacenews.com/eutelsat-awards-three-launches-to-arianespace-hints-at-ariane-6-interest/>.

[17]Op. cit.—"Ariane 5 down to two dozen launches before Ariane 6 takes over."

[18]"ArianeGroup to Start Production of the First Ariane 62." 18 Dec. 2017. ArianeGroup, 5 June 2018 <https://www.ariane.group/en/news/ariane6-production-en/>.

[19]Op. cit.—"ESA signs up as Ariane 6 inaugural customer with two Galileo missions."

[20]Op. cit.—"Eutelsat awards three launches to Arianespace, hints at Ariane 6 interest."

between CNES and the DLR, and its first launch is expected in 2020.[21] And looking further ahead, ArianeGroup agreed on a contract with ESA in July 2017 around the development of the Prometheus reusable engine demonstrator, envisioned for powering European launch vehicles from 2030.[22]

2.1.2 *The USA*

The first mission of NASA's heavy-lift Space Launch System (SLS) and Orion, known as Exploration Mission 1 (EM-1), was previously planned to occur by November 2018. However, it was announced in April 2017 that this timeframe would not be met, and in November 2017 it was reported that NASA is now aiming for December 2019.[23,24] However, the then acting NASA administrator Robert Lightfoot stated that "while the review of the possible manufacturing and production schedule risks indicate a launch date of June 2020, the agency is managing to December 2019". Furthermore, after studying the possibility of launching EM-1 as a crewed mission, NASA announced in May 2017 that although "technically feasible", the mission would proceed as a non-crewed mission as originally planned.[25] Concerning the European Service Module for the Orion spacecraft, its expected delivery was moved back from January to August 2017,[26] and it was ultimately completed in September 2018.[27]

Following a report in September 2017 that "technical and funding challenges" would necessitate the continued use of Russian RD-180 engines on ULA Atlas V

---

[21]Henry, Caleb. "France, Germany studying reusability with a subscale flyback booster." 8 Jan. 2018, SpaceNews, 6 Sep. 2018 <https://spacenews.com/france-germany-studying-reusability-with-a-subscale-flyback-booster/>.

[22]"ArianeGroup Signs a First Contract with ESA to Develop the Future Prometheus Engine." 21 June 2017. ArianeGroup, 5 June 2018 <https://www.ariane.group/en/news/arianegroup-signs-a-first-contract-with-esa-to-develop-the-future-prometheus-engine/>

[23]Jeff Foust "NASA plans to delay the first Orion/SLS mission to 2019" 27th April 2017. SpaceNews 6th June 2018 <https://spacenews.com/nasa-plans-to-delay-first-slsorion-mission-to-2019/>.

[24]Jeff Foust "NASA sets December 2019 date for first SLS launch" 8th November 2017. SpaceNews 6th June 2018 <https://spacenews.com/nasa-sets-december-2019-date-for-first-sls-launch/>.

[25]Jeff Foust "NASA decides not to place a crew on first SLS/Orion mission" 12th May 2017. SpaceNews 6th June 2018 <https://spacenews.com/nasa-decides-not-to-place-a-crew-on-first-slsorion-mission/>.

[26]Op. Cit "NASA plans to delay the first Orio/SLS mission to 2019" 27th April 2017. SpaceNews 6th June 2018 <https://spacenews.com/nasa-plans-to-delay-first-slsorion-mission-to-2019/>.

[27]"Orion first service module integration complete" 18th September 2018. ESA 26th September 2018 <https://www.esa.int/Our_Activities/Human_Spaceflight/Orion/Orion_s_first_Service_Module_integration_complete>.

rockets until the mid-2020s,[28] ULA President and CEO Tory Bruno announced the following month that this information was incorrect, stating the company is confident that development on ULA's next-generation Vulcan launcher will meet a 2022 deadline issued by the US Congress to cease orders for Russian engines.[29] As of May 2018, ULA was still evaluating the Blue Origin BE-4 engine as a replacement to the RD-180, along with Aerojet Rocketdyne's AR1 engine, although ULA and Blue Origin are in the process of developing a production contract for the BE-4.[30] This would be in line with the National Defense Authorization Act for Fiscal Year 201 (NDAA-18), which limits the use of NDAA-authorised funds used for the EELV programme to development of an American propulsion system "to replace non-allied space launch engines", among other items.[31]

2.1.3 Russia

As made public in government documents and announcements throughout 2017, Russia is continuing to pursue construction and operations of it spaceports with significant investments. As announced by Russian Prime Minister Dmitry Medvedev, the federal programme for the development of spaceports and their infrastructure over 2017–2025 is to be more than $5.9 billion.[32] For the Vostochny Cosmodrome in particular, which has been under construction since 2010, the Russian Finance Ministry is to provide $39 million, which is from funds classified in the budget for Outer Space Exploration and Use. This funding will be transferred specifically to Roscosmos for the operation of the spaceport's ground infrastructure and maintenance of supporting infrastructure.[33] A further $630 million will be used for the construction at Vostochny of the future Angara rocket's launch infrastructure, forecasted for completion in 2022.[34]

[28]Pasztor, Andy. "Pentagon Faces Delays in Shift Away from Russian Rocket Engines." 4 Sep. 2017. The Wall Street Journal, 20 June 2018 <https://www.wsj.com/articles/pentagon-faces-delays-in-shift-away-from-russian-rocket-engines-1504526402?mod=e2fb>.

[29]Bruno, Tony. "Building on a successful record in space to meet the challenges ahead." (Op-ed) 10. Oct. 2017. SpaceNews, 20 June 2018 <http://spacenews.com/op-ed-building-on-a-successful-record-in-space-to-meet-the-challenges-ahead/>.

[30]Foust, Jeff. "Blue Origin expects BE-4 qualification tests to be done by year's end." 3 May 2018. SpaceNews, 20 June 2018 <http://spacenews.com/blue-origin-expects-be-4-qualification-tests-to-be-done-by-years-end-2/>.

[31]Messier, Doug. "An Update on the Evolved Expendable Launch Vehicle Program." 6 June 2018. Parabolic Arc, 20 June 2018 <http://www.parabolicarc.com/2018/06/06/eelv-report/>.

[32]"More than $5.9 billion to be invested in spaceports development by 2025—Russian PM." 13 Sept. 2017. TASS, 11 July 2018 <http://tass.com/science/965379>.

[33]"Russian government allocates $39 mln for Vostochny spaceport operation." 18 Aug. 2017. TASS, 11 July 2018 <http://tass.com/science/960961>.

[34]"Russia to spend $630 mln on launch pad for Angara carrier rocket at Vostochny spaceport." 11 Aug. 2017. TASS, 11 July 2018 <http://tass.com/science/959991>.

Concerning launch systems, it was reported in July 2017 that "the first flight of a super-heavy-class carrier rocket is planned for 2028", which Russian TASS news agency indicates will be an Energiya-5V super-heavy rocket. Moreover, according to Roscosmos, the rocket's development and infrastructure (at Vostochny) will amount to a cost of 1.5 trillion rubles or $25 billion.[35] Regarding the Soyuz-5, the Energiya Rocket and Space Corporation stated in November 2017 that "The conceptual designing of the Soyuz-5 medium-class carrier rocket will be completed at the end of the current year". The model's debut launch is forecasted for 2021; it is intended to replace Zenit and will carry the Federatsiya spacecraft.[36] Finally, the Angara family is intended to replace Proton, and its third, fourth and fifth launches are set for 2019.[37]

2.1.4 *Japan*

The government of Japan through JAXA awarded a contract to Mitsubishi Heavy Industries in 2014 for the construction of the H3 rocket, Japan's next-generation launcher which will follow its H-2A and H-2B rockets. In this contract, JAXA stipulated that the launch price per kilogramme should be cut by 50% as compared with the H-2A, and it was announced in 2018 that Japan plans to adapt the Tanegashima spaceport's infrastructure to enable an increase in its launch rate with the new launcher.[38] Currently, Tanegashima hosts one pad for the H2-A and one for the H2-B, both of which MHI will adapt for the H3. Moreover, the H3 is scheduled to start launching in 2020 and is intended to boost the overall launch rate to ten per year from the current approximately four per year.[39] According to MHI, the H3 will be capable of delivering 10–20 satellites per launch and lift a maximum of 7 metric tonnes to a geostationary transfer orbit.

Following the successful second launch of the three-stage solid-fuel Epsilon launcher on 20 December 2016—carrying the "Exploration of energization and Radiation in Geospace" (ERG) scientific satellite on board—no new launches

[35] "First launch of Russia's super-heavy carrier rocket scheduled for 2028" 19th July 2017. Russian News Agency 6th June 2018 <http://tass.com/science/956957>.

[36] "Conceptual design of Russia's new Soyuz-5 carrier rocket to be ready by yearend" 10th of November. Russian News Agency, 6th June 2018 <http://tass.com/science/975015>.

[37] Howell, Elizabeth "Angara: Russian Rocket for LEO payloads" <https://www.space.com/40658-angara-rocket.html>.

[38] Henry, Caleb. "MHI says H3 rocket development on track for 2020." 26 June 2017. SpaceNews, 12 July 2018 <http://spacenews.com/mhi-says-h3-rocket-development-on-track-for-2020/>.

[39] Henry, Caleb. "Japan to modernize spaceport launch pads to support H3 rocket." 23 Mar. 2018. SpaceNews, 12 July 2018 <http://spacenews.com/japan-to-modernize-spaceport-launch-pads-to-support-h3-rocket/>.

occurred during the 2017 reporting period.[40] However, the Epsilon launched a third time on 18 January 2018, carrying the ASNARO-2 radar satellite.[41] The Epsilon launcher is also being developed to cut launcher costs, using the same solid-fuel strap-on booster as on the H-2A.[42] The Epsilon is developed by JAXA and is manufactured by IHI; its first launch took place in 2013, and it can deliver 450 kg to SSO and up to 1200 kg to LEO. It is designed for launching small satellites. [43]

2.1.5 China

China conducted the maiden launch of its two-stage Long March 7 on 25 June 2016, carrying a prototype version of its future next-generation crew vehicle (NGCV).[44] Launched from the newly constructed Wenchang Satellite Launch Centre on Hainan Island, the Long March 7 has a launch capacity of 13.5 metric tons to LEO and will eventually be used to launch taikonauts and cargo resupply missions to China's Tiangong space station.[45] During the 2017 reporting period, a second Long March 7 launched on the 20th of April again from Wenchang,[46] carrying the Tianzhou-1 cargo spacecraft which was destined for the Tiangong 2 space station (already launched in September 2016).[47] Furthermore, China's second launch of its Long March 5 took place 2 July 2017, also from Wenchang. The rocket was meant to launch the experimental Shijian 18 communication satellite, but suffered a failure.[48]

---

[40]"Success of Epsilon-2 Launch with ERG aboard." 20 Dec. 2016. JAXA 16 Aug. 2017 <http://global.jaxa.jp/press/2016/12/20161220_epsilon2.html>.

[41]"Epsilon Launch Vehicle" January 2018. JAXA 6th June 2018 <http://global.jaxa.jp/projects/rockets/epsilon/>.

[42]Matsuda, Shogo. "Japan's Epsilon rocket shoved aside?" 16 Jan. 2015. Nikkei Asian Review 21 Jan. 2016 <http://asia.nikkei.com/Tech-Science/Tech/Japan-s-Epsilon-rocket-shoved-aside>.

[43]"The Annual Compendium of Space Transportation: 2017" January 2017. FAA 6th of June 2018 <https://www.faa.gov/about/office_org/headquarters_offices/ast/media/2017_AST_Compendium.pdf>.

[44]Barbosa, Rui C. "China successfully debuts Long March 7—Recovers capsule." 25 June 2016. NASASpaceflight.com 16 Aug. 2017 <https://www.nasaspaceflight.com/2016/06/china-debuts-long-march-7-rocket/>.

[45]Clark, Stephen. "China's new Long March 7 rocket successful on first flight." 25 June 2016. Spaceflight Now 16 Aug. 2017 <https://spaceflightnow.com/2016/06/25/chinas-new-long-march-7-rocket-successful-on-first-flight/>.

[46]Clark Stephen "First supply ship for Chinese space station lifts off on test flight" 20th April 2017. Spaceflight 6th June 2018 <https://spaceflightnow.com/2017/04/20/tianzhou-1-launch/>.

[47]"China appears to be preparing to deorbit its Tiangong-2 space lab "20th June 2018. SpaceNews 27th June 2018 <https://spacenews.com/china-appears-to-be-preparing-to-deorbit-its-tiangong-2-space-lab/>.

[48]Barbarosa, Rui C. "Long March 5 suffers failure with Shijan-18 launch" 2nd July 2017. NASA Spaceflight 6th June 2018 <https://www.nasaspaceflight.com/2017/07/long-march-5-lofts-shijian-18/>.

2.1.6 India

Over 2017, the PSLV (Polar Satellite Launch Vehicle) launched three times, including missions C37, C38 and C39, the third of which failed.[49] Since its debut in 1993, the PSLV has launched a total of 41 times, with just one other unsuccessful mission.[50] Notably, the C37 mission of 15 February 2017 launched a record total of 104 satellites;[51] its primary payload was the Cartosat 2D remote sensing satellite.[52] The PSLV can lift 3250 kg to LEO, 1750 to SSO, and costs an estimated $21–$31 million per launch.[53] Furthermore, in recent years India's low-cost launch capacity, and India's refusal to sign a commercial launch agreement with the US government to ensure fair market-based pricing, is the reason US companies have been restricted from using the PSLV.[54]

The GSLV Mk III (Geosynchronous Satellite Launch Vehicle Mark III) is a heavy-lift, three-stage launch system using a core liquid booster, a cryogenic upper stage and two solid strap-ons. A first experimental launch took place in December 2014, followed by a developmental launch on 5 June 2017 deploying the satellite GSAT-19 into GTO. The GSLV Mk III can launch 8000 kg to LEO and 4000 kg to GTO.[55] The GSLV is manufactured by ISRO, with ISRO and Antrix as joint service providers. It costs an estimated $47 million per launch.[56]

ISRO successfully launched a Reusable Launch Vehicle (RLV) Technology Demonstration mission on 22 May 2016. First conceived by ISRO in 2009, the long delayed first flight test of the scramjet propulsion system took place at the Satish Dhawan Space Centre. Perched atop a HS9 solid booster, the 1750 kg technology demonstrator was launched to autonomously test the RLV's hypersonic aerodynamic properties, avionics, thermal protection and control systems and its mission

[49]"PSLV-C39 mission fails; IRNSS-H1 satellite trapped inside heat shield of the rocket" 1st of September 2017. The Time of India 6th June 2018 <https://timesofindia.indiatimes.com/india/pslv-c39-mission-fails-irnss-h1-satellite-trapped-inside-heat-shield-of-the-the-rocket/articleshow/60311488.cms>.

[50]"List of PSLV Launches". ISRO 6th June 2018 <https://www.isro.gov.in/launchers/list-of-pslv-launches>.

[51]Mathewson, Samantha "India Launches Record-Breaking 104 Satellites on Single Rocket" 15th February 2017. Space.com 6th June 2018 <https://www.space.com/35709-india-rocket-launches-record-104-satellites.html> .

[52]"Cartosat-2" 6th June 2018. Gunter's Space Page 7th June 2018 <https://space.skyrocket.de/doc_sdat/cartosat-2.htm>.

[53]"The Annual Compendium of Space Transportation: 2017" January 2017. FAA 6th of June 2018 <https://www.faa.gov/about/office_org/headquarters_offices/ast/media/2017_AST_Compendium.pdf>.

[54]SpaceNews Editor. "Getting the Cubesat Revolution Out of Low Gear." 24 Nov. 2015. SpaceNews 6 Jan. 2016 <http://spacenews.com/getting-the-cubesat-revolution-out-of-low-gear/>.

[55]"GSLV Mk III" . ISRO 6th June 2018 <https://www.isro.gov.in/launchers/gslv-mk-iii>.

[56]"The Annual Compendium of Space Transportation: 2017" January 2017. FAA 6th of June 2018 <https://www.faa.gov/about/office_org/headquarters_offices/ast/media/2017_AST_Compendium.pdf>.

management.[57] It reached an altitude of 65 km and then glided into the Bay of Bengal. The second mission is expected in 2019, when the spacecraft will be lifted via helicopter to an altitude of 3 km and detached, then gliding to an airstrip landing. A third test will involve a full launch to orbit followed by atmospheric re-entry and landing.[58]

2.2 Space Science and Exploration

The main scope of this section is to provide information to understand the forward steps conducted in human spaceflight activities and science exploration. Space science is mainly remote observation to make discoveries on the origin, evolution and future of the universe, its galaxies, our solar system and other celestial bodies, e.g. stars, exoplanets, comets and asteroids. Space exploration, on the other hand, involves human and robotic spaceflight missions. While traditional governmental space agencies dominate in both these fields, progress in the latter category can be seen with the development of exploration involving commercial players and with new space powers demonstrating the technology needed to carry out such missions.

2.2.1 Human Spaceflight Activities

The human spaceflight activities were focused on Low Earth Orbit (LEO), with the International Space Station (ISS) at centre stage, following its formal extension to 2024. Russia is one of the actors open to the extension of the partnership beyond the planned end of the programme so far, and it is the sole launch provider capable of transporting astronauts to the orbital laboratory.[59] The Soyuz-FG from Baikonur carried out the transfers of four expeditions, the first of which began in April 2017 and ended the 2nd of September of the same year (Expedition 51/52).

Fyodor Yurchikhin (Roscosmos) and Jack Fischer (NASA) were the flight engineers who reached the on-board crew of Thomas Pesquet (ESA) and Oleg Novitskiy (Roscosmos). Commanded by NASA's Peggy Whitson (launched 17 November 2016), the team used the space station robotic arm Canadarm 2 to take hold of the Cygnus after a 4-day delivery mission, the seventh Orbital ATK

---

[57]Gebhardt, Chris. "India successfully launches Reusable Launch Vehicle demonstrator mission." 22 May 2016. NASA Spaceflight.com 16 Aug. 2017 <https://www.nasaspaceflight.com/2016/05/india-launch-reusable-launch-vehicle-demonstrator-mission/>.

[58]https://timesofindia.indiatimes.com/india/manned-mission-isro-to-hold-2nd-test-of-reusable-launch-vehicle-next-year/articleshow/65377953.cms

[59]Stuart Clark "Global partners could consider extending life of ISS." 13th April 2017. The Guardian <https://www.theguardian.com/science/2017/apr/13/international-space-station-extension-russia-nasa> 19th September 2018.

cargo delivery on 18 April 2017. The Cygnus was launched from Cape Canaveral on a ULA Atlas V rocket.[60] ESA astronaut Thomas Pesquet began his 6-month Expedition 50/51 mission on the ISS under the mission banner "Proxima" on 19 November 2016. He is the first French astronaut to visit the ISS since the installation of the Columbus module in 2008, and he performed about 50 scientific experiments for ESA and CNES, taking part in many research activities for the other station partners.[61] In January 2017, Thomas Pesquet and NASA astronaut Shane Kimbrough performed two spacewalks on the ISS replacing older-technology batteries with newer lithium-ion designs.[62] During his 51st mission, in May, he assisted to the spacewalk of Peggy Wilson and Jack Fischer, a 4-h spacewalk to replace the supplying electricity avionic box, to repair the connecting point of the Japanese robotic arm and to install a protective shield on the Pressurized Mating Adapter-3.[63]

Also in May, the two NASA astronauts conducted an additional spacewalk of 2 h and 46 min, for the replacement of a computer relay box and to install two wireless communication antennas to permit communication during future Extravehicular Activities (EVA). The EVAs reached a total of 1250 h and 41 min until this 201st spacewalk, the 10th for Peggy Wilson and the second one for Fischer.[64] Moreover, the Commander Randy Bresnik of NASA, Sergey Ryazanskiy of Roscosmos and Italian astronaut Paolo Nespoli of ESA started their journey on the ISS on the 28th of July for the Expedition 52/53. More than 200 experiments were conducted, 11 of which selected by ASI for the mission named VITA (Vitality, Innovation, Technology, Ability). The VITA mission comes as a result of the bilateral agreement Multipurpose Pressurized Module (MPLM) between ASI and NASA enabling ASI to exploit the mission's permanent module.[65] On 17 August 2017, spacewalkers Fyodor Yurchikhin and Sergey Ryazanskiy manually deployed five nanosatellites, one of which aimed to test the effect of the LEO environment on the 3-D printed

[60]Garcia, Mark "Cygnus captured after four days delivery mission." 22nd April 2017. NASA Space Station <https://blogs.nasa.gov/spacestation/2017/04/22/cygnus-captured-after-four-day-delivery-mission/> 19 th September 2018.

[61]"ESA Astronaut Thomas Pesquet Arrives at the International Space Station." 19 Nov. 2016. ESA 8 May 2017 <http://www.esa.int/Our_Activities/Human_Spaceflight/Proxima/ESA_astronaut_Thomas_Pesquet_arrives_at_the_International_Space_Station>.

[62]"Spacewalk for Thomas Pasquet." 16 Dec. 2016. ESA 8 May 2017 <http://www.esa.int/Our_Activities/Human_Spaceflight/Proxima/Spacewalk_for_Thomas_Pesquet>.

[63]Garcia, Mark "200th Station Spacewalk Comes to an End" 12th May 2017. NASA Space Station <https://blogs.nasa.gov/spacestation/2017/05/12/200th-station-spacewalk-comes-to-an-end> 19th September 2018.

[64]Garcia, Mark "Short Spacewalk completed after successful installation work" 23rd May 2017. NASA Space Station <https://blogs.nasa.gov/spacestation/2017/05/23/short-spacewalk-complete-after-successful-installation-work/> 19th September 2018.

[65]Redazione ASI "Nespoli sulla stazione con VITA" 28th of July 2017. ASI NEWS <https://www.asi.it/it/news/nespoli-in-orbita-con-vita> 20th September 2018.

materials it was made of. It was, also, an occasion to collect residue samples from the Russian station segment.[66]

In September, the Soyuz MS-06 carried to the orbital laboratory Roscosmos' Aleksandr Misurkin and NASA's Mark T. Vande Hei and Joseph Acaba. Only 1 month later, three spacewalks were conducted during the Expedition 53, and each activity lasted more than 6 h, all of them commanded by Randy Bresnik. During the first spacewalk the 5th of October, the astronauts replaced one of the two Latching End Effectors (LEE) on the Canadarm 2 robotic arm.[67] During further extravehicular activity in October, respectively, the 10th, Expedition 53 Commander Randy Bresnik and flight engineer Mark Vande Hei accomplished the replacement of a camera system and adequate maintenance lubricating of LEE components they installed previously.[68] During the last spacewalk of 2017, conducted by the Commander Bresnik and Joe Acaba, a new radiator grapple bar was installed and one of two spare pump modules on separate stowage platforms were prepared, enabling easier access for future robotic replacement tasks. The 20 October EVA was the fifth for Bresnik and the third for Acaba for a total time of 53 days, 6 h and 25 min in outer space.[69] The Russian Soyuz MS-07 launched the last crew of the year in December 2017: transporting Anton Shkaplerov (Roscosmos), Scott Tingle (NASA) and Norishige Kanai (JAXA) for the ISS 54 Expedition. The astronaut researchers conducted experiments for manufacturing fibre optics in microgravity; they measured the amount of sunlight Earth receives and recollected data on space debris in low Earth orbit, monitoring the situation with the Space Debris Sensor (SDS), along with bacteria experiments.[70] On 4 June 2017, the Japan Aerospace Exploration Agency (JAXA) delivered to the JEM, via the US Dragon to the Kibo module, an Internal Ball Camera drone that can record video in the orbital module under remote control from the ground.[71]

With ESA's final Automated Transfer Vehicle (ATV) mission to the ISS completed on 15 February 2015, ESA will continue covering its dues for the ISS onward from 2017 by using the knowledge gained from the ATV programme to build the

---

[66]Whiting, Melanie "Spacewalks comes to a close" 17th August 2017. NASA Space Station <https://blogs.nasa.gov/spacestation/2017/08/17/spacewalk-comes-to-a-close/> 19th September 2018.

[67]Garcia, Mark "Spacewalkers wrap up robotic arm work" 5th October 2017. NASA Space Station <https://blogs.nasa.gov/spacestation/2017/10/05/spacewalkers-wrap-up-robotic-arm-work/> 19th September 2018.

[68]Garcia, Mark "Astronauts back inside the station after the second spacewalk" 10th October 2017. NASA Space Station <https://blogs.nasa.gov/spacestation/2017/10/10/astronauts-back-inside-station-after-second-spacewalk/> 19th September 2018.

[69]Whiting, Melanie "Expedition 53 spacewalk successfully comes to an end" 20th October 2017. NASA Space Station <https://blogs.nasa.gov/spacestation/2017/10/20/expedition-53-spacewalk-successfully-comes-to-an-end/> 20th September 2018.

[70]NASA Mission Summary Expedition 54 December 2017 <https://www.nasa.gov/sites/default/files/atoms/files/exp-54-summary_-_final.pdf> 20th September 2018.

[71]JAXA–ISS "First disclosure of images taken by the Kibo's internal drone "Int-Ball" 14th July 2017 <http://iss.jaxa.jp/en/kiboexp/news/170714_int_ball_en.html> 24th September 2018.

European Service Module for NASA's Orion spacecraft.[72] The Service Module will provide propulsion, electrical power, water and thermal control to the Orion spacecraft and will maintain the oxygen and nitrogen atmosphere for its crew. The first Service Module, built in Turin, Italy, initially underwent a comprehensive series of independent tests at Plum Brook in Ohio, USA, from November 2015 to December 2016; the Service Module will now undergo further testing at vehicle level in conjunction with NASA's Orion Crew Module Structural Test Article.[73] NASA will conduct the first launch of the Orion spacecraft and Service Module on its Space Launch System on a month-long un-crewed demonstration mission around the Moon. By 7 December 2016, the agencies had agreed to extend their collaboration in human space exploration, whereby ESA will provide a second Service Module to support NASA's first crewed Orion mission, slated to launch as early as 2021.[74]

In total, nine crew/cargo missions were conducted to the ISS and an additional one to the Chinese station Tiangong 2. SpaceX inaugurated the year on the 19th of February with the tenth Dragon Commercial Resupply Spacecraft (CRS-10) launched by a Falcon 9 v 1.2 from the Kennedy Space Centre. Three other SpaceX resupply cargo missions were launched to the ISS during the year, along with the US Cygnus OA-7 on an Atlas 5 (401) by ULA and Cygnus OA-8 on an Antares-230 by Orbital ATK. The Russian Progress-MS 04 was the first of four crewed capsules launched by the Europeanised Soyuz-FG, operated by Starsem.[75]

On the other side of orbital research, on 20 April 2017, China's Long March 7 (CZ-7) launch vehicle lifted off from the Wenchang Space Launch Centre.[76] The rocket was carrying the Tianzhou-1 unmanned spacecraft to the Tiangong 2 station, with 12,910 kg of mass at launch. The Tiangong 2 was refuelled by Tianzhou-1 after 5 days from the docking.[77] This cargo spacecraft mission provides an important technological basis to build the future Chinese space station; it has the capacity to carry six tons of goods and two tons of fuel, and it can fly unmanned for 3 months.[78]

---

[72]"Last ATV reentry leaves legacy for future space exploration." 15 Feb. 2015. ESA 30 May 2016 <http://m.esa.int/Our_Activities/Human_Spaceflight/ATV/Last_ATV_reentry_leaves_legacy_for_future_space_exploration>.

[73]"Testing Completed on Orion Service Module." 14 Dec. 2016. ESA 8 May 2017 <http://www.esa.int/Our_Activities/Human_Spaceflight/Orion/Testing_completed_on_Orion_service_module>.

[74]"ESA to Supply Service Module for First Crewed Orion Mission." 7 Dec. 2016. ESA 8 May 2017 <http://www.esa.int/Our_Activities/Human_Spaceflight/Orion/ESA_to_supply_Service_Module_for_first_crewed_Orion_mission>.

[75]ESPI database, 21st of September 2018.

[76]ESPI database, 24th of September 2018.

[77]Todd, David "China launches its first unmanned space freighter Tianzhou-1 to dock with Tiangong-2 mini-station" 20th April Of 2017. 24th September 2018. <https://www.seradata.com/china-launches-its-first-unmanned-space-freighter-tianzhou-1-to-dock-with-tiangong-2-mini-station/>.

[78]Reuters "Tiangong 2: China first cargo spacecraft docks with orbiting space labs" 22nd April 2017, The Guardian. 24th September 2018 <https://www.theguardian.com/science/2017/apr/22/tiangong-1-chinas-first-cargo-spacecraft-docks-with-orbiting-space-lab>.

The previous space lab, Tiangong 1, ended its mission in March 2016, after the station stopped sending telemetry data back to Earth.[79] The Tiangong 1—in sleep mode after its last crew departed in June 2013—was not planned to be a permanent orbital station; however, in September 2016, the Chinese government confirmed speculation that the dormant 8.5 ton space station would begin an uncontrolled re-entry into Earth's atmosphere, which finally occurred in April 2018.[80,81] Moving forward, China launched its Tiangong 2 space lab on 15 September 2016 on its Long March 2F rocket from the Jiuquan Satellite Launch Center. The Tiangong 2 uses the same basic module as the Tiangong 1 space lab but will be a test bed for the rendezvous, docking and life-support technologies intended for its future planned space station. The Shenzhou-11 spacecraft, launched on 16 October 2016, delivered the first two Taikonauts to the space lab for a 30-day visit.[82] Construction of China's space station will begin in 2018 with the launch of its experimental core module; China aims to complete construction of the space station by 2022.[83]

2.2.2 Lunar Science

Interest in the Moon is ongoing for both its science and exploration value, in addition to being the finish line for several private space companies competing to win the Google Lunar XPRIZE. Thanks to this kind of high-level worldwide competitions, landing on the Moon is not seen anymore only as a space agency prerogative, but it had shown how small teams and experts throughout the world could reach the goal.

Ultimately, 2017 was an unfruitful year in terms of winners, but the Google Lunar XPRIZE was a source of motivation for the private sector actors. Since 2007, the Google Lunar XPRIZE is a competition for a grand prize of $20 million and a second prize of $5 million for the first two privately funded teams to safely land on the Moon, travel at least 500 m across its surface and send high-definition video, images and data back to the Earth. In late 2013, the XPRIZE Foundation and Google announced a series of interim "milestone" prizes available to assist the competing teams in accessing finance at a critical point in their mission timeline and to raise

---

[79]Jones, Morris. "Has Tiangong 1 gone rogue." 30th Mar. 2016. SpaceDaily 24th May 2017 <http://www.spacedaily.com/reports/Has_Tiangong_1_gone_rogue_999.html>.

[80]Wall, Mike. "Tiangong-1 Space Lab Will Fall to Earth Next Year, China Says." 19 Sept. 2016. Space.com 23 May 2017 <http://www.space.com/34089-tiangong-1-fall-earth-2017.html>.

[81]Byrd, Deborah. "Tiangong-1's fiery death and ocean plunge." 2 April 2018. EarthSky, 10 Oct. 2018 <http://earthsky.org/space/china-tiangong-1-uncontrolled-re-entry-mar-apr-2018>.

[82]Reis, Michael. "Tiangong-2: China's Space Lab for Long Missions (Infographic)." 3 Nov. 2016. Space.com 24 May 2017 <http://www.space.com/34614-tiangong-2-china-space-lab-explained-infographic.html>.

[83]"China to Launch Space Station Core Module in 2018." 3 Mar. 2017. Chinese Academy of Sciences 24 May 2017 <http://english.cas.cn/newsroom/china_research/201703/t20170303_174538.shtml>.

public excitement and support for the teams. The Google Lunar XPRIZE competition was extended to the end of 2017, after two teams had met the XPRIZE Foundation's 16 December 2014 stipulation requiring at least one team to have made launch arrangements by the end of 2015.[84] Only five teams remained in the Google Lunar XPRIZE competition by the end of 2017, out of an initial 33 entrants, as competitors had until the end of the year to submit their own launch documentation to remain in the competition.[85] The Japanese Hakuto team stands among the finalists, along with the Indian Indus team and the Israeli team,[86] SpaceIL, the first to be confirmed by the foundation on 7 October 2015.[87] The American team Moon Express[88] and the international team Synergy Moon were the next teams to have their launch contracts verified by the XPRIZE Foundation, followed by the Indian team Indus on 1 December 2016.[89] The American team Astrobotic Technology separated from the Google Lunar XPRIZE competition by choosing not to secure a launch at the end of 2016; Astrobotic Technology intends to fly its first mission to the Moon in 2019.[90] The team Indus signed an agreement with the team Hakuto, planning to carry a rover built by the Japanese team, under the startup ispace that raised $90 million until December 2017.[91] The XPRIZE Foundation extended the deadline, once again, to 31 March 2018. After a consultation with the five finalist teams, they concluded that none of those teams could be in the right path to reach the Moon in the established time. In addition, the modality of the competition will change, seeking in the meantime a new sponsor after 10 years of Google support. The competition during the past years had a large commercial influence, guiding experts through the creation of the first commercial companies in India, Malaysia, Israel and Hungary. Pursuing to create a long-term business model, the XPRIZE

[84]Foust, Jeff. "Google Lunar X Prize Extends Competition Deadline." 17 Dec. 2014. SpaceNews 23 July 2015 <http://spacenews.com/google-lunar-x-prize-extends-competition-deadline/>.

[85]Foust, Jeff. "X Prize Verifies Moon Express Launch Contract." 8 Dec. 2015. SpaceNews 6 Jan. 2016 <http://spacenews.com/x-prize-verifies-moon-express-launch-contract/>.

[86]XPRIZE Competing Teams—Finalist Team. 25th of September 2018 <https://lunar.xprize.org/prizes/google-lunar/competing-teams>.

[87]Foust, Jeff. "Israeli X Prize Team Announces Launch Contract for Lunar Mission." 7 Oct. 2015. SpaceNews 5 Jan. 2016 <http://spacenews.com/israeli-x-prize-team-announces-launch-contract-for-lunar-mission/>.

[88]Foust, Jeff. "Moon Express Buys Rocket Lab Launches for Lunar Missions." 1 Oct. 2015. SpaceNews 5 Jan. 2016 <http://spacenews.com/moon-express-buys-rocket-lab-launches-for-lunar-missions/>.

[89]Foust, Jeff. "Indian X Prize team secures launch contract with ISRO." 2 Dec. 2016. SpaceNews 4 Jan. 2017 <http://spacenews.com/indian-x-prize-team-secures-launch-contract-with-isro/>.

[90]Thornton, John. "Graduating from the Google Lunar X Prize." 19 Dec. 2016. SpaceNews Magazine 14 Aug. 2017 <http://www.spacenewsmag.com/commentary/graduating-from-the-google-lunar-x-prize>.

[91]Foust, Jeff "Google Lunar XPRIZE to end without winner" 23rd January 2018. SpaceNews 25th of September 2018 <https://spacenews.com/google-lunar-x-prize-to-end-without-winner/>.

Foundation will continue to lead future competitions for a Moon-shot, acquiring different modalities, rules and awards too.[92]

NASA's Exploration Systems Mission Directorate (ESMD) operated the Lunar Reconnaissance Orbiter LRO spacecraft and its instruments during the 1-year mission phase, started in June 2009, and it should last only for 1 year. Scouting the Moon in preparation for future lunar exploration, including finding landing sites and locating resources such as water, ice and hydrogen, it is tremendously important also to investigate the long-term effects of the lunar environment. The LRO completed its second 2-year extended science mission in September 2016 and was extended for a third 2-year extended science mission, which will run through October 2018. This new "Cornerstone Mission" will focus on three questions relevant to the US Decadal Survey goals between 2013 and 2022, including (1) Volatiles and the Space Environment, (2) Volcanism and Interior Processes and Impacts and (3) Regolith Evolution.[93] On 23 March 2015, new research published in the journal *Nature* provided evidence that the spin axis of the Moon shifted by about five degrees roughly 3 billion years ago, by examining the distribution of ice at each of the lunar poles. While ice can exist in permanently shadowed areas on the Moon, direct exposure to sunlight would make it evaporate into space; researchers were able to find that the path of the ice that survived this axis shift matched models predicting where the ice could remain stable.[94] On 29 April 2016, data from the LRO allowed researchers to develop models published in various journals to explain how "lunar swirls" are formed, i.e. patterns which extend tens of kilometres and are peppered across the Moon's surface in areas where ancient bits of magnetic field are embedded in the lunar crust and appear less weathered than their surroundings. While the swirls were first thought to have formed from plumes of material ejected by comet impacts or by fine dust particles lofted by micrometeorite impacts on the Moon's surface, a third theory which seems to be supported by LRO data is that the less-weathered areas are protected by magnetic field shields embedded in the Moon's surface which create strong electric fields that are able to deflect some of the slower moving charged solar wind particles.[95] On 13 October 2016, new observations by the LRO determined that the Moon's surface experiences a heavier bombardment by small asteroids than previous models had predicted, which implies that a future lunar base may have to be made sturdier than anticipated to withstand secondary debris

---

[92]Diamandis, Peter H. "An important update from Lunar XPRIZE"23 Jan 2018. XPRIZE Articles, 25th September 2018 <https://lunar.xprize.org/prizes/google-lunar/articles/an-important-update-from-google-lunar-xprize>.

[93]Keller, J. W. and N. E. Petro. "The Lunar reconnaissance orbiter cornerstone mission: a synergistic study of fundamental solar system processes." 11 July 2016. Annual Meeting of the Lunar Exploration Analysis Group (2016) 12 Aug. 2017 <https://www.hou.usra.edu/meetings/leag2016/pdf/5036.pdf>.

[94]"Ancient Polar Ice Reveals Tilting of Earth's Moon." 23 Mar. 2016. NASA 12 Aug. 2017 <https://www.nasa.gov/ames/feature/lunar-polar-ice-reveals-tilting-axis-of-earth-s-moon>.

[95]Steigerwald, Bill. "NASA Research Gives New Insights into How the Moon Got 'Inked'." 29 Apr. 2016. NASA 13 Aug. 2017 <https://www.nasa.gov/feature/goddard/2016/lunar-swirls>.

impacts moving at up to 500 m/s.[96] After 9 years of lunar operation, there is evidence that the inertial measurement unit (IMU) is close to its end of life, a reason why the LRO team wants to reserve the remaining power for critical occasions such as lunar eclipses.[97]

China's Chang'e 5-TI test capsule returned to lunar orbit in the week of 12 January 2015.[98] Launched on 23 October 2014, the prototype sample return capsule reached the Moon within a day, circling it before returning to eject its sample capsule at a higher than average velocity into Earth's atmosphere. Following the successful release of the capsule on 1 November 2015, Chang'e 5-TI began making its way to the Earth-Moon Lagrange (L2) point on the opposite side of the Moon. Reaching L2 by late November 2015, the service module then completed three circles around that point prior to returning to lunar orbit.[99] In addition to testing critical manoeuvres, the Chang'e 5-TI carries a camera system that will help to identify future landing sites for the Chang'e 5 robotic sample return mission planned for launch, at first, in the second half of 2017.[100] This later Chang'e 5 mission will involve a soft landing on the Moon and the collection of 200 g of samples prior to bringing them to Earth. However, officials from the Chinese space programme announced there will be a delay, stating that the lunar probe Chang'e 5 will be launched in 2019, mainly due to a launch failure of the Long March in July 2017. It was also disclosed that the lunar probe is expected to land in the Mons Rumker region. The Chang'e 4 mission will be launched with a Long March 4C, and, originally, it should have been launched after Chang'e 5. It will be a two-phase mission to explore the dark side of the Moon utilising an orbiter, a lander and a rover. Carrying 4 international payloads over 11,[101] the Chang'e 4 mission is expected to start at the end of 2018.[102] In this view, Liu Jizhong, director of the China Lunar Exploration and Space Engineering Center of the China National Space Administration (CNSA), proposed a joint exploration with ESA of the lunar polar region and

[96]Jones, Nancy and Bill Steigerwald. "Earth's Moon Hit by Surprising Number of Meteoroids." 13 Oct. 2016. NASA 13 Aug. 2017 <https://www.nasa.gov/press-release/goddard/2016/lro-lunar-cratering>.

[97]LRO Mission Status Report. 7 May 2018. NASA Official 25th September 2018 <https://www.nasa.gov/feature/goddard/2018/lro-mission-status-report>.

[98]David, Leonard. "Chinese Spacecraft Enters Orbit around the Moon." 20 Jan. 2015. SpaceNews 8 Jan. 2016 <http://spacenews.com/chinese-spacecraft-enters-orbit-around-the-moon/>.

[99]"Service Module of Chinese Probe Successfully Enters Lunar Orbit." 11 Jan. 2015. Sputnik News 22 Jan. 2016 <http://sputniknews.com/science/20150111/1016776437.html>.

[100]Jones, Andrew. "China's Chang'e-5 Moon sample return mission on course for 2017 launch." 24 Oct. 2016. GBTimes 13 Aug. 2017 <http://gbtimes.com/china/chinas-change-5-moon-sample-return-mission-course-2017-launch>.

[101]Xinhua "China discloses Chang'e 5 lunar probe landing side" 6th July 2017. ChinaDaily.com.cn 25th September 2018 <http://www.xinhuanet.com/english/2017-06/07/c_136346632.htm>.

[102]Deyana Goh "China's Chang'e 5 mission delayed to 2019, Chang'e 4 on schedule." 26th of April 2018. SpaceTech 25th September 2018 <http://www.spacetechasia.com/chinas-change-5-lunar-mission-delayed-to-2019-change-4-on-schedule/>.

the construction of a scientific research station as a guide for the international Moon village or station. Moreover, he also proposed the creation of an open cooperation platform in accordance with the principle of "sharing the risks and achievements".[103]

India is also running for its second mission to the Moon, Chandrayaan-2, which is composed of an orbiter, a lander and a rover. The mission will carry a six-wheeled rover which will be mostly commanded by the ground station, while the instruments will collect lunar soil characteristics. Chandrayaan-2 has a total mass of 3290 kg, and it will collect scientific information on lunar topography, mineralogy, elemental abundance, the lunar exosphere and signatures of hydroxyl and water-ice from the second half of 2018.[104] India's first lunar mission was launched in 2008 and operated successfully until August 2009, when a failure occurred and the communication was lost after 3400 orbits.[105] Chandrayaan-1 gathered data on the chemical, mineralogical and geologic composition of the Moon, orbiting at 100 km from the surface and reaching 200 km of altitude; the lunar surface was mapped in visible, near-infrared, X-ray and low-energy gamma ray for a high remote sensing resolution.[106] Chandrayaan-1 was an example of international cooperation, as the spacecraft carried 11 scientific instruments built in the UK, the USA, Sweden, Bulgaria, Germany and India.[107]

In pursuing further international cooperation, NASA and Roscosmos announced on 27 September 2017 at the IAC Congress in Adelaide a partnership for NASA's Deep Space Gateway, a mission architecture that intends to send astronauts into cislunar space by the 2020s. The agreement had the purpose to show the affordability and sustainability of this kind of exploration architecture, inviting the private sector to participate and contributing to the realisation of the ambitious project. Bigelow Aerospace, Orbital ATK and Lockheed Martin[108] have already signed awarded contracts for the design of the Deep Space habitat.[109] Under Next Space Technologies for Exploration Partnerships (NextSTEP),[110] they

---

[103]Xinhua"China discloses Chang'e 5lunar probe landing side" 6th July 2017. ChinaDaily.com.cn 25th September 2018 <http://www.xinhuanet.com/english/2017-06/07/c_136346632.htm>.

[104]Department of Space ISRO website. 25th September 2018 <https://www.isro.gov.in/gslv-f10-chandrayaan-2-mission>.

[105]"Chandrayaan-1" 22nd of October 2008. ISRO 25th September 2018 <https://www.isro.gov.in/Spacecraft/chandrayaan-1>.

[106]"Chandrayaan-1 MIP" 11th of December 2017. Gunter's Space Page 25th of September 2018. <https://space.skyrocket.de/doc_sdat/chandrayaan-1.htm>.

[107]"Chandrayaan-1" 22nd of October 2008. ISRO 25th September 2018 <https://www.isro.gov.in/Spacecraft/chandrayaan-1>.

[108]See also Lockheed Martin website: <https://news.lockheedmartin.com/2017-07-20-Living-in-Deep-Space-Lockheed-Martin-to-Build-Full-Scale-Prototype-of-NASA-Cislunar-Habitat>

[109]Weitering, Hanneke "NASA and Russia partner up for crewed deep-space missions" 27th September 2017. Space.com 25th September <https://www.space.com/38287-nasa-russia-deep-space-gateway-partnership.html>.

[110]Foust, Jeff "NASA moving ahead with plans for cislunar human outpost" 10th March 2017. SpaceNews 1st October 2018 <https://spacenews.com/nasa-moving-ahead-with-plans-for-cislunar-human-outpost/Y>.

advanced habitation concepts and prototypes improving the capabilities for the mission.[111] The ultimate goal of the mission architecture is to build a strategic capability for advancing and sustaining human space exploration and to test the system for future challenging human exploration missions such as to Mars. Moreover, the lunar environment is ideal for gaining experience for further missions into the solar system.[112] The mission concept is divided into four segmented steps, and, so far, activities are approaching the phase 1 from phase 0, concerning the resolution of mission challenges through system testing on the ISS.[113] The Deep Space Gateway is still in its concept formulation, but the joint statement reflects the common vision of the two agencies, already partnering with concrete results in the ISS programme.[114] The Space Launch System (SLS) was elected for the launch, along with the Orion spacecraft for the EM-1 (Exploration Mission), originally scheduled for the end of 2018, but now delayed.[115] The Orion spacecraft will be powered by the European Service Module (ESM), assembled by Airbus in Bremen, for the first mission since 1972 to take humans beyond LEO (EM-2). Prevented for the end of 2018, the first Orion mission will be unmanned, and the month-long flight will prove the spacecraft for the manned mission to come, with both ESMs built under the lead of Airbus Defence and Space by 11 companies, improving the existing technology from ESA's Automated Transfer Vehicle, which has flown five times to the ISS.[116] The model is awaiting extensive testing by NASA, ESA and Airbus, powered by 21 engines, including the Shuttle Orbital Manoeuvring System (OMS) engine, 8 auxiliary thrusters and 12 smaller thrusters.[117] Using currently developed technologies for the first phase of exploration in the Deep Space Gateway mission, it will confirm in the second phase of the mission the capabilities necessary for human long-duration missions beyond

[111]Loff , Sarah "NASA, Roscosmos sign joint state statement on researching, exploring deep space" 27th September 2018. NASA-Moon to Mars—26th of September 2018. <https://www.nasa.gov/feature/nasa-roscosmos-sign-joint-statement-on-researching-exploring-deep-space>.

[112]Hambleton, Kathryn "Deep Space Gateway to Open Opportunities for Distant Destinations" 28th of March 2017. NASA—Moon to Mars—26th of September 2018 <https://www.nasa.gov/feature/deep-space-gateway-to-open-opportunities-for-distant-destinations>.

[113]NASA Advisory Council Human Exploration & Operations Committee "Cislunar Habitation & Environmental Control & Life Support Systems" 29th March 2017. 1st of October 2017 <https://www.nasa.gov/sites/default/files/atoms/files/20170329-nacheoc-crusan-gatens-hab-eclss-v5b.pdf>.

[114]Loff, Sarah "NASA, Roscosmos sign joint state statement on researching, exploring deep space" 27th September 2018. NASA—Moon to Mars—26th of September 2018. <https://www.nasa.gov/feature/nasa-roscosmos-sign-joint-statement-on-researching-exploring-deep-space>.

[115]Jeff Foust "NASA moving ahead with plans for cislunar human outpost" 10th March 2017. SpaceNews 01st of October 2018 <https://spacenews.com/nasa-moving-ahead-with-plans-for-cislunar-human-outpost/>.

[116]ESA "Building Orion service module for the first astronaut mission" 17 February 2017. ESA – Orion—1st October 2018 <https://www.esa.int/Our_Activities/Human_Spaceflight/Orion/Building_Orion_service_module_for_first_astronaut_mission>.

[117]ESA space in images "Waiting for the burn" copyright by NASA, 27th February 2017. ESA blog 1st of October 2018 <https://www.esa.int/spaceinimages/Images/2017/02/Waiting_for_the_burn>.

the Moon.[118] JAXA has proposed its intention of collaboration in view of the International Space Exploration Forum in March 2018. And earlier, in November 2017, Masazumi Miyake, director of the International Relations Department of JAXA, shared via the Global Partnership for Space Exploration High Level Forum the agency's intention to cooperate in this ambitious project: Through conducting an inspection of the lunar south pole in order to assess the possibility of utilising water for fuel, Japan would participate in the US-led Deep Space Gateway programme with key technologies, envisioning to send Japanese astronauts to deep space and elaborating key technologies for the international Human Lunar Surface Exploration programme, announcing a preparatory mission around 2025. Under an international collaboration, there is willingness to construct a fuel plant at the lunar south pole, using both human and robotic capabilities. Lastly, the common principles of international space exploration would lead the mission for peaceful purposes, the promotion of international collaboration, public engagement and, consequently, economic expansion in continuity for environmental stewardship.[119]

2.2.3 Mars Science

For decades the focus for Mars science has been the investigation of the planet's habitability, in a search for the presence of water. The collected data continues to suggest that Mars was once partially covered by large oceans and that life could have been possible in many locations on the planet's surface.

ESA's Mars Express is one of the agency's most scientifically productive missions, being one of few that has allowed experts to publish more than 1000 papers for information releases,[120] and is considered a "flexible mission" of the ESA New Horizon programme.[121] Launched in June 2003 via a Soyuz-Fregat, the Mars Express orbiter has conducted a high-resolution imaging mission of Mars, including the mapping of its mineral composition and atmosphere, determining the structure of the subsurface and studying the effect of the atmosphere on the surface and the interaction of the atmosphere with the solar winds. The images taken by the high-resolution stereo camera on Mars Express in mid-2015 helped to reveal in

[118]Hambleton, Kathryn "Deep Space Gateway to Open Opportunities for Distant Destinations" 28th of March 2017. NASA—Moon to Mars—26th of September 2018 <https://www.nasa.gov/feature/deep-space-gateway-to-open-opportunities-for-distant-destinations>.

[119]Miyake, Masazumi, JAXA "Global partnership for Space Exploration High Level Forum" 7th of November 2017. UNOOSA, 1st October 2018 <http://www.unoosa.org/documents/pdf/hlf/HLF2017/presentations/Day2/Session_6/Presentation4.pdf>.

[120]EGU General Assembly "Mars Express recent finding and future plans"—Geophysical Research Abstracts—Vol. 19, EGU2017-15392, 2017.

[121]"Mars Express" missioni e progetti. ASI 1st October 2018 <https://www.asi.it/it/attivita/esplorare-lo-spazio/esplorazione-del-sistema-solare/mars-express>.

extraordinary detail complex features such as the Noctis Labyrinthus region on the western edge of Valles Marineris, an entire network of plateaus and fractures spanning around 1200 km that suggests many episodes of tectonic stretching and volcanic activity in Mars' Tharsis region.[122] Moreover, images of the Arda Valles, north of the Holden crater and Ladon Valles, revealed a dendritic drainage system carved by vast volumes of water that once flowed from the southern highlands.[123] Images from Mars' Colles Nili region showed the erosional remnants of a former plateau, whose layered deposits gently sloped away from the sides of hills and series of ridges and troughs found around the mounds and inside some of the impact craters on the channel floors. These are thought to be associated with buried ice that has since been covered over by wind-blown dust and local debris.[124] And images of ridges and troughs in the western part of Acheron Fossae, 1000 km north of Olympus Mons and other volcanic giants in the Tharsis bulge, suggest a complex history, as its pattern of cross-cutting faults implies that the region experienced stresses from different directions over time; as Acheron Fossae has been likened to Earth's continental rift system and is associated with plate tectonics, rifts are important for studies of the general evolution of the crust as well as the thermal evolution of the deeper subsurface.[125] Additionally, plasma and solar wind measurements from Mars Express' Analyser for Space Plasmas and Energetic Atoms (ASPERA-3) plasma instrument suite and the Mars Advanced Radar for Subsurface and Ionospheric Sounding (MARSIS) instrument revealed that the cloud-like plume that reached an altitude of 250 km in 2012 was likely the result of a coronal mass ejection (CME) from the Sun that was large enough to impact Mars and increase the escape of plasma from the planet's atmosphere, thus adding an important angle on the potential role of space weather in how Mars may have lost much of its atmosphere in the past.[126] Subsequently, scientists have discovered that the results related to the structure of the ionosphere detected by the MARSIS radar and ASPERA-3 observations were significantly affected by many space factors, such as solar activity, the influx of cometary dust and the magnetic field of the crust. The scientific observations have furthermore revealed the presence of a denser sunlit ionosphere

---

[122] "Martian Labyrinth." 28 Jan. 2016. ESA 24 May 2017 <http://www.esa.int/Our_Activities/Space_Science/Mars_Express/Martian_labyrinth>.

[123] "Footprints of a Martian Flood." 18 Feb. 2016. ESA 24 May 2017 <http://www.esa.int/Our_Activities/Space_Science/Mars_Express/Footprints_of_a_martian_flood>.

[124] "Buried Glaciers on Mars." 13 Oct. 2016. ESA 24 May 2017 <http://www.esa.int/Our_Activities/Space_Science/Mars_Express/Buried_glaciers_on_Mars>.

[125] "A Record of Ancient Tectonic Stress on Mars." 3 Nov. 2016. ESA 24 May 2017 <http://www.esa.int/Our_Activities/Space_Science/Mars_Express/A_record_of_ancient_tectonic_stress_on_Mars>.

[126] "Are Mystery Mars Plumes Caused by Space Weather?" 23 May 2016. ESA 24 May 2017 <http://www.esa.int/Our_Activities/Space_Science/Mars_Express/Are_mystery_Mars_plumes_caused_by_space_weather>.

over regions characterised by robust crustal fields.[127] Moreover, in March, Mars Express captured the Kasei Valles channel, a 3000 km system of outflow channels, sourcing from the Echus Chasma region to the plains of Chyse Planitia.[128] Further, ancient volcanic activity on the red planet was confirmed by fractures originating in the crust that divide the Tharsis region into two blocks.[129] The objectives proposed until 2016 were fully accomplished, and the mission has been extended until 2018, awaiting for the scientific case mission extension approval.[130]

The ExoMars programme—Exobiology on Mars—is a joint endeavour between ESA and the Russian Space Agency, Roscosmos, to address whether life has ever existed on Mars. The programme is comprised of two missions: the first began in 2016 with the launch of ESA's Trace Gas Orbiter (TGO) and the Entry, Descent and Landing Demonstrator Module (EDM) "Schiaparelli"; the second mission will take place in 2020 and comprises a rover and surface science platform. Both missions will be launched to Mars using Roscosmos' Proton launcher.[131] The first ExoMars mission began its 7-month journey to Mars on 14 March 2016; on reaching a distance of 900,000 km from the planet on 16 October 2016, the TGO and Schiaparelli separated, with the TGO moving into a 4-day elliptical orbit around Mars, while Schiaparelli veered into the atmosphere to descend to Mars' surface.[132] After separation, the TGO began a series of complex aerobraking manoeuvres that will continue over the course of 13 months to lower its elliptical orbit of 250 km by 98,000 km to a circular orbit of 400 km before its main scientific mission to analyse rare gases in the atmosphere begins. It will also act as a data relay for surface rovers, providing two to three flyovers of each Mars rover every day to send signals back to Earth. Despite being placed into safe mode for a brief period after faulty configuration of the TGO's main engine caused a temporary glitch during preliminary testing, the TGO was performing nominally by the end of 2016.[133] Nevertheless, on 19 October 2016, Schiaparelli experienced an anomaly as it descended into Mars's atmosphere; while its radar Doppler altimeter functioned correctly and the measurements were included in the guidance, navigation and control system, a saturation of

---

[127]EGU General Assembly "Mars Express recent finding and future plans"—Geophysical Research Abstracts—Vol. 19, EGU2017-15392, 2017.

[128]"Remnants of a Mega-Flood in Mars" 2nd March 2017. ESA 1st of October 2018 <https://www.esa.int/Our_Activities/Space_Science/Mars_Express/Remnants_of_a_mega-flood_on_Mars>.

[129]"Fracture Swarms on Mars" 16th November 2017. ESA 1st October 2018 <https://www.esa.int/Our_Activities/Space_Science/Mars_Express/Fracture_swarms_on_Mars>.

[130]EGU General Assembly "Mars Express recent finding and future plans"—Geophysical Research Abstracts—Vol. 19, EGU2017-15392, 2017.

[131]"What is ExoMars?" 29 Mar. 2017. ESA 26 May 2017 <http://www.esa.int/Our_Activities/Space_Science/ExoMars/What_is_ExoMars>.

[132]ExoMars on its way to solve the Red Planet's mysteries." 14 Mar. 2016. ESA 26 May 2017 <http://www.esa.int/Our_Activities/Space_Science/ExoMars/ExoMars_on_its_way_to_solve_the_Red_Planct_s_mysteries>.

[133]"Skimming an alien atmosphere." 16 Dec. 2016. ESA 26 May 2017 <http://www.esa.int/Our_Activities/Operations/Skimming_an_alien_atmosphere>.

the inertial measurement unit (IMU) occurred shortly after the parachute deployed, causing its navigation system to estimate that it was already below ground level and triggering the final landing stages, while the vehicle was at an altitude of around 3.7 km. ESA has called for an external independent inquiry board to conduct a technical investigation into the anomaly.[134] In May 2017, ESA provided a full report concluding that the on-board computer generated conflicting information, causing the premature descent of the module Schiaparelli.[135] In the report, the Inquiry Board underlined that Schiaparelli was in free fall from 3.7 km altitude roughly at a speed of 540 km/h, that overall it had been very close to a successful landing and that most demonstration objectives had been achieved.[136]

With the primary science goal to collect data to evaluate the effective conductibility of human life, NASA's Mars Odyssey—Mars Global Mapper (MGM)—mission was launched on 7 April 2001 with a Delta-7925,[137] resulting in it being the longest-operating spacecraft sent to Mars. The Mars Odyssey mission marked a turning point for NASA in Mars exploration, after the failure of two preceding missions in 1999 prompted an overhaul of NASA's exploration plans. While its prime mission to make the first global map of the amount and distribution of numerous chemical elements and minerals on Mars' surface was completed in August 2004, its mission had been extended for 13 additional years by 2017; further, Mars Odyssey is also a communications relay for rovers and landers on the planet.[138] In March 2016, NASA published its most detailed gravity map of Mars, derived from Doppler and range tracking data collected by NASA's Mars Odyssey, Mars Global Surveyor and Mars Reconnaissance Orbiter spacecraft missions. The improved resolution allows researchers to better interpret how the crust of the planet has changed over Mars' history in many regions. Moreover, by analysing tides in the crust and mantle caused by the gravitational pull of the sun and the two moons of Mars, researchers confirmed that Mars has a liquid outer core of molten rock. Researchers also determined that during the winter season for both hemispheres within Mars' 11-year orbit around the Sun, approximately 3–4 trillion tons of carbon dioxide, i.e. 12–16% of the mass of the entire Martian atmosphere, freezes out of the

[134]"Schiaparelli landing investigation makes progress." 23rd Nov. 2016. ESA 26th May 2017 <http://www.esa.int/Our_Activities/Space_Science/ExoMars/Schiaparelli_landing_investigation_makes_progress>.

[135]"Schiaparelli Landing Investigation Completed" 24th of May 2017. ESA Robotic Exploration of Mars 1st of October 2018 <http://exploration.esa.int/mars/59184-schiaparelli-landing-investigation-completed/>.

[136]ESA Report "EXOMARS 2016—Schiaparelli Anomaly Inquiry" issued: 18th May 2017. ESA ExoMars 1st October 2018 <http://exploration.esa.int/mars/59176-exomars-2016-schiaparelli-anomaly-inquiry/#>.

[137]"2001 MARS Odyssey (MGM)" 11th December 2017. Gunter's space page 2nd October 2018 <https://space.skyrocket.de/doc_sdat/mars_odyssey_2001.htm>.

[138]"2001 Mars Odyssey." NASA 29 May 2017 <https://mars.jpl.nasa.gov/programmissions/missions/present/odyssey/>.

atmosphere onto the northern and southern polar caps, respectively.[139] On 26 December 2016, the Mars Odyssey spacecraft briefly put itself into safe mode due to uncertainty about its orientation with regard to the Earth and the Sun—similar to a fault experienced in December 2013; following a reset of the inertial measurement unit and the circuit card that serves as the interface between that sensor, the flight software and the star tracker, for determining spacecraft attitude, the orbiter's knowledge of its orientation was restored. By 30 December 2016, Mars Odyssey had resumed communication relay assistance to Martian rovers, with science observations of Mars following shortly afterward.[140] In September 2017, Odyssey captured its first look at the Martian Moon Phobos, with the Thermal Emission Imaging System (THEMIS) infrared camera. Combining infrared and visible wavelength data, THEMIS will be helpful for the evaluation of Phobos as a possible future astronaut outpost, likewise for the smallest Mars Moon Deimos.[141]

NASA's Mars Reconnaissance Orbiter (MRO) continued its scientific operations in 2017, providing data for the purpose of determining whether life has ever existed on Mars, characterising the climate and geology and preparing for future human exploration. In the decade following its insertion into Mars' orbit in March 2006, the MRO completed 45,000 orbits of Mars[142]; while its primary science mission ended in November 2008, the MRO is currently in its third extended mission, which began in October 2012.[143] In March 2016, new evidence from the MRO's Compact Reconnaissance Imaging Spectrometer for Mars (CRISM) instrument suggested that Mars' Sisyphi Montes region, studded with flat-topped mountains, formed when volcanoes erupted beneath an ice sheet billions of years ago; the research helps show that there was extensive ice on ancient Mars and adds information about an environment combining heat and moisture that might have provided favourable conditions for microbial life.[144] Moreover, "radargrams" produced with data from MRO's Shallow Subsurface Radar (SHARAD) instrument confirmed previous 2003 and 2007 models that indicate that Mars' most recent ice age ended about 400,000 years ago, as the poles began to cool and thicken relative to Mars' equator.[145] Other

[139]"New Gravity Map Gives Best View Yet Inside Mars." 21 Mar. 2016. NASA 29 May 2017 <https://mars.jpl.nasa.gov/news/new-gravity-map-gives-best-view-yet-inside-mars>.

[140]"NASA Mars Odyssey Orbiter Resumes Full Operations." 28 Dec. 2016. NASA 29 May 2017 <https://mars.jpl.nasa.gov/news/2016/mars-odyssey-mission-status-report-orbiter-recovering-from-precautionary-pause-in-activity>.

[141]"Examining Mars's Moon Phobos in a Different Light" 4th of October 2017. NASA Science, Mars Exploration Program 2nd of October 2018 <https://mars.nasa.gov/news/examining-mars-moon-phobos-in-a-different-light/>.

[142]"Ten Years of Discovery by Mars Reconnaissance Orbiter." 9 Mar. 2016. NASA 30 May 2017 <https://mars.nasa.gov/news/ten-years-of-discovery-by-mars-reconnaissance-orbiter>.

[143]"Mission Timeline." NASA 30 May 2017 <https://mars.nasa.gov/mro/mission/timeline/>.

[144]"Found: Clues about Volcanoes Under Ice on Ancient Mars." 3 May 2016. NASA 30 May 2017 <https://mars.nasa.gov/news/found-clues-about-volcanoes-under-ice-on-ancient-mars>.

[145]"NASA Radar Finds Ice Age Record in Mars' Polar Cap." 26 May 2016. NASA 30 May 2017 <https://mars.nasa.gov/news/nasa-radar-finds-ice-age-record-in-mars-polar-cap>.

clues about Mars' climate history were revealed when images of the northern Arabia Terra region from the Context Camera and High-Resolution Imaging Science Experiment camera showed the formation of runoff lakes and streams that appeared roughly a billion years after an earlier era of wet conditions on ancient Mars.[146] After a decade of observation, in March 2017 the Context Camera (CTX) performed an almost total coverage of the planet at high resolution, collecting 90,000 images since 2006.[147] Moreover, the MRO also observed ancient seafloor hydrothermal deposits in southern Mars, representing an important discovery that may offer a hypothesis of the Earth's origin of life. The Compact Reconnaissance Spectrometer for Mars contributed to furnish information for the identification of minerals in the large deposits within Eridania basin, which is estimated to have held 210,000 cubic kilometres of water, representing the combined volume of all of North America's Great Lakes.[148]

On 5 August 2012, NASA's Mars Science Laboratory (MSL) rover "Curiosity" began its mission to address whether Mars ever had the right environmental conditions to support microbial life. It seeks to determine whether life ever arose on Mars and characterise its climate and geology to help prepare for human exploration. Its biological objectives are to determine the nature and inventory of organic carbon compounds, conduct an inventory of the chemical building blocks of life and identify features that may represent the effects of biological processes; its geological and geochemical objectives are to investigate the chemical, isotopic and mineralogical composition of Martian geological materials and to interpret the processes that have formed and modified rocks and soils; its planetary process objectives include assessing 4-billion-year timescale atmospheric evolution processes and determining the present state, distribution and cycling of water and carbon dioxide; finally its surface radiation objective is to characterise the broad spectrum of surface radiation, including galactic cosmic radiation, solar proton events and secondary neutrons.[149] Curiosity began its 3rd Martian year on 11 May 2016, having recorded environmental patterns through two full cycles of Martian seasons while crossing the Gale Crater's rugged terrain. Each year on Mars lasts 668.6 sols, with each sol lasting 39.6 min longer than an Earth day.[150] The rover continues to redefine researchers'

[146]"Some Ancient Mars Lakes Came Long After Others." 15 Sept. 2016. NASA 30 May 2017 <https://mars.nasa.gov/news/some-ancient-mars-lakes-came-long-after-others>.

[147]"Mars Global coverage by context and camera on MRO" 29th March 2017. NASA Science, Mars Exploration program 2nd of October 2018 <https://mars.nasa.gov/resources/mars-global-coverage-by-context-camera-on-mro/>.

[148]"Mars Study Yields Clues to Possible Cradle of Life" 6th of October 2017. NASA Science, Mars Exploration Program 3rd October 2018 <https://mars.nasa.gov/news/mars-study-yields-clues-to-possible-cradle-of-life/>.

[149]"Mars Science Laboratory: Mission Objectives." NASA 31 May 2017 <https://mars.nasa.gov/msl/mission/science/objectives/>.

[150]"Second Cycle of Martian Seasons Completing for Curiosity Rover." 11 May 2016. NASA 1 June 2017 <https://mars.nasa.gov/news/second-cycle-of-martian-seasons-completing-for-curiosity-rover>.

assumptions of Mars' volcanic history after an analysis of a rock sample using an X-ray diffraction instrument detected significant amounts of tridymite, created when large amounts of silica are heated at extremely high temperatures; the finding suggests that Mars once had explosive volcanoes that led to the presence of the mineral.[151] Another surprise came when a study of active sand dunes at the Bagnold Dunes of Mars' Mount Sharp showed sinuous crest lines of ripples similar to underwater ripples on Earth, with each ripple steeper than the face on the other side, instead of impact ripples formed by wind-carried sand grains colliding with other sand grains along the ground; this led researchers to believe that the mechanism forming ripples on Mars was its past thicker atmosphere which worked like a fluid to form smaller wind-drag ripples.[152] On 2 July 2016, Curiosity placed itself into safe mode as a precaution after experiencing an unexpected mismatch between camera software and data processing software in the main computer; the rover had also gone into safe mode three times in 2013. The rover was taken out of safe mode in the following week and resumed full operation on 11 July 2016.[153] Curiosity began its second 2-year mission extension on 1 October 2016, to explore key sites on lower Mount Sharp including a ridge capped with material rich in the iron-oxide mineral hematite and an exposure of clay-rich bedrock. Furthermore, additional extensions for exploring farther up Mount Sharp appear on the horizon.[154] Creating however a core problem for NASA JPL, Curiosity's drill mechanism faulted unexpectedly, and the rover team is testing possible ways to drill without using stabilisers.[155] Lastly, in June 2017, NASA developed the first AI system on a distant spacecraft, announcing a software upgrade that allows Curiosity to autonomously choose its targets.[156] The Autonomous Exploration for Gathering Increased Science (AEGIS) appears to be 93% accurate compared to the anticipated 24% "without intelligent targeting", permitting also an acceleration in data gathering. The software has been used 54 times from sol 1343 to sol 1663, approximately between 13 May 2016 and 7 April 2017 on Earth.[157]

---

[151] "NASA Scientists Discover Unexpected Mineral on Mars." 22 June 2016. NASA 1 June 2017 <https://mars.nasa.gov/news/nasa-scientists-discover-unexpected-mineral-on-mars>.

[152] "NASA Rover's Sand-Dune Studies Yield Surprise." 30 June 2016. NASA 1 June 2017 <https://mars.nasa.gov/news/nasa-rovers-sand-dune-studies-yield-surprise>.

[153] "Curiosity Mars Rover Resumes Full Operations." 11 July 2016. NASA 6 June 2017 <https://mars.nasa.gov/news/curiosity-mars-rover-resumes-full-operations>.

[154] "NASA's Curiosity Rover Begins Next Mars Chapter." 3 Oct. 2016. NASA 6 June 2017 <https://mars.nasa.gov/news/nasas-curiosity-rover-begins-next-mars-chapter>.

[155] Martin Perez "Curiosity Mars Rover Begins Study of Ridge Destination" 7th August 2017. NASA Mars Curiosity 2nd October 2018 <https://www.nasa.gov/feature/jpl/curiosity-mars-rover-begins-study-of-ridge-destination>.

[156] Bob Yirka "Software on Mars rover allows it to pick research targets autonomously" 22nd of June 2017. Phys.org 2nd of October 2018 <https://phys.org/news/2017-06-software-mars-rover-autonomously.html>.

[157] "AEGIS autonomous targeting for ChemCam on Mars Science Laboratory: Deployment and results of initial science team use" 21st of June 2017. Science Robotic Vol 2, 2nd of October 2018 <http://robotics.sciencemag.org/content/2/7/eaan4582>.

A mission designed to explore the planet's upper atmosphere, ionosphere and interactions with the Sun and solar wind is NASA's Mars Atmosphere and Volatile Evolution (MAVEN), which will be used to determine the role that the loss to space of volatile compounds from the Mars atmosphere has played in the history of Mars' habitability.[158] In March 2016, new research published in the journal *Geophysical Research Letters* showed how the Comet Siding Spring plunged Mars' magnetic field into chaos as it passed within a distance of 140,000 km in October 2014. While sensitive equipment aboard MAVEN was turned off during the flyby, its magnetometer remained on to observe how the charged plasma in Mars upper atmosphere interacted with Siding Spring's charged coma as it washed over the planet, flooding it with additional charged particles and temporarily merging the comet's magnetic field with Mars' weak field generated in its upper atmosphere; the effect of the plasma tide was similar to that of a strong but short-lived solar storm and likely fuelled a temporary surge in the amount of gas escaping from Mars' upper atmosphere.[159] New images from MAVEN's Imaging Ultraviolet Spectrograph (IUVS) published on 19 October 2016 provided an unprecedented ultraviolet view of Mars' nightside atmosphere verifying the presence of a "nightglow" emitted from nitric oxide; the phenomenon begins with chemical reactions on Mars' dayside as ultraviolet light from the Sun breaks down carbon dioxide and nitrogen in the upper atmosphere that are then blown around the planet and eventually descend to lower altitudes on the nightside where nitrogen and oxygen collide to form the nitric oxide molecules that generate the ultraviolet glow from the release of additional energy. Prior to these images, the nitric oxide glow on Mars was predicted by researchers, and its presence had been detected in prior missions.[160] In September 2017, MAVEN detected the energetic particles from a large solar storm both in Mars' orbit and on its surface[161] and, consequently, the appearance of a bright aurora. Through the Imaging Ultraviolet Spectrograph instrument (IUVS), scientists observed that the aurora was occurring because Mars' atmosphere was injected by energetic particles from the solar storm, giving birth to gases. Differently from planet

[158]"MAVEN." University of Colorado Boulder 14 Jan. 2013. <http://lasp.colorado.edu/home/maven/>.

[159]"Close Comet Flyby Threw Mars' Magnetic Field Into Chaos." 9 Mar. 2016. NASA 7 June 2017 <https://www.nasa.gov/feature/goddard/2016/close-comet-flyby-threw-mars-magnetic-field-into-chaos>.

[160]"NASA's MAVEN Mission Gives Unprecedented Ultraviolet View of Mars." 17 Oct. 2016. NASA 7 June 2017 <https://mars.nasa.gov/news/2016/nasas-maven-mission-gives-unprecedented-ultraviolet-view-of-mars>.

[161]"Solar storm radiation at Martian orbit and surface" 27th September 2017.University of Colorado Boulder, laboratory for Atmospheric and Science physic 2nd October 2018 <http://lasp.colorado.edu/home/maven/solar-storms-radiation-at-martian-orbit-and-surface/>.

Earth, the Martian aurora is not located in the polar regions because of the weaker magnetic field.[162]

ISRO's Mars Orbiter Mission (MOM) "Mangalayaan" is the first Asian interplanetary space probe to reach the planet Mars, joining the ranks of ESA, NASA and Roscosmos; ISRO is also the first space agency to have been successful in its maiden attempt and at the relatively low total cost of ₹450 crores ($73 million).[163] MOM lifted into space on 5 November 2013 and inserted in Mars' orbit on 24 September 2014. While MOM's primary mission is intended as a technology demonstration, its secondary mission is to study Mars' surface features, morphology, mineralogy and atmosphere with its suite of five indigenous scientific instruments: the Mars Colour Camera, the Methane Sensor for Mars, the Thermal Infrared Imaging Spectrometer, the Mars Exospheric Neutral Composition Analyser and the Lyman Alpha Photometer.[164] While ISRO announced the first science results of the mission on 11 November 2015, on 8 December 2016, it was made known that the highly anticipated measurements of atmospheric methane (a biomarker whose presence indicates the current or historical presence of life) from MOM's Methane Sensor for Mars (MSM) instrument might never arrive due to a flaw in the sensor design of the Fabry-Pérot interferometer, which didn't allow for methane data alone to be extracted from carbon dioxide and other gases in Mars' atmosphere.[165] Nevertheless, the MSM is still effective as an albedo mapper and in measuring reflected sunlight.[166] Subsequent science data collected by all instruments of the spacecraft is still being studied and prepared for publication. MOM was launched on a PSLV-C25, an XL variant of the PSLV launcher with 1350 kg of mass at launch,[167] the same model that launched the Moon mission Chandrayaan in 2008.[168] Lastly, plans for a second MOM spacecraft are already underway with

[162]"Sun storm triggers whole-planet aurora at Mars" 27th September 2017.University of Colorado Boulder, laboratory for Atmospheric and Science physic 2nd October 2018 <http://lasp.colorado.edu/home/maven/sun-storm-triggers-whole-planet-aurora-at-mars-video/>.

[163]"Mars Mission: India creates history as Mangalyaan successfully enters Mars orbit in first attempt." 24 Sept. 2014. The Economic Times 22 Mar. 2015 <http://economictimes.indiatimes.com/news/science/mars-mission-india-creates-history-as-mangalyaan-successfully-enters-mars-orbit-in-first-attempt/articleshow/43299562.cms>.

[164]Laxman, Srinivas. "Mars Orbiter Mission activates all science instruments as NASA, ISRO form joint Mars working group." 1 Oct. 2014. Planetary.org 22 Mar. 2015 <http://www.planetary.org/blogs/guest-blogs/2014/10010914-mars-orbiter-mission.html>.

[165]Mukunth, Vasudevan. "ISRO Mars Orbiter Mission's Methane Instrument Has a Glitch." 15 Dec. 2016. The Wire 8 June 2017 <https://thewire.in/85859/methane-isro-msm-spectroscopy/>.

[166]Klotz, Irene. "India's Mars Orbiter Mission Has a Methane Problem." 8 Dec. 2016 Space.com 8 June 2017 <http://www.space.com/34943-india-mars-orbiter-mission-methane-detector-flaw-red-planet.html>.

[167]"Mars Orbiter Mission" 11th December 2017. Gunter's Space Page 2nd of October 2018 <https://space.skyrocket.de/doc_sdat/mars-orbiter-mission.htm>.

[168]"PSLV-C25" 5th of November 2013. ISRO 2nd of October 2018 <https://www.isro.gov.in/launcher/pslv-c25>.

ISRO's "announcement of opportunity"; the follow-up mission is expected to be launched in March 2018.[169]

2.2.4 Mercury Science

Mercury is considered as one of the most mysterious planets of the solar system, hard to reach due to its proximity to the Sun. Further, given the Sun's glare, telescopic analysis is likewise difficult to conduct, and overall there have only been two successful missions that orbited the planet in the past: the NASA Mariner 10 (1970) and Messenger (2011). The BepiColombo programme is a joint endeavour between ESA and the Japanese space agency, JAXA, to measure and understand the composition, geophysics, atmosphere, magnetosphere and history of the planet Mercury. BepiColombo will be Europe's first mission to Mercury and is based on two spacecraft: the ESA-led Mercury Planetary Orbiter (MPO), built by Airbus as prime contractor,[170] and the JAXA-led Mercury Magnetospheric Orbiter (MMO). The MPO will use a suite of 11 experiments and instruments to study the planet's geology, composition, inner structure and exosphere, while the MMO has five experiments and instruments to study the planet's magnetic field, atmosphere, magnetosphere and inner interplanetary space.[171] The mission was expected to launch in October 2018, having been delayed by 6 months from its earlier intended April 2018 window, after a major electrical problem in the mission's Mercury Transfer Module (MTM) was encountered prior to a thermal test. While the postponement will have no impact on the science return of the mission, it will extend the flight time to 7.2 years, arriving in December 2025—1 year later than previously anticipated.[172] BepiColombo will follow in the footsteps of NASA's MErcury Surface, Space ENvironment, GEochemistry and Ranging (MESSENGER) mission that launched in August 2004 and orbited Mercury from March 2011 until it impacted the surface on 30 April 2015.[173] In June 2017, the final tests for launch

[169]Mukunth, Vasudevan. "ISRO Plans Return to Mars with Mangalyaan 2.0." 12 Aug. 2016. The Wire 8 June 2017 <https://thewire.in/58415/mars-orbiter-isro-planetary/>.

[170]"Mercury: a special challenge for visiting spacecraft" 12th June 2017. Airbus 3rd of October 2018. <https://www.airbus.com/newsroom/topics-in-focus/BepiColombo-Mission-to-Mercury.html>.

[171]"Factsheet." 1 Dec. 2016. ESA 12 June 2017 <http://sci.esa.int/bepicolombo/47346-fact-sheet/>.

[172]"BepiColombo Launch Rescheduled for October 2018." 25 Nov. 2016. ESA 12 June 2017 <http://sci.esa.int/bepicolombo/58591-bepicolombo-launch-rescheduled-for-october-2018/>.

[173]Nowakowski, Tomasz. "BepiColombo mission to Mercury on track for April 2018 launch." 21 July 2016. Spaceflight Insider 12 June 2017 <http://www.spaceflightinsider.com/missions/solar-system/bepicolombo-mission-mercury-april-2018-launch/>.

configuration were conducted, and afterward the thermal vacuum chamber test was conducted, checking the extremely high temperatures of the route to Mercury.[174]

2.2.5 *Jupiter Science*

NASA's Juno mission aims to answer questions about Jupiter's origins by observing its gravity and magnetic fields, atmospheric dynamics and composition and evolution. Launched on 5 August 2011, Juno inserted into an elliptical polar orbit around Jupiter on 4 July 2016; its original aim was to conduct 37 science orbits of Jupiter over the course of 20 months, before ending its mission in February 2018.[175] Its first close flyby took place on 27 August 2016, with Juno reaching a distance of 4200 km from Jupiter's clouds and with its suite of eight science instruments activated to transmit data over the following days.[176] The next close flyby took place in safe mode on 19 October 2016 after the spacecraft's telemetry indicated that two important helium check valves in its main engine had not operated as expected, opening for several minutes rather than a few seconds as intended, during the execution of a period reduction manoeuvre (PRM); while Juno rebooted successfully in the following week and returned to healthy operation, no science data was collected on that flyby.[177] The next flyby took place on 11 December 2016, with data collected from seven of Juno's eight science instruments—not including its Jovian Infrared Auroral Mapper (JIRAM) which was receiving a software patch at the time.[178] NASA had furthermore decided to also forego the PRM in that third flyby, finding that it would not have affected the quality of the science collected by Juno during its closest approach; however, as the manoeuvre would have reduced Juno's orbital period from 53.4 to 14 days, maintaining its current path will result in around 12 science orbits instead of 37. The latest studies of May 2017 demonstrated that Jupiter's gravitational field is askew, differently shaped in the southern and northern poles; this suggests to scientists the presence of asymmetrical hydrogen-rich gas, relevant for evaluating the rotation of Jupiter's interior and in determining whether it is rotating as a single solid body or having different layers spinning

[174]"Preparing for Mercury: BepiColombo Staks Complete Testing" 6th of July 2017. ESA Science & Technology, 3rd of October 2018 <http://sci.esa.int/bepicolombo/59294-preparing-for-mercury-bepicolombo-stack-completes-testing/>.

[175]"Mission to Jupiter—Juno—About the Mission." NASA 12th June 2017 <https://www.jpl.nasa.gov/missions/juno/>.

[176]"NASA's Juno Successfully Completes Jupiter Flyby." 27 Aug. 2016. NASA 12 June 2017 <https://www.missionjuno.swri.edu/news/juno-completes-jupiter-flyby>.

[177]"Juno Spacecraft in Safe Mode for Latest Jupiter Flyby. Scientists Intrigued by Data from First Flyby." 19 Oct. 2016. NASA 12 June 2017 <https://www.missionjuno.swri.edu/news/mission-status-report-10-19-2016>.

[178]"NASA Juno Mission Prepares for December 11 Jupiter Flyby." 9 Dec. 2016. NASA 12 June 2017 <https://www.missionjuno.swri.edu/news/dec11-jupiter-flyby>.

separately.[179] The extension of the $1 billion mission has been approved until July 2021, providing additional months to achieve NASA's science objectives.[180]

The next spacecraft to orbit Jupiter will be ESA's Jupiter Icy moon Explorer (JUICE) in development under ESA's Cosmic Vision 2015–2025 plan. Its foreseen launch date is in 2022 aboard an Ariane 5 launcher, and it should arrive at the Jovian planet in 2030. The nearly 5500 kg spacecraft will make a careful investigation of Jupiter's three biggest moons, using the gravity of Jupiter to initiate a series of close flybys around Callisto, Ganymede and Europa, and then finally settle into an orbit around Ganymede for an 8-month study.[181] As all three moons are suspected of having oceans of water beneath their icy crusts, scientists are trying to understand whether there is any possibility that these moons could host microbial life.[182] On 16 July 2015, ESA selected Airbus D&S for a 350.8 million € ($374 million) contract to build the JUICE Orbiter. The contract was formalised on 9 December 2015.[183] JUICE will spend the first 2 years realising 30 observation overflights of the 3 moons, and it will carry 10 instruments, such as an ice-penetrating radar, a sensor to monitor the magnetic field and cameras. The radar sound instrument, RIME, will penetrate below the Galilean icy moons' surface; the radar antennas were, moreover, tested by Airbus in a helicopter flight campaign.[184]

2.2.6 *Saturn Science*

The Cassini-Huygens mission, a joint NASA, ESA and ASI mission, was launched in 1997. Reaching Saturn in 2004, Cassini went on to drop the Huygens probe onto Saturn's moon Titan. From there, the Cassini Solstice Mission was supposed to end in June 2008; however, funding was provided to allow continued operation to

[179] Alexandra Witze, "Juno probe discovers surprising activity in Jupiter's interior" 24th of October 2017. Phys.org 3rd of October 2017 <https://phys.org/news/2017-10-juno-probe-jupiter-interior.html>.

[180] JPL "NASA re-plans Juno Jupiter mission" 6th June of 2018. NASA Jet Propulsion Laboratory 3rd of October 2018 <https://www.jpl.nasa.gov/news/news.php?feature=7153>.

[181] De Selding, Peter B. "Airbus to Build ESA's Jupiter-bound Juice Orbiter." 17 July 2015. SpaceNews 11 Jan. 2016 <http://spacenews.com/airbus-to-build-esas-jupiter-bound-juice-orbiter/>.

[182] "ESA Selects 1bn-Euro JUICE Probe to Jupiter." 2 May 2012. BBC News 27 Mar. 2014 <http://www.bbc.com/news/science-environment-17917102>.

[183] De Selding, Peter B. "ESA, Airbus Formalize Jupiter Icy Moons Contract." 11 Dec. 2015. SpaceNews 11 Jan. 2016 <http://spacenews.com/esa-airbus-formalize-jupiter-icy-moons-contract/>.

[184] "JUICE radar antenna tested in helicopter flight campaign" 26th September 2017. Airbus 3rd of October 2017 <https://www.airbus.com/newsroom/press-releases/en/2017/09/JUICE-radar-antenna.html>.

provide new insights on Saturn and its moons.[185] The Cassini mission began its final year on 14 September 2016, after more than 12 years of studying Saturn, its rings and moons.[186] On 23 January 2016, Cassini took its second of five large propulsive manoeuvres using gravity assists from Titan to take the spacecraft from an equatorial orbit around Saturn to an inclination above its poles for Cassini's "Grand Finale" that would involve passes through Saturn's rings and end with a plunge into its atmosphere.[187] Cassini's final mission phase began shortly before its first close dive through the centre of Saturn's outermost F ring on 4 December 2016; the spacecraft made 20 of these passes until 22 April 2017 and then transitioned between Saturn's innermost ring and atmosphere for an additional 22 passes.[188] A 2016 study published in the *Journal of Geophysical Research* independently confirmed that Ligeia Mare, the second largest sea on Titan, is rich in methane rather than ethane, which is produced in abundance in Titan's atmosphere. A possible explanation for the sea's nearly pure methane composition could be that the sea is regularly replenished with fresh methane rainfall or that the denser ethane somehow drains into the larger adjacent sea, Kraken Mare. Using radar data collected from a 2013 experiment that bounced radio signals off Ligeia Mare, researchers were also able to find that its seabed is likely covered by a sludge layer of organic-rich compounds.[189] Another recent paper in the journal *Geophysical Research Letters* described how researchers using data from Cassini's radar imager were able to find the first direct evidence of liquid-filled channels on Titan, along with the first observation of canyons hundreds of metres deep, indicating that it was formed from a process that was active for a long time or that the channels had eroded down much faster than other areas on Titan's surface.[190] ASI and NASA had signed a Memorandum of Understanding in 1995 whereby ASI was commissioned to develop the high-gain antenna, incorporated with the low-gain counterpart, and the radar in addition to the VIMS spectrometer and Radio Science Subsystem (RSIS). Cassini's mission ended 15 September 2017 under wide media coverage; only 1 week before its final day, it

---

[185] Mason, Betsy. "Cassini Gets Life Extension to Explore Saturn Until 2017." 3 Feb. 2010. WIRED 18 Dec. 2012 <http://www.wired.com/wiredscience/2010/02/cassini-life-extension-2017/>.

[186] "Cassini Begins Epic Final Year at Saturn." 14 Sept. 2016. NASA 8 June 2017 <https://saturn.jpl.nasa.gov/news/2939/cassini-begins-epic-final-year-at-saturn/>.

[187] "Cassini Heads for 'Higher Ground' at Saturn." 24 Jan. 2016. NASA 8 June 2017 <https://saturn.jpl.nasa.gov/news/2801/cassini-heads-for-higher-ground-at-saturn/>.

[188] Dyches, Preston. "Cassini Makes First Ring-Grazing Plunge." 4 Dec. 2016. NASA 9 June 2017 <https://saturn.jpl.nasa.gov/news/2974/cassini-makes-first-ring-grazing-plunge/>.

[189] Dyches, Preston and Markus Bauer. "Cassini Explores a Methane Sea on Titan." 25 Apr. 2016. NASA 9 June 2017 <https://saturn.jpl.nasa.gov/news/2897/cassini-explores-a-methane-sea-on-titan/>.

[190] Dyches, Preston. "Cassini Finds Flooded Canyons on Titan." 9 Aug. 2016. NASA 9 June 2017 <https://saturn.jpl.nasa.gov/news/2927/cassini-finds-flooded-canyons-on-titan/>.

was regularly monitoring Saturn's most mysterious moon Enceladus,[191] revealing that it hosts elements necessary for life.[192] ESA ground stations supported the plunge of the spacecraft into the planet, receiving until the very last minute data in almost real time until the antenna didn't point toward Earth anymore.[193]

2.2.7 Solar Observation

The monitoring of space weather involves long-term surveillance of solar phenomena. Continued observation of the Sun's external activity has the benefit of improving our understanding of its interior, its corona, the monitoring of solar wind and its consequences on Earth and its neighbouring planets. Coronal mass ejections (CMEs) from the Sun emit surges of charged particles in directions that may cross Earth's path and can damage satellites, impede space-based services and affect the terrestrial electrical infrastructure.

Launched in November 2009, ESA's PRoject for OnBoard Autonomy (PROBA)-2 microsatellite mission tracks spikes in CMEs ejecting from the Sun that have previously been seen to just skim Earth, typically bringing with them a burst of radio energy.[194] Building on nearly 8 years of successful PROBA-1 experience, PROBA-2 continues as part of ESA's In-orbit Technology Demonstration Programme, whose missions are dedicated to the demonstration of innovative technologies; its payload carries 17 new technological developments and 4 scientific experiments.[195] The lifetime mission of PROBA-2 was once again extended on 22 November 2016, and its third operational mission will continue from 1 January 2017 to 31 December 2018.[196] On 9 March 2016, PROBA-2 was able to observe a partial solar eclipse with its SWAP imager from its 800-km-altitude polar orbit; the SWAP views the solar disc at extreme ultraviolet wavelengths to capture the turbulent surface of the

[191]"Cassini, ultime immagini aspettando il gran finale." 6th September 2017. ASI NEWS 3rd October 2018 <https://www.asi.it/it/news/cassini-ultime-immagini-aspettando-il-grand-finale> <https://www.asi.it/en/activity/solar-system-exploration/cassini-huygens>.

[192]"What's next? " Cassini. NASA Science and Exploration 3rd of October 2018 <https://solarsystem.nasa.gov/missions/cassini/mission/grand-finale/whats-next/>.

[193]"CASSINI-HUYGENS" 15th September 2017. ESA Science 3rd of October 2018 <https://www.esa.int/Our_Activities/Space_Science/CassiniHuygens/Cassini_concludes_pioneering_mission_at_Saturn>.

[194]"Small Sun-Watcher Proba-2 Offers Detailed View of Massive Solar Eruption." 9 June 2011. ESA 17 Apr. 2013 <http://www.esa.int/Our_Activities/Technology/Small_Sun-watcher_Proba-2_offers_detailed_view_of_massive_solar_eruption>.

[195]"About PROBA-2." 17 Dec. 2012. ESA 13 June 2017 <http://www.esa.int/Our_Activities/Space_Engineering_Technology/Proba_Missions/About_Proba-2>.

[196]"Two-year extensions confirmed for ESA's science missions." 22 Nov. 2016. ESA 12 June 2017 <http://sci.esa.int/director-desk/58589-two-year-extensions-confirmed-for-esa-s-science-missions/>.

Sun and its swirling corona.[197] As the Sun's 11-year activity cycle drew closer to its minimum, where the number of sun spots, active regions, solar flares and eruptions diminish, PROBA-2 was in the right position to view an annular solar eclipse on 1 September 2016. Orbiting the Earth once every 90 min, PROBA-2's SWAP imager and its LYRA radiometer, which measures solar irradiance, managed to view four partial eclipses.[198] ESA is also developing PROBA-3, as a pair of satellites maintaining a fixed configuration to form a 150-m long solar chronograph to study the Sun's faint corona closer to the solar rim than previously achieved.[199] The two satellites, a coronagraph spacecraft and an occulter, will be launched together in late 2020 into a highly elliptical tandem orbit, repetitively demonstrating acquisition, rendezvous, proximity operations, formation flying, coronagraph operations, separation and convoy flying.[200] Aiming to produce artificial total solar eclipses, PROBA-3 will position the small satellite between the Sun and the visual field of its counterpart, creating a shadowed area firstly allowing the view of the corona, being the source of solar wind that could affect satellites, and considering also events such as coronal mass ejections. Using new technologies and autonomous drive, the main challenge will be to keep the satellites in a correct and safe position.[201] Beside the scientific aims, the pair will be a precise tool for the achievement of perfect positioning as ESA's first precision formation flying technologies.[202]

The Sun's activity is evaluated in ten different wavelengths revealing a range of solar features[203] from NASA's Solar Dynamics Observatory (SDO), a mission that seeks to determine how the Sun's magnetic field is generated and structured and how this stored magnetic energy is released in the form of solar wind, energetic particles and variations in the solar irradiance. The spacecraft is comprised of three scientific experiments: the Atmospheric Imaging Assembly (AIA), the EUV Variability Experiment (EVE) and the Helioseismic and Magnetic Imager (HMI).[204] Located in a geosynchronous orbit around the Earth, SDO's global view of the Sun facilitates

---

[197] "Pacific solar eclipse seen from Proba-2." 9 Mar. 2016. ESA 13 June 2017 <http://www.esa.int/spaceinvideos/Videos/2016/03/Pacific_solar_eclipse_seen_from_Proba-2>.

[198] D'Huys, Elke. "Another eclipse!" 13 June 2016. ESA/ROB 13 June 2017 <http://proba2.sidc.be/node/316>.

[199] "About PROBA3." 19 Nov. 2012. ESA 1 Apr. 2014 <http://www.esa.int/Our_Activities/Technology/Proba_Missions/About_Proba-3>.

[200] "Fact Sheet." ESA 13 June 2017 <http://esamultimedia.esa.int/docs/Proba/Proba-3_fact-sheet_final.pdf>.

[201] "ESA's Proba-3 will create artificial solar eclipses" 18th August 2017, provided by European Space Agency. Phys.org 4th of October 2018 <https://phys.org/news/2017-08-esa-proba-artificial-solar-eclipses.html>.

[202] ESA Fact Sheet Proba-3. ESA Multimedia 5th October 2018 <https://esamultimedia.esa.int/docs/Proba/Proba-3_fact-sheet_final.pdf>.

[203] SDO Gallery 7th of August 2017. NASA SDO Solar Mission <https://sdo.gsfc.nasa.gov/gallery/potw/item/932>.

[204] "SDO | Solar Dynamics Observatory." NASA Goddard Space Flight Center 4 Mar. 2013 <http://sdo.gsfc.nasa.gov/>.

research that focuses on the previously unrecorded real fine structure of the star.[205] SDO continuously observed the Sun's activity throughout 2016, capturing images of coronal holes and solar flares and occasionally catching the transit of a planet passing in front of the star. On 9 May 2016, SDO (along with the NASA/ESA SOHO mission and the JAXA/NASA/UK Hinode mission) studied the planet Mercury as it transited between the Earth and the Sun—an event that happens 13 times or so in a century. The transit helped SDO to both align and calibrate its space instruments, enabling researchers to mitigate the effect of stray light in SDO's instruments that should otherwise have viewed the profile of Mercury transiting the Sun without a glow surrounding its circumference. SOHO used the transit opportunity to measure the Sun's rotation axis.[206] On 2 August 2016, SDO entered inertial mode as the Moon transited the Sun; while its HMI and EVE instruments came back online within 2 days of the transit, a temporary glitch in SDO's AIA instrument delayed full science mode operations until 10 August 2016.[207] And in research published on 11 October 2016, scientists using SDO and IRIS data were able to observe certain frequencies of solar seismic waves channelling upwards through the chromosphere and corona atmospheric layers into the Sun's photosphere. The technique gives scientists a new tool to understand the Sun's lower atmosphere and also might help to address a long-standing question in solar physics regarding excess heat in the Sun's corona, which is about 100 times hotter than the chromosphere below.[208] In September 2017, the SDO imaged two intense solar flares emanating from the solar region labelled AR 2673, and NOAA's Space Weather Prediction Center denoted and evaluated the negative repercussions of the powerful eruptions.[209]

The SOlar and Heliospheric Observatory (SOHO) mission is a joint collaboration between ESA and NASA to study the Sun from its deep core to the outer corona as well as solar wind. Launched in December 1995 for a 3-year mission that was meant to end in 1998,[210] its success prompted several mission extensions with operations

[205] SpaceNews Staff. "NASA Boasts Big Results from 5-minute Spaceflight." 28 Jan. 2013 SpaceNews 4 Mar. 2013 <http://www.spacenews.com/article/nasa-boasts-big-results-from-5-minute-spaceflight>.

[206] "Satellites to See Mercury Enter Spotlight on May 9." 3 May 2016. NASA 13 June 2017 <https://www.nasa.gov/feature/goddard/2016/satellites-to-see-mercury-enter-spotlight-on-may-9>.

[207] "SDO Status Update—Aug. 10, 2016." 10 Aug. 2016. NASA 13 June 2017 <https://www.nasa.gov/feature/goddard/2016/sdo-status>.

[208] Frazier, Sarah. "Tracking Waves from Sunspots Gives New Solar Insight." 20 Oct. 2016. NASA 13 June 2017 <https://www.nasa.gov/feature/goddard/2016/tracking-waves-from-sunspots-gives-new-solar-insight>.

[209] "Two significant solar flares imaged by NASA's SDO" 6th of September 2017. Phys.org 5th of October 2018 <https://phys.org/news/2017-09-significant-solar-flares-imaged-nasa.html>.

[210] "SOHO Overview." 5 June 2013. ESA 13 June 2017 <http://www.esa.int/Our_Activities/Space_Science/SOHO_overview2>.

continuing through 2018, confirmed by ESA's Science Programme Committee.[211] SOHO orbits around the Sun in step with the Earth, at a distance of 1.5 million kilometres from Earth, enabling an uninterrupted view of the star.[212] SOHO's mission was once again extended on 22 November 2016 and will continue from 1 January 2017 to 31 December 2018.[213] In anticipation of Mercury's 9 May 2016 transit of the Sun, 2 of SOHO's 12 instruments, the Extreme ultraviolet Imaging Telescope (EIT) and the Michelson Doppler Imager (MDI), were restored to full operation, after a 5-year dormancy, to take measurements of the event.[214] On 4 August 2016, SOHO caught the demise of a Kreutz-type comet that plunged too close to the Sun in its highly elliptical orbit.[215] Further, a paper appearing in *Astronomy and Astrophysics* on 6 June 2016 summarised research on a new model, based on data collected from SOHO and STEREO, to map out where solar energetic particles (SEPs) might be found as they spread out and travel away from the Sun; the model takes into consideration the fact that turbulence in solar material can cause magnetic field lines to wander, showing SEPs taking a much wider path than previous models predicted and explaining how SEPs can reach the far side of the Sun.[216] Recent findings made via SOHO, announced by ESA in August 2017, furthermore suggest that the core of the Sun is rotating four times faster than the surface. Moreover, 16 years of data collection of SOHO's Global Oscillations and Low Frequency (GOLF) instrument revealed the functionality of the solar interior by studying pressure and gravity waves.[217]

On 25 October, SOHO's lower right corner field of view was intersected by the comet 96P, which it has detected four times since 1996. And in a rare occurrence, NASA's STEREO mission also observed the comet from its vantage point located on

[211]"Two Years Extensions Confirmed for ESA'S Science Missions" 22th November 2016. ESA's Director Desk 5th October 2018 <http://sci.esa.int/director-desk/58589-two-year-extensions-confirmed-for-esa-s-science-missions/>.

[212]About the SOHO Mission. "SOHO Fact Sheet." SOHO—Solar and Heliospheric Observatory 4 Mar. 2013 <http://sohowww.nascom.nasa.gov/about/docs/SOHO_Fact_Sheet.pdf>.

[213]"Two-year extensions confirmed for ESA's science missions." 22 Nov. 2016. ESA 12 June 2017 <http://sci.esa.int/director-desk/58589-two-year-extensions-confirmed-for-esa-s-science-missions/>.

[214]"Satellites to See Mercury Enter Spotlight on May 9." 3 May 2016. NASA 13 June 2017 <https://www.nasa.gov/feature/goddard/2016/satellites-to-see-mercury-enter-spotlight-on-may-9>.

[215]"ESA, NASA's SOHO Sees Bright Sungrazer Comet." 4 Aug. 2016. NASA 14 June 2017 <https://www.nasa.gov/image-feature/goddard/2016/esa-nasa-s-soho-sees-bright-sungrazer-comet>.

[216]Tran, Lina. "Wayward Field Lines Challenge Solar Radiation Models." 17 Oct. 2016. NASA 14 June 2017 <https://www.nasa.gov/feature/goddard/2016/wayward-field-lines-challenge-solar-radiation-models>.

[217]"Gravity Waves Detected in Sun's Interior Reveal Rapidly Rotating Core" 1st of August 2017. ESA Science and Technology, SOHO 5th of October 2018 <http://sci.esa.int/soho/59362-gravity-waves-detected-in-suns-interior-reveal-rapidly-rotating-core/>.

the other side of the Earth orbit. This incident will help researchers gain knowledge on the composition of the comet and its interaction with solar wind.[218]

NASA's Solar TErrestrial RElations Observatory (STEREO) mission, launched on 26 October 2006, provides the first-ever stereoscopic measurements of the Sun and its CMEs. Made up of two space-based observatories, i.e. STEREO-A travelling in a smaller and faster orbit (ahead of Earth's orbit) and STEREO-B trailing behind with a larger and slower orbit, the mission aims to understand the causes and mechanisms of CME initiation, characterise the propagation of CMEs through the heliosphere, discover the mechanisms and sites of energetic particle acceleration in the low corona and the interplanetary medium and improve the determination of the structure of the ambient solar wind.[219] While NASA mission operations had lost communication with STEREO-B on 1 October 2014, during a test of the spacecraft's command loss timer as it neared solar conjunction, NASA managed to re-established contact with STEREO-B on 21 August 2016, following several months of attempts to contact the spacecraft without the Sun's interference.[220] Despite intermittent contact with STEREO-B since that time, the operations team is still in the process of assessing the spacecraft's health, re-establishing attitude control and evaluating all subsystems and instruments, which could take months or even years.[221] STEREO-A continued collecting data, having resumed normal science operations on 17 November 2015, after undergoing side lobe repointing operations on 20 August and 1 December 2014 which had the spacecraft transmit lower resolution data for most of 2015.[222]

JAXA's Hinode mission, formerly Solar-B, is a joint collaboration with NASA and the UK to measure solar magnetic fields; study the generation, transport and dissipation of magnetic energy from the photosphere to the corona; and record how energy stored in the Sun's magnetic field is released as the field rises into the Sun's outer atmosphere.[223] Launched in September 2006 for a 3-year mission, the polar, Sun-synchronous orbiting spacecraft has allowed scientists to study solar phenomena, from solar explosions to the delicate motion of solar spicules, in great detail for

---

[218]"Return of Comet: 96P Spotted by ESA/NASA satellites" 3rd of November 2017. NASA Comets, 5th of October 2018 <https://www.nasa.gov/feature/goddard/2017/return-of-the-comet-96p-spotted-by-esa-nasa-satellites>.

[219]"About the STEREO Mission." 9 Feb. 2013. NASA 14 June 2017 <https://stereo.gsfc.nasa.gov/mission/mission.shtml>.

[220]"NASA Establishes Contact with STEREO Mission." 22 Aug. 2016. NASA 14 June 2017 <https://www.nasa.gov/feature/goddard/2016/nasa-establishes-contact-with-stereo-mission>.

[221]"STEREO: 10 Years of Revolutionary Solar Views." 25 Oct. 2016. NASA 14 June 2017 <https://www.nasa.gov/feature/goddard/2016/stereo-10-years-of-revolutionary-solar-views>.

[222]"Q & A for Operations of STEREO during Superior Conjunction." 3 July 2014 NASA 14 June 2017 <https://www.nasa.gov/content/goddard/q-a-for-stereo-during-superior-conjunction>.

[223]"Hinode: Mission to the Sun." 31 July 2015. NASA 19 June 2017 <https://www.nasa.gov/mission_pages/hinode/mission.html>.

more than 10 years, continuing beyond 2016.[224] In a research published on 19 April 2016, scientists using Hinode, SDO and STEREO-A observations' from a December 2013 solar flare were provided with speed, temperature, density and size measurements that strengthened scientists' understanding that the electromagnetic phenomenon called a "current sheet" is the result of magnetic reconnection on the Sun. A current sheet is a very fast, very flat flow of electrically charged material that forms when two oppositely aligned magnetic fields come in close contact, creating very high magnetic pressure that is unstable and can lead to new configurations, with the heat and light from the transformation producing a solar flare. As current sheets and magnetic reconnection are so closely associated, such detailed observations bolster the idea that magnetic reconnection is the force behind solar flares.[225]

NASA's Interface Region Imaging Spectrograph (IRIS) satellite is a Small Explorer mission to observe how solar material moves, gathers energy and heats up as it travels through the Sun's lower atmosphere. Its mission, launched in June 2013 and operating in a polar, Sun-synchronous orbit, complements the SDO and the Hinode missions to explore the Sun's variable atmosphere and how it impacts Earth; that is, while SDO and Hinode monitor the photosphere (solar surface) and corona (outer atmosphere), IRIS observes the chromosphere and transition region between. In addition to being where most of the Sun's ultraviolet emission is generated, this region powers the Sun's million-degree atmosphere and drives the solar wind. Like with the SDO spacecraft, IRIS used the transit of Mercury in front of the Sun to help recalibrate its telescope to correct any changes that might have occurred during its launch into orbit.[226] IRIS observed a mid-level solar flare on 24 July 2016, capturing how large amounts of magnetic energy are released, heating the sun's atmosphere and releasing energised particles out into space, which in turn drives post-flare loops of plasma (i.e. coronal rain) to puzzlingly rapidly cool from millions down to a few tens of thousands of kelvins while descending to the photosphere.[227] While IRIS's prime mission was for 2 years, it has been extended through September 2018, with the possibility of additional extensions afterward.[228]

NASA's Reuven Ramaty High-Energy Solar Spectroscopic Imager (RHESSI) focuses on the basic physics of particle acceleration and energy release in solar

---

[224]Frazier, Sarah. "Highlights from a Decade of JAXA and NASA's Hinode Solar Observatory." 22 Sept. 2016. NASA 19 June 2017 <https://www.nasa.gov/feature/goddard/2016/highlights-from-a-decade-of-jaxa-and-nasa-s-hinode-solar-observatory>.

[225]Frazier, Sarah. "Seeing Double: NASA Missions Measure Solar Flare from 2 Spots in Space." 19 Apr. 2016. NASA 19 June 2017 <https://www.nasa.gov/feature/goddard/2016/seeing-double-nasa-missions-measure-solar-flare-from-2-spots-in-space>.

[226]Fox, Karen C. "IRIS Releases New Imagery of Mercury Transit." 18 May 2016. NASA 14 June 2017 <https://www.nasa.gov/feature/goddard/2016/iris-releases-new-imagery-of-mercury-transit>.

[227]Tran, Lina. "IRIS Spots Plasma Rain on Sun's Surface." 5 Aug. 2016. NASA 14 June 2017 <https://www.nasa.gov/feature/goddard/2016/iris-spots-plasma-rain-on-suns-surface>.

[228]Tran, Lina. "NASA's Sun-Observing IRIS Mission." 1 Dec. 2016. NASA 14 June 2017 <https://www.nasa.gov/feature/goddard/2016/nasas-sun-observing-iris-mission>.

flares. Launched in February 2002 for a 2-year mission, RHESSI has operated for more than 14 years and underwent its 5th month-long detector anneal from 23 February to 29 April 2016. An annual procedure involves heating up detectors to rejuvenate them from accumulated radiation damage and then cooling them back down to operating temperatures for the spacecraft to resume collecting solar X-ray and gamma ray data.[229] RHESSI has provided diagnostic observations of high-energy processes in solar flares for over 15 years since its launch in February 2002. These observations address the key Heliophysics goal of understanding the fundamental processes of energy release and particle acceleration in solar eruptions, both flares and coronal mass ejections (CMEs). Designed for imaging spectroscopy of hard X-ray (HXR), RHESSI is one of the essential composing elements of the Heliophysics System Observatory (HSO), and its re-entry is estimated for December 2022, modified from the previous prediction of October 2021.[230] Due to the interests in space weather and, in particular, to solar flares, NASA is proceeding with the selection of two over five missions to improve the knowledge about interferences with radio and satellite communication in proximity to Earth. One of the proposals is the third mission of the Focusing Optics X-ray Solar Imager (FOXSI-3) rocket, a collaboration between the University of Minnesota, University of California Berkeley, NASA/Marshall, NASA/Goddard, University of Tokyo/Kavli IPMU, Nagoya University, Tokyo University of Science, JAXA/ISAS and the National Astronomical Observatory of Japan[231]; moreover, from September 2018 it will implement RHESSI as the observer of thousands of soft and higher-energy gamma rays.[232]

2.2.8 *Solar System Science*

NASA's Dawn mission studies the asteroid Vesta and dwarf planet Ceres, the two largest bodies in the asteroid belt between Mars and Jupiter (with average diameters of 525 km and 950 km, respectively), to characterise the early solar system and the processes that dominated its formation. Launched on 27 September 2007, Dawn inserted into orbit around Ceres on 6 March 2015.[233] In the following months, Dawn mapped Ceres with increasing detail as it lowered its orbit from 4400 km on 7 June

---

[229]"RHESSI STATUS." 29 Apr. 2016. NASA 19 June 2017 <https://hesperia.gsfc.nasa.gov/rhessi3/news-and-resources/status/index.html>.

[230]Samuel Krucker "Heliophysics review—The Reuven Ramaty High Energy Solar Spectroscopic Imager" 27th February 2017. NASA 5th of October 2018 <https://hesperia.gsfc.nasa.gov/senior_review/2017/senior_review_proposal_2017.pdf>.

[231]"Foxsi sounding rocket project". 5th of October 2018 <http://foxsi.umn.edu/>.

[232]"Proposed NASA Mission Would Investigate Where Space Weather Begins" 14th of November 2017. NASA SpaceTech 5th of October 2018 <https://www.nasa.gov/feature/goddard/2017/proposed-nasa-mission-would-investigate-where-space-weather-begins>.

[233]"NASA Spacecraft Becomes First to Orbit a Dwarf Planet." 6 Mar. 2015. NASA 24 May 2016 <http://dawn.jpl.nasa.gov/news/news-detail.html?id=4503>.

2015 to 1470 km on 25 August 2015, to its final orbit of 380 km on 26 October 2015.[234] Capturing images at a resolution of 35 m per pixel, Dawn's visible and infrared mapping spectrometer (VIR) enables scientists to identify specific minerals by their reflected wavelength, while data from its gamma ray and neutron detector (GRaND) provides details on Ceres' composition and on the abundances of elements in its surface.[235] Observations of the domed 3 km tall Ahuna Mons in Ceres' bright Occator Crater revealed an abundance of bright material on some of its slopes, and less on others.[236] While ice has been located near the surface and in cold traps in permanent shadow on Ceres, the bright material in the crater is likely to be highly reflective salts.[237] The intricate geometry of the crater's interior and of Ahuna Mons suggests that geologic activity occurred in the recent past; moreover, studies of Ahuna Mons have led researchers to conclude the domed mountain to be the first known example of a cryovolcano that erupts a liquid made of volatiles such as water in the form of a salty mud mix, instead of silicates. Researchers have also found evidence that Ceres might once have had a weak temporary atmosphere.[238] Dawns' prime mission ended on 30 June 2016; its extended mission phase began in the following month with an elliptical orbit of 7200 km to view the dwarf planet from a higher vantage.[239] In February 2017, the first organic material traces were found on Ceres and organic evidence native to the dwarf planet provided for a hypothesis of chemical activity.[240]

Prioritising the collection of data with Dawn's gamma ray and neutron spectrometer, NASA authorised one more extension of the mission at Ceres during which the spacecraft will slope to 200 km to the surface to study the most relevant object in the asteroid belt between Mars and Jupiter. Thus, it will be an additional opportunity to analyse the dwarf planet in its perihelion moment in April 2018. Further, the DLR, ASI, the Italian National Astrophysical Institute and the Max Planck Institute for Solar Research are additional partners of the mission team.[241]

[234]"Dawn Heads Toward Final Orbit." 26 Oct. 2015. NASA 24 May 2016 <http://dawn.jpl.nasa.gov/news/news-detail.html?id=4751>.

[235]"New Details on Ceres Seen in Dawn Images." 12 Jan. 2016. NASA 19 June 2017 <https://dawn.jpl.nasa.gov/news/news-detail.html?id=4817>.

[236]"Dawn's First Year at Ceres: A Mountain Emerges." 7 Mar. 2016. NASA 19 June 2017 <https://dawn.jpl.nasa.gov/news/news-detail.html?id=5745>.

[237]"Where is the Ice on Ceres? New NASA Dawn Findings." 15 Dec. 2016. NASA 19 June 2017 <https://dawn.jpl.nasa.gov/news/news-detail.html?id=6703>.

[238]"Ceres' Geological Activity, Ice Revealed in New Research." 1 Sept. 2016. NASA 19 June 2017 <https://dawn.jpl.nasa.gov/news/news-detail.html?id=6611>.

[239]"Where is the Ice on Ceres? New NASA Dawn Findings." 15 Dec. 2016. NASA 19 June 2017 <https://dawn.jpl.nasa.gov/news/news-detail.html?id=6703>.

[240]"Dawn Discovers Evidence for Organic Material on Ceres" 17th of February 2017. NASA 5th of October 2018 <https://www.nasa.gov/feature/jpl/dawn-discovers-evidence-for-organic-material-on-ceres>.

[241]"Dawn Mission Extended at Ceres" 19th October of 2017. NASA JPL California Institute 5th of October 2018 <https://www.jpl.nasa.gov/news/news.php?feature=6980>.

ESA's Rosetta mission to perform a detailed study of the comet 67P/Churyumov-Gerasimenko concluded on 30 September 2016. Launched on 2 March 2004 on Europe's Ariane 5 launcher, the Rosetta orbiter and its Philae lander had reached the comet after a decade-long journey and became the first mission to successfully orbit a comet on 6 August 2014.[242] On 12 November 2014, Philae accomplished the first successful landing on a comet, albeit bouncing twice due to a misfire of its harpoons, and landing in the shadow of a cliff. Philae spent nearly 57 h performing its science objectives, managing to transmit all of the results from its final sequence of measurements before its battery ended.[243] From there, contact was established sporadically before the comet's closest approach to the Sun in August 2015. In the following months, ESA and the DLR extended Rosetta's mission into September 2016 to monitor the comet's evolution as it journeyed deeper into the solar system. In a study published in the journal *Nature*, researchers using data from Rosetta's RSI instrument confirmed that the comet's low density was likely due to a porous mixture of dust particles and ice rather than a cavernous interior, consistent with earlier results from Rosetta's CONSERT radar experiment.[244] Additional studies published throughout 2016 addressed the detection of magnetic field-free cavities and its association with outgassing[245]: how the colour of the comet changes following a pass around the Sun, as surface dust is ejected—lifted by the sublimated ice beneath it.[246] Prior to the end of its mission, on 2 September 2016, Rosetta's OSIRIS narrow-angle camera had captured images of the Philae lander at 5 cm per pixel, providing the precise location and a contextual conclusion to the data generated in Philae's journey.[247] In the following weeks, the Rosetta spacecraft was set on a collision course with the Ma'at region of the comet's small lobe which plays an important role in the comet's activity, to study the comet's gas, dust and plasma environment near its surface and for very high-resolution images.[248] Rosetta provided a detailed telemetry until the very last moment before the mission conclusion

---

[242]Algar, Jim. "Rosetta's lander Philae snaps selfie with comet." 4 Oct. 2014. Tech Times 20 June 2017 <http://www.techtimes.com/articles/17887/20141014/rosetta-s-lander-philae-snaps-selfie-with-comet.htm>.

[243]Beatty, Kelly. "Philae Wins Race to Return Comet Findings." 15 Nov. 2014. Sky & Telescope 20 June 2017 <http://www.skyandtelescope.com/astronomy-news/philae-lander-success-11152014/>.

[244]"Inside Rosetta's comet." 4 Feb. 2016. ESA 20 June 2016 <http://m.esa.int/Our_Activities/Space_Science/Rosetta/Inside_Rosetta_s_comet>.

[245]"Rosetta finds magnetic field-free bubble at comet." 11 Mar. 2016. ESA 20 June 2017 <http://m.esa.int/Our_Activities/Space_Science/Rosetta/Rosetta_finds_magnetic_field-free_bubble_at_comet>.

[246]"The colour-changing comet." 7 Apr. 2016. ESA 20 June 2017 <http://m.esa.int/Our_Activities/Space_Science/Rosetta/The_colour-changing_comet>.

[247]"Philae found!" 5 Sept. 2016. ESA 20 June 2017 <http://m.esa.int/Our_Activities/Space_Science/Rosetta/Philae_found>.

[248]"Mission Complete: Rosetta's Journey Ends in Daring Descent to Comet." 30 Sept. 2016. ESA 20 June 2017 <http://www.esa.int/Our_Activities/Space_Science/Rosetta/Mission_complete_Rosetta_s_journey_ends_in_daring_descent_to_comet>.

on Comet 67P. Before the shutdown at Sais, the impact site,[249] it returned its final data, sent via its navigation star trackers, concerning a "large object" in its visible field, being the comet's horizon.[250] Furthermore, and although not immediately recognised by the processing software, Rosetta's last transmission sent data concerning the gas outflow, dust and plasma along with temperature variations of the comet.[251]

NASA's reboot of the Wide-field Infrared Survey Explorer (WISE) mission, this time to discover and characterise near-Earth objects (NEOs) with infrared light, was reactivated in September 2013 for 3 additional years of service.[252] The original WISE mission was completed early in 2011 and was placed in 31 months of hibernation after surveying the whole sky twice in infrared light.[253] Now dubbed NEOWISE, soon after the mission began its 3rd year of operation at the start of 2016, NASA released its latest data from the spacecraft, having detected and characterised 72 NEOs (eight of which being potentially hazardous asteroids (PHA)), bringing the total number of known NEOs to 439.[254] The NEOWISE mission continued its search through 2016, spotting a comet C/2016 U1 (NEOWISE) on 21 October 2016 and another "2016 WF9" NEO likely with cometary origins on 27 November 2016—the trajectories of both objects appear not to threaten Earth in the near future.[255] In January 2016, a new study published in the *Astrophysical Journal Letters* added to findings on the most luminous galaxy, W2246-0526, which belongs to a new class of extremely luminous infrared galaxies (ELIRG). Discovered using WISE data in 2015, W2246-0526 shines with the light of more than 300 trillion suns, some 12.4 billion light years from Earth, and is thought to have a behemoth black hole at its centre that is heating surrounding gasses to temperatures of millions of degrees and blasting out high-energy, visible, ultraviolet and X-ray light, as it sucks in matter. The researchers studying W2246-0526 found large amounts of ionised carbon in a very turbulent state throughout the entire

---

[249] "Rosetta impact site named Sais" 3rd of October 2016. ESA Rosetta Blog 5th October 2018 <http://blogs.esa.int/rosetta/2016/10/03/rosetta-impact-site-named-sais/>.

[250] "Rosetta's last words: science descending to a comet" 15th December 2016. ESA Science—Rosetta 5th of October 2018 <http://www.esa.int/Our_Activities/Space_Science/Rosetta/Rosetta_s_last_words_science_descending_to_a_comet>.

[251] "Rosetta Impact Site named Sais" 3rd of October 2016. ESA Rosetta Blog 5th of October 2018 <http://blogs.esa.int/rosetta/2016/10/03/rosetta-impact-site-named-sais/> <http://www.esa.int/Our_Activities/Space_Science/Rosetta/Rosetta_s_last_words_science_descending_to_a_comet>.

[252] "NASA Spacecraft Reactivated to Hunt for Asteroids." 21 Aug. 2013. NASA 2 Apr. 2014 <http://www.jpl.nasa.gov/news/news.php?release=2013-257>.

[253] "NASA's Asteroid Hunter Spacecraft Returns First Images after Reactivation." 19 Dec. 2013. NASA 9 Mar. 2015 <http://www.jpl.nasa.gov/wise/newsfeatures.cfm?release=2013-373>.

[254] Zolfagharifard, Ellie. "All of the asteroids near Earth in one video: Nasa releases latest data from Neowise 'hunter' spacecraft." 7 Apr. 2016. MailOnline 21 June 2017 <http://www.dailymail.co.uk/sciencetech/article-3528830/All-asteroids-near-Earth-one-video-Nasa-releases-latest-data-Neowise-hunter-spacecraft.html>.

[255] "NASA's NEOWISE Mission Spies One Comet, Maybe Two." 29 Dec. 2016. NASA 21 June 2017 <https://www.jpl.nasa.gov/wise/newsfeatures.cfm?release=2016-328>.

galaxy, instead of flowing in specific directions from the black hole's accretion disc, suggesting that the momentum and energy of the particles of light deposited in the gas are so great that they are pushing the gas out in all directions; should all the gas and dust eventually blow out of the Hot, Dust-Obscured Galaxy (Hot DOG) the quasar surrounding the black hole would likely be visible.[256] Following NASA's reactivation of the mission with a primary goal of scanning NEOs, the statistics of the 3rd year of data collection were announced in June 2017 with the final result of 97 unknown celestial objects being discovered, of which 28 were NEOs; as such, 693 is the total number of NEOs characterised since the NEOWISE mission restarted in 2013.[257] Moreover, in February 2017, a NASA-funded "Backyard Worlds: Planet 9" citizen project[258] bolstered the search for new planets in the solar system through WISE[259] data resulting in the discovery of a new substellar object, as the *Astrophysical Journal Letters* reported, utilising this new volunteer methodology of research whereby sets of animated images are uploaded online for classification.[260]

Commercial interest in NEOs has increased in recent years with several private US companies already investing millions of dollars to be the first to mine asteroids. As an initial step, Planetary Resources and Deep Space Industries (DSI) business models focus on developing prospecting spacecraft that will image and characterise promising asteroids. On 16 July 2015, Planetary Resources launched its Arkyd 3 Reflight (A3R) spacecraft from the ISS to validate several core technologies that will be incorporated into a future fleet of Arkyd spacecraft that will be launched into the solar system.[261] Similarly, DSI is developing its FireFly spacecraft; but rather than testing its technologies in Earth orbit, the first generation spacecraft will be launched directly toward promising asteroids. The main focus in asteroid mining is in the potential availability of rare metals that could be returned to Earth or water that could be converted into propellant and oxygen for deeper space exploration. While initial estimates of the value of some known asteroids are upwards of $100 trillion, to reach these caches, both companies will need to invest substantial resources in developing the necessary mining technology, in addition to successfully

[256] "Most Luminous Galaxy Is Ripping Itself Apart." 15 Jan. 2016. NASA 21 June 2017 <https://www.jpl.nasa.gov/wise/newsfeatures.cfm?release=2016-011>.

[257] "NASA's Asteroid-Hunting Spacecraft a Discovery Machine" 5th of June 2017. NASA JPL 5th of October 2018 <https://www.jpl.nasa.gov/wise/newsfeatures.cfm?release=2017-159>.

[258] "Backyard Worlds: Planet 9." Zooniverse, 11 Oct. 2018 <https://www.zooniverse.org/projects/marckuchner/backyard-worlds-planet-9>

[259] "WISE mission overview". Wide-field infrared Survey Explorer 5th of October 2018 <https://www.nasa.gov/mission_pages/WISE/mission/index.html>.

[260] "The First Brown Dwarf Discovered by the Backyard Worlds: Planet 9 Citizen Science Project" 1st of June 2017. The Astrophysical journal Letter 5ht of October 2018 <http://iopscience.iop.org/article/10.3847/2041-8213/aa7200/pdf>.

[261] "Planetary Resources' First Spacecraft Successfully Deployed, Testing Asteroid Prospecting Technology in Orbit." 16 July 2015. Planetary Resources 27 Nov. 2015 <http://www.planetaryresources.com/2015/07/planetary-resources-first-spacecraft-deployed/>.

rendezvousing and securing a spacecraft onto an asteroid's surface.[262] On the occasion of the 2017 Paris Air Show in Le Bourget, Luxembourg and ESA enhanced cooperation in this field by signing a statement on future asteroids, related technologies and space resource utilisation and exploration. The agreement contributes to further evaluations over feasibility and technical maturity of asteroid utilisation for in situ extraction and operations.[263] Further, 1 month later, the Luxembourg Parliament voted in favour of a space resource law, rendering the nation the first European country to adopt a legal framework granting private companies the legal means to operate in space resource exploitation. Similar to the US Commercial Space Launch Competitiveness Act of 2015, Luxembourg's law declares that space resources can be appropriated,[264] although resource ownership is granted only after extraction, thus sidestepping a conflict in terms with the Outer Space Treaty.[265]

2.2.9 *Science Beyond the Solar System*

ESA's Gaia mission is measuring the positions and motions of more than 1 billion stars of the roughly 100 billion stars in our galaxy to create the most accurate map yet of the Milky Way. Launched on 19 December 2013, and operating from the L2 Lagrange point, the spacecraft completed the 2nd year of a 5-year survey on 16 August 2016.[266] On 14 September 2016, ESA published the first catalogue based on data collected during its first 14 months of science operations, which feature the density of stars across the entire sky and the parallax, i.e. the apparent motion of a star against a distant background, of more than two million stars. While the image contained stripes and other artefacts, these will gradually fade in subsequent scans as the trace amounts of ice deposits which remained in the spacecraft following its commissioning will be outgassed using heaters beneath Gaia's

[262]Calandrelli, Emily. "The Potential $100 Trillion Market for Space Mining." 9 July 2015. Techcrunch.com 26 Nov. 2015 <http://techcrunch.com/2015/07/09/the-potential-100-trillion-market-for-space-mining/>.

[263]Ministry of economy of the Grand Duchy of Luxembourg, Press release of the 20th June 2017. Space Resources Luxembourg 5th October 2018 <https://spaceresources.public.lu/damassets/pressrelease/2017/2017_06_21%20Press%20Release%20ESA%20LeBourget.pdf>.

[264]"Loi du 20 juillet 2017 sur l'exploration et l'utilisation des ressources de l'espace" 20th July 2017 Journal Officiel de Luxembourg. 8th October 2018 <http://legilux.public.lu/eli/etat/leg/loi/2017/07/20/a674/jo>.

[265]Jeff Foust "Luxembourg adopts space resources law" 17th of July 2017. SpaceNews 5th of October 2018 <https://spacenews.com/luxembourg-adopts-space-resources-law/>.

[266]"Gaia's Second Anniversary Marked by Successes and Challenges." 16 Aug. 2016. ESA 21 June 2017 <http://sci.esa.int/gaia/58135-gaia-s-second-anniversary-marked-by-successes-and-challenges/>.

mirrors.[267] In the lead up to the publication, ESA released a sonification, converting astronomical data into sound, to portray the status of astrometric catalogues prior to the advent of Gaia and demonstrate the remarkable progress that was being made in the field of astrometry.[268] Researchers scrutinising Gaia data have detected over a thousand transient bright astronomical sources, due to stars undergoing a major outburst or supernova, but in July and August 2016, they observed two rare instances of gravitational microlensing, where the gravity of a massive object between a star and observer causes the path of its light to distort. The first instance, classified as Gaia16aua, observed a faint star of magnitude 19 suddenly brighten by two magnitudes, while the second instance, classified as Gaia16aye, observed the anomalous peaks and troughs in brightness of a magnitude 14.5 star; by pairing Gaia measurements with ground-based data, researchers will be able to estimate the position and mass of objects (e.g. a star or black hole) causing the gravitational microlensing with high precision.[269] In the Milky Way, most of the stars are located on the flat Galactic disc, which is remote and difficult to analyse; however, as reported in June 2017, Gaia identified six stars travelling at "hypervelocity" speeds away from the centre of the Galaxy, a finding which will "yield crucial information about the gravitational field of the Milky Way from the centre to its outskirts".[270] Looking ahead, in April 2018 a second data collection is to be released covering motions and distances. This further data release is expected to denote a new astrometry era, allowing more accurate scientific study of the past and future positions of stars.[271] And on 7 December 2017, the ESA's Science Programme Committee approved a mission extension of Gaia from July 2019 to the end of 2020.[272]

NASA's Kepler mission to discover hundreds of Earth size and smaller planets in or near the habitable zone of more than 150,000 stars and determine the fraction of the hundreds of billions of stars in our galaxy that might have such planets began in May 2009. By May 2013, the loss of two of the four reaction wheels on the spacecraft brought an end to Kepler's 4-year science mission as the spacecraft had lost its ability to precisely point at the original field of view; but by May 2014,

[267] "Gaia's Billion-Star Map Hints at Treasures to come." 14 Sept. 2016. ESA 22 June 2017 <http://www.esa.int/Our_Activities/Space_Science/Gaia/Gaia_s_billion-star_map_hints_at_treasures_to_come>.

[268] "From Hipparchus to Hipparcos: A Sonification of Stellar Catalogues." 22 Sept. 2016. ESA 22 June 2017 <http://sci.esa.int/gaia/58311-from-hipparchus-to-hipparcos-a-sonification-of-stellar-catalogues/>.

[269] "Gaia Spies Two Temporarily Magnified Stars." 27 Oct. 2016. ESA 22 June 2017 <http://sci.esa.int/gaia/58546-gaia-spies-two-temporarily-magnified-stars/>.

[270] "Artificial brain helps Gaia catch speeding stars" 26th June 2017. ESA GAIA 5th of October 2018 <http://sci.esa.int/gaia/59263-artificial-brain-helps-gaia-catch-speeding-stars/>.

[271] "Two million stars on the move" 12th April 2017. ESA Gaia 5th of October 2018 <http://sci.esa.int/gaia/59004-two-million-stars-on-the-move/>.

[272] "Green light for continued operations for ESA science missions" 7th December 2017. ESA Director's Desk 8th October 2018 <http://sci.esa.int/director-desk/59839-green-light-for-continued-operations-of-esa-science-missions/>.

NASA had approved the Kepler 2 (K2) community-driven mission which repurposed Kepler to accurately point at target sky fields along the ecliptic plane of Earth's orbit using the Sun to maintain its stability. The K2 mission is expected to continue operations into 2018.[273] By the end of 2016, continued analyses of Kepler data had revealed more than 5100 planet candidates, with more than 2500 verified planets[274]; K2 added another 447 candidates, with 154 verified planets by mid--year.[275] As at 24 August 2016, 22 planets were known to be Earth-sized and orbiting within the habitable zone of their nearest stars.[276] In May 2016, a new study of the Kepler-223 star system showed its four planets to have the same configuration as that of Jupiter, Saturn, Uranus and Neptune in the early history of our Sun's solar system; but whereas our the orbit of planets in the solar system have evolved since its birth 4.6 billion years ago, the much older Kepler-223 system appears to have maintained a single orbital configuration resonating around its star for far longer. As resonances are extremely fragile, it's possible that interactions with numerous asteroids and planetesimals may have dislodged the Sun's giants from their own resonance.[277] Another recent study published in the *Astrophysical Journal* measured the orbits of 19 heartbeat star systems that were identified by the Kepler mission; because these binary systems are in elongated elliptical orbits, the diameters of the stars tend to rapidly fluctuate at the point of their closest encounter due to the tidal forces caused by each star's gravitational pull. While the tidal stretching of these heartbeat stars should have quickly caused their systems to evolve into circular orbits, the researchers postulate that third or fourth stars might exist in these systems that have gone undetected, which may be maintaining these highly stretched-out, elliptical orbits.[278] As reported in June 2017, mission observations have added ten near-Earth size planets in the so-called habitable zone out of 219 new candidates

[273]"Mission overview." 8 Feb. 2017. NASA 23 June 2017 <https://www.nasa.gov/mission_pages/kepler/overview/index.html>.

[274]"NASA Kepler Visionary Honoured by American Association for the Advancement of Science." 21 Nov. 2016. NASA 23 June 2017 <https://www.nasa.gov/feature/kepler/nasa-kepler-visionary-honored-by-american-association-for-the-advancement-of-science>.

[275]"Mission Manager Update: K2 Marches On." 9 June 2016. NASA 23 June 2017 <https://www.nasa.gov/feature/ames/kepler/mission-manager-update-k2-marches-on>; and "NASA's Kepler Confirms 100+ Exoplanets during Its K2 Mission." 18 July 2016. NASA 23 June 2017 <https://www.nasa.gov/feature/ames/kepler/nasa-s-kepler-confirms-100-exoplanets-during-its-k2-mission>.

[276]"Kepler's Small Habitable Zone Planets." 11 May 2016. NASA 23 June 2017 <https://www.nasa.gov/mission_pages/kepler/multimedia/images/keplers-small-habitable-zone-planets>; "ESO Discovers Earth-Size Planet in Habitable Zone of Nearest Star." 24 Aug. 2016. NASA 23 June 2017 <https://www.nasa.gov/feature/jpl/eso-discovers-earth-size-planet-in-habitable-zone-of-nearest-star>.

[277]"Kepler-223 System: Clues to Planetary Migration." 17 May 2016. NASA 23 June 2017 <https://www.nasa.gov/feature/jpl/kepler-223-system-clues-to-planetary-migration>.

[278]" 'Heartbeat Stars' Unlocked in New Study." 21 Oct. 2016. NASA 23 June 2017 <https://www.nasa.gov/feature/jpl/heartbeat-stars-unlocked-in-new-study>.

overall,[279] while in December, NASA announced that Kepler, collaborating with Google's AI, discovered the eighth planet of the star system Kepler-90, previously known as a seven planet system.[280]

NASA's Spitzer Space Telescope, launched in August 2003, studied the early universe, young galaxies and forming stars and was used to detect dust discs around stars.[281] After running out of the coolant needed to chill its longer-wavelength instruments in 2009, Spitzer was repurposed to track exoplanets around stars with the use of infrared light. As exoplanets cross in front of their stars, they block out a fraction of the light, allowing the size of the planet to be revealed, in addition to giving clues about the planet's atmosphere by the infrared light that they also emit.[282] In March 2016, Spitzer data helped researchers to create the first temperature map of an Earth-like exoplanet that is double in size and tidally locked to its star; rather than having a thick atmosphere with winds moving heat around the planet as previously thought, the exoplanet was seen to inefficiently transport heat, and lava flows likely warmed the nightside of the planet.[283] Another planet-finding technique used by Spitzer is called microlensing, which occurs when the light of a distant star is magnified and brightened by the gravity of another star that passes in its foreground. Should the closer star have a planet it its orbit, the planet might cause a blip in the magnification. By mid-2015, Spitzer had viewed a total of 142 microlensing events[284]; moreover, in late 2016 NASA's Spitzer and Swift space telescopes were used to observe a microlensing event of a newly discovered brown dwarf, OGLE-2015-BLG-1319, marking the first time two space telescopes have collaborated to observe a microlensing event, rather than pairing observations with a ground observatory.[285] Spitzer's mission was extended for 2½ years beginning on 1 October

---

[279]"NASA's Kepler telescope finds 10 Earth-like planets: 'We are not alone'" 20th June 2017. The Guardian 8th October 2018 <https://www.theguardian.com/science/2017/jun/20/we-are-not-alone-nasa-telescope-finds-10-earth-like-planets>.

[280]"Why NASA Kepler mission in toast" 15th December 2017. Forbes 8th October2018 <https://www.forbes.com/sites/startswithabang/2017/12/15/why-nasas-kepler-mission-is-toast/#321c09401da0>

[281]"Mission Overview." 30 July 2008. NASA 23 June 2017 <https://www.nasa.gov/centers/jpl/missions/spitzer.html>.

[282]"How Engineers Revamped Spitzer to Probe Exoplanets." 24 Sept. 2013. NASA 3 Apr. 2014 <http://www.spitzer.caltech.edu/news/1560-feature13-07-How-Engineers-Revamped-Spitzer-to-Probe-Exoplanets>.

[283]"NASA's Spitzer Maps Climate Patterns on a Super-Earth." 30 Mar. 2016. NASA 23 June 2017 <http://www.spitzer.caltech.edu/news/1869-ssc2016-01-NASA-s-Spitzer-Maps-Climate-Patterns-on-a-Super-Earth>.

[284]"NASA's Spitzer Spots Planet Deep Within Our Galaxy." 14 Apr. 2015. NASA 22 May 2016 <http://www.spitzer.caltech.edu/news/1746-feature15-05-NASA-s-Spitzer-Spots-Planet-Deep-Within-Our-Galaxy>.

[285]"NASA Space Telescopes Pinpoint Elusive Brown Dwarf." 10 Nov. 2016. NASA 26 June 2017 <http://www.spitzer.caltech.edu/news/1918-feature16-18-NASA-Space-Telescopes-Pinpoint-Elusive-Brown-Dwarf>.

2016, and [286] this "Beyond" phase will explore a wide range of topics in astronomy and cosmology, as well as planetary bodies in and out of our solar system.[287] In May 2016, researchers employing the Transit Planets and Planetesimals Small Telescope (TRAPPIST) announced the existence of three exoplanets, two of which were confirmed by Spitzer. Spitzer also identified another five exoplanets in the TRAPPIST-1 system, which was thus known to have seven overall, and thereby holds a record for the largest number of planets in the habitable zone identified around a star outside the solar system; the system's name pays tribute to the TRAPPIST telescope.[288]

In conjunction with the Chandra X-Ray Observatory, the Compton Gamma Ray Observatory and the Hubble Space Telescope, the Spitzer Telescope forms a part of the Great Observatories. Each telescope analyses a specific wavelength of the electromagnetic spectrum, starting from the visible and near-ultraviolet with Hubble, observing gamma rays with Compton's four different gamma detectors, the soft X-rays with Chandra (CXO) and the infrared with Spitzer (SST).[289] NASA planned to continue funding the Spitzer Space Telescope until 2019, which it declared to be open for private partnership proposals after the mission end. Parenthetically, NASA had a previous unique model agreement experience related to the GALEX astronomy satellite, which it continued operating with support from Caltech private funds, and it was the first governmental-owned science probe business model.[290] While the Compton Gamma Ray Observatory was launched aboard the Space Shuttle Atlantis in April 1991 and had a safely returned to Earth in June 2000,[291] Chandra and Hubble are still operational from, respectively, 1999 and 1990. Chandra with its X-Ray resolution captured images of the centre of the Milky Way, observing its supermassive black hole, other than supernovas and exploded stars. It contributed to researching the separation of dark matter in galaxy cluster collision.[292] In January 2017, NASA released an image from Chandra's Deep Field-South Catalogue showing the highest concentration of black holes, amounting to roughly the same

[286] "Spitzer Operations to Continue into 2019." 9 June 2016. NASA 23 June 2017 <http://www.spitzer.caltech.edu/news/1882-feature16-09-Spitzer-Operations-to-Continue-into-2019>.

[287] "Spitzer Space Telescope Begins 'Beyond' Phase." 25 Aug. 2016. NASA 26 June 2017 <https://www.nasa.gov/feature/jpl/spitzer-space-telescope-begins-beyond-phase>.

[288] "NASA Telescope reveals Largest batch of Earth-sized habitable zone planet around a single star" 22nd of February 2017. NASA JPL 8th October 2018 <http://www.spitzer.caltech.edu/news/1924-ssc2017-01-NASA-Telescope-Reveals-Largest-Batch-of-Earth-Size-Habitable-Zone-Planets-Around-Single-Star>.

[289] "NASA Great Observatories" 3rd of November 2004. NASA 17th October 2018 <https://www.nasa.gov/audience/forstudents/postsecondary/features/F_NASA_Great_Observatories.html>.

[290] "NASA might privatize one of its great observatories" 22nd July 2017. AstronomyNow 18th October 2018 <https://astronomynow.com/2017/07/22/nasa-might-privatise-one-of-its-great-observatories/>.

[291] "Launch of the Compton Gamma-Ray Observatory" 4th April 2014. NASA 18th October 2018 <https://www.nasa.gov/centers/marshall/history/cgro140404.html>.

[292] "NASA's Chandra X-Ray Observatory Celebrates 15th anniversary" 22nd of July 2014. NASA 18th October 2018 <https://www.nasa.gov/chandra/news/chandra-15th-anniversary.html>.

dimensions as the Moon; to achieve the deepest X-Ray image ever obtained, Chandra used over 7 million seconds of observing time, supporting astronomers conducting further studies of the moment immediately after the Big Bang.[293] Hubble (HST) aided both ESA and NASA in studying the ageing of the universe and its expansion rate, which is accelerating and yet not constant. Launched on the Shuttle Discovery in April 1990 with a defected lens, it was successfully repaired in 1993 with the shuttle mission Endeavour, serviced in total five times. The Wide Field Planetary Camera, the Faint Object Camera, the Goddard High-Resolution Spectrograph (GHRS) and the Faint Object Spectrograph (FOS) with the High Speed Photometer were the instruments composing the Hubble telescope, which collected and transferred information through the Tracking and Data Relay Satellite System (TDRSS),[294] elaborated from the Space Telescope Science Institute (STScI) and NASA's Goddard Space Flight Centre (GSFC).[295] The James Webb Space Telescope (JWST) is to be Hubble's successor; it possesses different capabilities and will observe in infrared, compared with the ultraviolet and optical capacities of Hubble. The location of observation also represents a major difference between the two missions: whereas Hubble was located in LEO, the JWST will be 1.5 million kilometres from Earth at the second Lagrange point. It is programmed to be launched in 2021.[296]

China's Dark Matter Particle Explorer (DAMPE) mission was successfully launched to a 500 km Sun-synchronous orbit on 17 December 2015. Its 3-year mission aims to shed new light on the nature of dark matter—a hypothetical kind of matter (along with dark energy) that could explain where the remaining estimated 85% of the total mass-energy in the universe exists. The DAMPE mission searches for dark matter by measuring the properties of particles that annihilate or decay in space, as with experiments conducted inter alia by the Alpha Magnetic Spectrometer (AMS) on board the ISS, or the CALorimetric Electron Telescope (CALET) recently attached to Japan's Kibo module. However, DAMPE extends the search into the multi-TeV region with an energy resolution of 1.5% at 100 GeV (i.e. at least three times higher than international peers) and will also take precise measurements of the flux of nuclei with a spectrum up to above 100 TeV (i.e. nine times wider than the AMS), to gain insight into the origin and propagation of high-energy cosmic rays.[297] In the initial 2 years of its mission, DAMPE will scan space in all directions, followed by another year or more that will focus on areas where the potential

[293]"Chandra Deep Field-South: Deepest X-Ray Image Ever" NASA-Harvard Chandra X-Rays Observatory 18th October 2017 <http://chandra.harvard.edu/photo/2017/cdfs/>.

[294]"Hubble telescope: pictures, facts and histories" 14th December 2017. Space.com 18th October 2018 <https://www.space.com/15892-hubble-space-telescope.html>.

[295]"Operating Hubble" 26th October 2007. ESA Hubble 18th October 2018 <http://sci.esa.int/hubble/34035-operating-hubble/>.

[296]"Webb vs Hubble". NASA James Webb Space Telescope 18th October 2018 <https://jwst.nasa.gov/comparison_about.html>.

[297]"DAMPE joins the search for dark matter in space." 12 Feb. 2016. CERN Courier 26 June 2017 <http://cerncourier.com/cws/article/cern/63969>.

signatures of dark matter are most likely to be observed.[298] By 19 December 2016, the spacecraft had collected 1.8 billion cosmic rays, with more than one million high-energy electrons among the collected particles. The mission's first results were published in November 2017 in *Nature*, showing the international outcomes of Chinese, Swiss and Italian scientists. During this time, DAMPE sensed 3.5 million cosmic rays events aiming to answer to a deviation curve that could be evidence of dark matter.[299] The astrophysicist Chang Jin, leader of the collaboration at the Chinese Academy of Science's (CAS) Purple Mountain Observatory (PMO), expected DAMPE to be able to collect over ten billion cosmic ray events and to extend the mission to 2022, aiming to clarify the hypothetical connection between the atypical signal and the dark matter extinction.[300]

ISRO's Astrosat mission, launched on 28 September 2015, is India's first astronomy satellite to provide optical, ultraviolet and X-ray images of black holes and other related phenomena over the course of 5 years.[301] Its first scientific results and the future scope were presented on 29 September 2016 at the Inter-University Centre for Astronomy and Astrophysics (IUCAA), Pune, India, with the spacecraft having orbited the Earth more than 5400 times, executing 343 individual pointings to 141 different cosmic sources.[302] Included among the findings, Astrosat's Large Area X-ray Proportional Counter (LAXPC) instrument observed for the very first time rapid variability of high-energy (particularly >20 keV) X-ray emissions from the enigmatic "GRS 1915+105" black hole system. These quasi-period oscillations are thought to occur because the inner part of the disc surrounding the black hole wobbles as the spinning black hole drags the space-time fabric around it, as predicted by Einstein's General theory of relativity. While these oscillations have been studied in the past in low-energy X-rays, observing phenomenon also in high-energy X-rays from higher-energy photons that are emitted closer to the black hole allows researchers to measure the arrival time difference between the energy bands and provides clues to the geometry and dynamic behaviour of the gas swirling around a

---

[298] Yue, Huang. "China Launches Country's First Dark Matter Satellite." 17 Dec. 2015. CRI English News 22 Jan. 2016 <http://english.cri.cn/12394/2015/12/17/4201s908753.htm>.

[299] "First finding of China's DAMPE may shed light on dark matter research" 30th November 2017, Chinese Academy of Science. Phys.org 8th October 2018 <https://phys.org/news/2017-11-china-dampe-dark.html>.

[300] Dennis Normile "China's dark matter space probe detects tantalizing signal" 29th November 2017. Science Magazine 8th October 2018 <https://www.sciencemag.org/news/2017/11/china-s-dark-matter-space-probe-detects-tantalizing-signal>.

[301] De Selding, Peter B. "PSLV Rocket Launches India's 1st Astronomy Satellite, 4 Spire Cubesats." 28 Sept. 2015. SpaceNews 5 Jan. 2016 <http://spacenews.com/pslv-rocket-launches-indias-1st-astronomy-satellite/>.

[302] "A science meet to commemorate one year of AstroSat in orbit." 29 Sept. 2016. ISRO 27 June 2017 <http://www.isro.gov.in/update/29-sep-2016/science-meet-to-commemorate-one-year-of-astrosat-orbit>.

black hole.[303] The spacecraft began its Open Phase observations based on proposals by numerous institutions on 1 October 2016.

ESA is planning to launch the first mission for exoplanetary transit, CHEOPS (CHaracterising ExOPlanet Satellite), equipped for ultrahigh precision photometry. Selected in 2012 as the first S-class mission in Cosmic Vision 2015–2025, it will be ready for launch at the end of 2018. The 250 kg smallsat deployed with a single medium-sized telescope will pursue the goal of measuring the density of so-called super-Earths. It will operate in Sun-synchronous orbit at an altitude of 700 km[304] and will permit an exact determination of the radii of transiting planets, as well as radii of newly discovered planets identified by next-generation ground-based transit surveys, in particular those that are Neptune sized and/or smaller.[305]

A further mission planned for 2026 by ESA is PLATO (Planetary Transits and Oscillations of stars), which will operate from L2 and focus on the discovery and analysis of Earth-sized planets and super-Earths orbiting Sun-like stars in the zone considered habitable, focusing also on seismic activity of these stars and collecting information over star evolution.[306] The DLR Institute of Planetary Research in Berlin will lead the consortium of numerous European research institutions in charge of providing the scientific instrument that consists of 26 telescopes, providing novel accuracy.[307] The complementary strategies utilised to address different science cases include a long-duration observation as well as a shorter exposure of different targets. Current observations are based on a long-duration observation phase (LOP) of two sky fields, each for a 2-year time lap. Alternatively, an LOP lasting 3 years with a step-and-stare (SOP) phase lasting 1 year of duration could be utilised.[308]

---

[303]"AstroSat observes the high energy X-ray variability of a black hole system." 31 Aug. 2016. ISRO 27 June 2017 <http://www.isro.gov.in/pslv-c30-astrosat-mission/astrosat-observes-high-energy-x-ray-variability-of-black-hole-system>.

[304]"Cheops characterising exoplanet satellite—overview". ESA CHEOPS 8th October 2018 <https://www.cosmos.esa.int/web/cheops>.

[305]"Cheops at a glance" Cheops in Italy. 8th October 2018 <http://www.oact.inaf.it/cheops-it/Home.html>

[306]"Gravitational Wave Mission Selected, Planet-Hunting Mission Moves Forward" 20th June 2017. ESA Cosmic vision 8th October 2017 <http://sci.esa.int/cosmic-vision/59243-gravitational-wave-mission-selected-planet-hunting-mission-moves-forward/>.

[307]"Green light for PLATO mission" 20th June 2017. DLR News 8th October 2018 <https://www.dlr.de/dlr/en/desktopdefault.aspx/tabid-10081/151_read-22858/#/gallery/27241>.

[308]"Mission observation strategy" PLATO mission. 8th October <https://platomission.com/2018/05/20/mission-profile/>.

Chapter 3
ESPI Thought Papers on Selected 2017 and 2018 Developments

3.1 China's 2016 White Paper on Space: An Analysis

3.1.1 Introduction to China's Space White Papers

On 27 December 2016, the Information Office of China' State Council released the new version of the country's 5-year policy guidelines for national space activities. Published in the form of a government White Paper, the document highlights the major developments of the past 5 years (2011–2015) and provides a comprehensive description of the programmatic intentions for the sector over the period 2016–2020.

In this respect, it is important to highlight that China's space activities are planned and executed within the Framework of China's Five-Year Plans (*Zhongguo Wunian Jihua*). Discussed and adopted by the Central Committee of the Chinese Communist Party (CCP) and subsequently ratified by the National People's Congress since 1953, these plans provide a grand blueprint of the overall objectives and goals related to national social and economic growth and industrial planning in key sectors and regions.

The 2016 White Paper hence represents the specific plan for the space sector during the 13th Five-Year Plan, which governs the period 2016–2020. Remarkably, the 2016 White Paper is the fourth policy document of this kind, the other three having been released in 2001, 2006 and 2011, in conjunction with the 10th, 11th and 12th Five-Year Plans, respectively.

3.1.2 What's Inside and What's Left Out

Like the three previous versions, "China's Space Activities in 2016" is organised around five major sections. In the first one, the document spells out the purpose, visions and principles of development for the country's space programme. It then

E. Burger, G. Bordacchini, *Yearbook on Space Policy 2017*, Yearbook on Space Policy, https://doi.org/10.1007/978-3-030-05417-5_3

hails the remarkable accomplishments and breakthroughs realised during the 12th Five-Year Plan (2011–2015) and enunciates the plans for the next 5 years. Finally, a set of policy measures designed to assist the realisation of the programmatic targets and the international cooperation policies are presented.

In line with the 2011 version, the new White Paper contains a quite extensive range of technical and operational information, which signal a greater level of confidence in the country's space capabilities and technological prowess as well as the desire to "enable the world community to better understand China's space industry". A stated objective for this increased level of transparency is to support the principle of *open development* through international cooperation, which is one of the four guiding principles to which China's space developments adhere, the others being *innovative*, *coordinated* and *peaceful* development. Consistent with these four guiding principles, the principal axes around which China's space programme is stated to operate can be summarised as follows:

- Implementing major space science and technology projects by pursuing independent innovation as core development strategy of the space industry
- Allocating various resources in a rational and coordinated manner to promote a comprehensive development of space science, space technology and space applications and improve the quality and efficiency of the overall space sector
- Adhering to the peaceful and clean utilisation of outer space by opposing the weaponisation and any arms race in space and taking effective measures to protect the space environment
- Combining self-reliance with opening to the outside world by actively engaging in international cooperation based on equality, mutual benefit, peaceful utilisation and inclusive development

Thanks to its open policy statements, China's White Paper marks several steps forward in enhancing the degree of transparency and in providing foreign audiences with an effective tool of communication on the purpose, visions and development principles of its space programme. Where the White Paper is, however, much less communicative—not to say silent—is in providing a comprehensive and clear picture of the uses China intends to make of its space programme, which is an equally important means for reducing ambiguities and potential tensions, while building confidence internationally.

Predictably, the document's focus is on peaceful development of space activities to support scientific and technological development and spur social and economic growth. The security- and military-related goals, policies and activities of programme are barely mentioned.[1] Admittedly, the document acknowledges that the purpose of the space programme is to meet the demands of national security and toward this "to build China into a space power in all respects". But apart from this and some other generic statements, the White Paper carefully omits to provide any

[1]In this 11,000 words document, the word "security" is mentioned only three times, while the word "military" or "dual use" finds no trace at all.

account of the security-related uses of its satellite systems, the majority of which is known to serve dual-use purposes.

Similarly, no indication is given about the resource allocations and the stakeholders involved in the management of the programme. This is not surprising. Chinese official documents have rarely released figures on the country's overall space spending, nor have they provided clear guidance to navigate the ocean of its complex organisational structures. In the White Paper, mention is made of the China National Space Administration (CNSA) which, however, is little more than a clearing house acting as the public interface of the Chinese space programme on the international stage. Arguably, this reticence about releasing information on its policy processes, capabilities and expenditures, while consistent with China's strategic thinking, continues to create a certain degree of ambiguity that undermines the stated goal of making China's space programme more transparent so as to promote deeper international cooperation.

3.1.3 China's Programmatic Targets for 2016–2020

Looking at the specific targets set out by the White Paper, the documents contains little that is new, though the full spectrum of activities is remarkably covered. The major targets fixed for the 2016–2020 period can be grouped under the following major headings: space transportation, space infrastructure, human spaceflight and space exploration, space applications, space science and technologies and ground infrastructures.

In the area of access to space, the new policy document pledges some efforts to improve the reliability of the new *Chang Zheng* (Long March—LM) rocket family, namely, the LM-5, the LM-6 and the LM-7, the development of which was completed during the 12th Five-Year Plan. In parallel, the document anticipates some major efforts and breakthroughs in the development of a new heavy-lift launch vehicle (unofficially referred to as LM-9) and, interestingly, "research into the technologies for low-cost launch vehicles, new upper stage and a reusable space transportation system".

In the area of satellite systems, which is identified as a key priority area during the 13th Five-Year Plan, efforts are pledged in the full range of China's applications satellites, including Earth observation, communications and navigation. Specifically, the new policy document calls for (a) the further development of three series of multifunctional, high-resolution satellites for observing the land, oceans and atmosphere and the creation of networks of satellites integrating them;[2] (b) the expansion

[2]China already operates five dedicated series of EO satellites, namely, the *Fengyun* (wind and cloud) series for meteorology; the *Haiyang* (ocean) series for oceanography and maritime observation; the *Ziyuan* (resources) series for natural resource monitoring; the *Gaofen* (high resolution) series for near-real-time geographical mapping, climate change monitoring, environmental

of a "comprehensive [satellite] system capable of providing broadband communications, fixed communications, direct-broadcast television, mobile communications and mobile multimedia broadcast services"; and (c) enhancing the capacities of the Beidou-2 GNSS constellation. In this respect, China plans "to start providing basic services to countries along the Silk Road Economic Belt and 21st-century Maritime Silk Road in 2018, form a network consisting of 35 satellites for global services by 2020, and provide all clients with more accurate and more reliable services through advancing the ground-based and satellite-based augmentation systems in an integrated way".

With regard to human spaceflight and space exploration, the key focus is understandably on the activities and technologies required for building and operating the future Chinese Space Station (CSS), including the launch of the *Tianzhou* (Heavenly Vessel) cargo spacecraft to dock with *Tiangong*-2. Interestingly, the assertions contained in the 2011 White Paper about the conduct of studies "on the preliminary plan for a human lunar landing" are not reaffirmed in the new White Paper, which limits to state that China will "strive to acquire key technologies [...] to raise manned spaceflight capacity and laying the foundation for exploring and developing cislunar space". Concerning robotic exploration, the White Paper confirms the three-step strategy of "orbiting, landing and returning" for its lunar exploration project by anticipating the launch of *Chang'e-5* sample return mission in 2017 and the launch of the *Chang'e-4* lunar probe in 2018 to achieve mankind's first soft landing on the far side of the Moon. Finally, indications are given on the launch of a new Mars probe by 2020, a mission that was put aside following the failure of Phobos-Grunt in 2012.

In terms of infrastructure development, the document emphasises the need to improve "the existing space launch sites by raising the reliability and IT application level and conducting adaptive improvements to ground facilities and equipment". The document also pledges to enhance China's TT&C systems—i.e. by launching and operating a second-generation data relay satellite system and building "a space-ground integrated TT&C network featuring security, reliability, quick response, flexible access, efficient operations and diverse services"—and to improve China's capability in characterising the space environment and associated risks through an advanced and more effective SSA system.

With regard to space science and technology development, key pioneering experiments are planned in the field of space astronomy and space physics, quantum mechanics, biology, life sciences, medicine and materials in space through dedicated space missions and various activities on-board the *Tiangong* space laboratory. Also, the document states that within the next 5 years, China is to develop and launch technology experiment satellites (including three new *Shijian* missions)[3] and to

surveying and precision agriculture; and the *Yaogan* (remote sensing) series for disaster management and surveillance operations combining optical, radar and electronic intelligence signals.

[3] *Shijian* (literally, "practice") is a scientific and technology demonstrator satellite series started in the 1970s.

conduct experiments on key technologies for new electric propulsion, laser communications and common platforms of new-generation communications satellites with the goal of further supporting the maturation of its space industry.

Finally, in the area of space applications, consideration is given to the need of raising the scale, operational standards, industrialisation and commercialisation levels of China's space applications so as to serve national security objectives and foster new growth points for the national economy. The document in particular puts the spotlight on ensuring an extensive provision of integrated space-based services for meeting the needs of industries, regions and the general public.

3.1.4 Between the Lines

The 2016 White Paper not only serves as a celebration of China's past achievements and a compendium of programmatic targets for the next 5 years. Although the statements are, as mentioned, rather scant, in reading between the lines and in comparing the new version with the one of 2011, a whole set of underlying policy goals and priorities comes into the fore.

To begin with, it does not go unnoticed that whereas the formula deployed to enunciate goals and working principles of China's space policy is essentially the same as the 2011 White Paper,[4] the new version adds the principle of *coordinated development*, an inclusion that, together with the ever-growing stress on international cooperation, seems to acknowledge the difficulties faced by the Chinese space community in balancing its involvement in so many different programmes and its pursuit of so many different objectives. While no new trailblazing programme was revealed in the paper, it should not be neglected that, in addition to the programmes already underway, the White Paper envisages many new efforts in the whole plethora of space activities during the 2016–2020 period. Building a space station is already an extremely costly undertaking, and the *soft-landing* scenario envisaged for China's economic performance over the next years recommends—indeed requires—China's space officials to be more pragmatic and effective in managing the ever-growing resources demand and policy goals. And this is evident in the set of policy measures spelled out in the fourth section to assist the realisation of the abovementioned programmatic targets and, more specifically, in increased interest toward private sector's participation in the space programme.[5]

[4]The operative language in the 2016 document reads as follows, "China's space industry is subject to and serves the national overall development strategy, and adheres to the principles of innovative, coordinated, peaceful and open development", while in the 2011 paper reads as follows: "China's space industry is subject to and serves the national overall development strategy, and adheres to the principles of independent, peaceful, innovative, and open development".

[5]Interestingly, these measures call for rationally arranging national space activities; enhancing space-related innovation ecosystem; upgrading space industry capacity; accelerating satellite application industry; improving a system of diverse funding for space activities by, i.e. increasing

A careful reading of the White Paper also shows how China intends to leverage its space programme to support objectives that are at once commercial, diplomatic and strategic. To illustrate, in the section "Key areas for future cooperation", China announces the construction of a "Belt and Road Initiative Space Information Corridor" (encompassing Earth observation, communications, navigation and positioning and satellites development; ground and application system construction; and application product development). This initiative is an integral part of the development strategy and framework known as *One Belt, One Road* to enable China to take a bigger role in global affairs by improving connectivity and strategic cooperation among the countries located along the land-based "Silk Road Economic Belt" and oceangoing "Maritime Silk Road". Similarly, the construction of the BRICS remote sensing satellite constellation and the construction of the APSCO Joint Small Multi-Mission Satellite Constellation Programme serve to position China as a strategic hub alternative to the USA on the geopolitical chessboard of twenty-first-century international relations.

Finally, the assertion that the space programme is intended to "protect China's national rights and interests and build up its overall strength" can be taken as a clear indication of the further development of military space capabilities to enhance the country's *Zhonghe Guoli* (Comprehensive National Power)[6] and avert behaviours that might limit its freedom of action, be it in space or on Earth.

3.1.5 Concluding Remarks

After 5 years of highly remarkable achievements, the vision announced by the 2016 White Paper on Space Activities may appear rather modest for a fast-rising space power like China: no ground-breaking plans were announced, and the majority of them were already known to the international space community. But the quantity of projects China will embark upon in the next 5 years is, in fact, impressive: new satellite systems for Earth observation, communications and navigation, new space science missions and robotic exploration missions to the Moon and Mars manned space missions to pave the way for the construction of the *Tiangong* space station, consolidation of the current launcher fleet and R&D for new launch vehicles, key technologies experiments, comprehensive development of space application for a variety of uses and strengthening of the ground infrastructure. All these undertakings will certainly contribute to making China a mature spacefaring nation as well as one of the most prominent powers in the international (space) hierarchy of the 2020s.

---

cooperation with private investors; strengthening the training of professionals for the space industry; and disseminating knowledge about space science.

[6]*Zhonghe Guoli* refers to the total sum of strengths of a country in economy, military affairs, science and technology, education and resource and its influence.

The very importance of this government White Paper should not be dismissed either. In outlining its policy posture in the open, China makes significant efforts to avoid ambiguities and speculations about its space programme and enable the international space community to increase understanding of the country's vision and goals in space. While this increased level transparency is certainly much appreciable, it is still far from being an effective tool for confidence-building and for unlocking the full potential of China's international cooperation efforts.

3.2 Trump's Administration Plans for Space: Implications for Europe

3.2.1 Introduction

In contrast with previous campaigns, the 2016 US Presidential election race paid little attention to the space domain. Yet, with no doubt major changes can be expected in space policy, on which the President-elect Donald Trump has expressed bold and sometimes controversial positions—which could have deep and lasting strategic and diplomatic implications across the world.

This notwithstanding, in light of this anticipated upheaval, several factors of stability still remain regarding US space policy. The first and foremost is that the overarching objective of the space dominance is not in question; on the contrary, it can be reasonably expected to be strongly reaffirmed. Moreover, it is buttressed by continued strong and deep bipartisan general support in the US Congress regarding contribution of national space assets to security and defence.

Nevertheless, some major changes appear unavoidable, primarily because of a number of external factors that are evolving or perceived differently by the ingoing administration: a shift from Russia to China as the US main competitor, a—so far—limited sensibility on the topic of climate change and an increasing sensitivity to the criticality of US space infrastructure and even more so of the vulnerability of such space assets vis-à-vis potential threats from foreign countries.

Furthermore, the US space sector continues to witness a momentous rise of private actors. The way their activities and efforts will be regulated in the US legal framework, both regarding the exploitation of LEO and, in the longer run, in space exploration and exploitation-related activities, will undoubtedly shape future US and global space endeavours.

3.2.2 What Might Be the Main Priorities of the New Administration?

There are little indications on what the priorities of the new administration might be. As no white paper nor official policy document has been released by the President-elect campaign or transition team so far, the global space sector is left to wait in suspense. Yet, statements released by space policy advisers to the campaign, chiefly by Robert S. Walker,[7] reveal that the next administration might work on updating the National Space Strategy document.[8] In Walker's words, the new space policy should be based on "vision, disruption, coordination and resilience".[9]

It thus is very likely that the new administration will seek to ensure US strategy global leadership in space, renewing strategic investments to support national security with dedicated programmes; enabling robust commercial space actors in LEO, space exploration and resource exploitation; and further refocusing NASA activities.

In this respect, Walker's statements underline the need to update organisational structures, contract procedures and bureaucracies that President-elect Trump considers as outdated. Under the authority of the next NASA administrator, who at this stage is still undecided, a reorganisation of the Agency will most likely be undertaken, i.e. by reducing or shifting NASA portfolio elements—namely, Earth science—focusing NASA mostly on space science and exploration missions. Walker also suggested along this restructuring that a National Space Policy Council might be reinstituted, likely headed by the vice president, with the aim of ensuring a better coordination between DARPA (DoD), NASA and the private sector. This would ensure that "the USA remains well ahead of the technology curve". Furthermore, under such a framework "every federal government agency [might] develop a plan for utilising space assets in the fulfilment of their mission", and a business-oriented approach would be implemented, to reduce costs and avoid duplications, with "an increased reliance on the private sector to provide solutions" even in very sensitive fields (e.g. military space activities).[10]

In terms of technology and hardware developments, statements from policy advisors and President-elect Trump himself put the spotlight on two major developments which have a strong dual-use approach. The first is smaller, new-generation and more robust constellations of satellites, chiefly to improve resiliency of US military capabilities. The second is new technologies that could cause an upheaval in

[7]Robert S. Walker is former chairman of the US House Science, Space and Technology Committee, former chairman of the Commission on the Future of the US Aerospace Industry and senior policy adviser to the Trump campaign.

[8]The last official US National Space Policy paper was released in 2010.

[9]Robert Walker intervention during a panel discussion at a meeting of the FAA Commercial Space Transportation Advisory Committee (COMSTAC), 26 Oct. 2016

[10]Foust, Jeff. "Next steps for space policy." 14 Nov. 2016. TheSpaceReview. Accessed 16 Jan. 2017. Web: <http://www.thespacereview.com/article/3105/1>.

warfare, including hypersonic weapons, referring again to the perceived threats from foreign space powers. Lastly, in line with a more protectionist view of the global economy, it seems unlikely that ITAR regulations would be subject to a relaxation.

3.2.3 Implications and Opportunities for Europe

The magnitude of the potential shifts in the space policy of the world's foremost space power undoubtedly will reverberate in Europe as well, revealing—as it is often the case—both opportunities and risks at the same time.

For example, one of the consequences of the eventual restructuring of NASA Earth science activities might be the opportunity for Europe to take the lead in areas such as environment monitoring and climate change where a US backtracking might leave a gap. This includes also disaster-risk reduction-related activities and energy-related businesses, where the role of space assets can be instrumental for the full exploitation of renewable resources. In this context, Europe can develop its capabilities and increase its geopolitical influence, particularly considering the fact that the effects of climate change have implications for all countries on the planet, well recognised as a political priority both in the context of COP21 and of the UN Agenda 2030 on Sustainable Development.

Conversely, Europe is still heavily reliant on the USA for a large amount of space-related electronic components as well as space-based information. Regarding the latter, if, for instance, space surveillance data dissemination is transferred to the Federal Aviation Administration (FAA), will European users still have free and unrestricted access to this vital information for operating their space assets? More broadly this could be put into perspective together with statements related to NATO as made by President-elect Trump. These two challenges—among others—should be indeed taken seriously by European policy-makers and with appropriate responses to be given in a short time frame. This is needed to guarantee full autonomy of Europe in the fields of space-critical technologies, SSA-related data and ultimately European Security and Defence. Last but not least, safeguarding these aspects would allow the European industry to remain an active player in all the expected private-led initiatives to come.

Furthermore, it is worth noting that the Europe-USA collaboration in human space activities and exploration has strong structuring effects on the crafting of the European policy in these matters. Since Europe has been a long-lasting, privileged partner of the USA in manned space exploration, it remains to be seen whether this partnership will be challenged by the expected change in the US international posture.

Regarding the legal framework applicable to the private exploitation initiatives of space resources following the US Commercial Space Launch Competitiveness Act (CSLCA), it is expected that it will be retained and possibly reinforced by the new administration. This might trigger strong resistance from international partners, pressing Europe to define its position in these matters. Similarly, a major policy change of the new US administration might involve further development of US antisatellite weapons and/or systems, along with the principles of "freedom of action in space" in order to secure US space assets. As said above, this decision would undoubtedly generate intense discussion—and it is unclear how Europe could respond in this regard, but quite clear to anticipate the reaction of Moscow and Beijing.

To conclude, despite these (and others, as of yet undefined) anticipated changes in US space policies, and aside from the often scathing rhetoric, the European leadership must ensure that this successful, long-lasting transatlantic partnership will prevail. In this regard, a key opportunity for Europe could be represented by a more active involvement of European stakeholders in the debate and positioning surrounding a potential new US national space policy, ultimately maybe to reframe a new basis for an improved space transatlantic partnership. At the same time, there is the perhaps historic opportunity for Europe to achieve a higher degree of autonomy (in space affairs, and beyond) and consolidate cohesion among European partners, at times of greater and greater worldwide uncertainties.

3.3 Europe: Out of the Box

The European space industry is at a disadvantage to its international competitors that are backed by massive institutional programmes and R&T funding, are less dependent on commercial sales and have formed robust synergies between civil and defence sectors. What's more, those competitors are bringing new innovation and commercial challenges, by applying aggressive strategies in both manufacturing processes and business models, which are changing the nature of competition. If European space sector hopes to maintain its industry's competitiveness, it will need to have the right framework to meet these new challenges, reallocate the shared risks between public and private actors and form a long-term vision on the drivers in space activity to remain ahead of the curve in space technologies and services.

3.3.1 Europe's Technology Policy

For Europe to become a leader in space innovation, it needs to develop a mechanism for technology maturation; i.e. it needs to have the right framework in place enabling for a sufficient maturation of technologies, in terms of having the right capability, level of performance and nondependence in components at affordable economic conditions. Many of ESA or EC's Space Research and Technology programmes such as TRP[11] are good steps in that direction, but they are most often focused on developing preliminary concepts. In order for those technologies to reach maturity, they first must be brought within an application or operational programme. What is missing is an alternative way for Europe's space industry to demonstrate new technologies in orbit to address new and emerging markets.

Europe would benefit from having an institutionally funded agency to expand the frontiers of technology and science, similar to how the US Defense Advanced Research Projects Agency (DARPA) helps to incubate technologies that are later picked up by NASA when it suits their mission objectives. DARPA, founded in response to the launch of Sputnik in 1957 to ensure that the USA would be the initiator and not the victim of strategic technological surprises, is a well-established mechanism for maturing technologies. Funded by the US Department of Defense (DoD), DARPA acts as an institutional anchor customer for the development of game-changing technologies whose applications might extend beyond military uses, such as with its development of the ARPANET platform—one of the precursors to the internet. Such an institution would likewise aide the EU's Global Strategy on Foreign and Security Policy, where issues of space and security are converging.

3.3.2 Risk-Sharing Schemes Between Public and Private Institutions

European institutions have shied away from sharing market risk with industry, preferring instead to invest in the development of space systems that bring quick returns on investment. Paradoxically, beyond feasibility studies and demonstrations within the ESA ARTES programme, commercial operations are not facilitated. To enter the commercial market, European companies often must turn to the private sector to help shoulder the risk; here, many European investors and venture capitalists are still shaking off the consequences of the last financial crisis.

That barrier to entry is much lower in the USA because its public actors have a larger appetite for amortising the investments and sharing risks with private companies to the point where they can be successful in the market. For instance, the US DoD has increasingly used microsatellites to conduct research and test new technologies, while in Europe its development is relatively absent. Another poignant example comes from SpaceX's Falcon 9 launcher whose continued US institutional support has insulated the company from market forces. While reusable launch capability is in no way beyond the means of Europe's space industry, it has consistently determined that there is no reason to expect that reusable launchers will bring a cost benefit especially when considering the level of institutional

[11]"About the Technology Research Programme (TRP)." 9 Feb. 2017. ESA 15 Feb. 2017. <http://www.esa.int/Our_Activities/Space_Engineering_Technology/Shaping_the_Future/About_the_Technology_Research_Programme_TRP>.

spending in Europe. Rather than share the risk with industry only in the development phase, Europe's public sector should take further steps in helping to share market risks with industry by creating a stronger institutional market.

3.3.3 Projecting Ourselves in a Long-Term Vision

Steps to better position Europe's space sector globally are being considered by the European Union in its recently published Space Strategy. Yet, Europe is also in need of a long-term vision for its space sector; otherwise, the new technologies and capabilities its space industry creates could be exploited by strong international competitors before they can be enjoyed by Europeans. For instance, while Europe considers how to reap the benefits of the data and signals it makes available through its Copernicus and Galileo systems in its Horizon 2020 Space Work Programme 2016–2017, technology giants Google, Amazon, Facebook, Apple (GAFAs) are ready to commercialise that data based on demand.

While Europe's space sector focuses on the early 2020s, the USA and other large space actors are looking into the distant horizon with human missions to Mars and space commercialisation. If Europe hopes to maintain its competitiveness beyond the 2030s, it needs to form its own vision for the types of markets that will exist at that time and to create a European-wide mechanism to enable these technologies to bloom. Otherwise technologies for deep space exploration, space resource exploitation and developing space structures on the Moon will remain just concepts for Europe's space industry and a fully realised capability for Europe's competitors.

3.4 Italy-China Collaboration

On 22 February 2017, the Italian Space Agency (ASI) and the China Manned Space Agency (CMSA) signed an agreement of cooperation between the two Agencies on a range of human spaceflight activities. This agreement is 1 of 13 signed recently by Italy and China in the perspective of a scientific and industrial collaboration between the 2 countries.

The scientific collaboration set up in the agreement between ASI and the CMSA includes biomedical research and physiological experiments, on the consequences of long-duration space missions and on the technology needs for human spaceflight. The agreement also considers joint work on scientific payloads, data sharing and the use of technical facilities. Overall, these 13 agreements represent an important step for Italy, with the aim to recreate a new "Silk Road", with space playing a fundamental role, as stated by Professor Roberto Battiston, President of ASI. This accord on ItalianSino space activity helps to highlight the growing importance of the globalisation of space, in which technology and science play a relevant part.

This decision will have a strong impact on the current and future activities of both countries, and it will possibly reshuffle future relations between Europe and China along with the future developments of related manned space activities.

On the Italian side, the partnership with the CMSA would reinforce an already existing relationship with China. Furthermore, in a broader sense, this scientific collaboration can be helpful for the whole Italian R&D sector, combining Italian know-how and the availability of Chinese space infrastructures.

On the other side, China is changing its attitude toward international cooperation. It has been keen to promote cooperation in space and has worked both with Russia's space programme, ESA, and others, such as France and Germany. The increased cooperation with western partners helps to present a friendlier and open image of China in space and strengthens its foreign policy efforts to enable more international engagement.

3.4.1 The Future of the ISS

The consequences of the Italy-China agreement could be relevant considering the leading position that Italy has maintained in the field of human spaceflight, particularly in the biomedical field and in the construction of the ISS modules.

In recent years China has developed a promising human spaceflight programme, especially with its development of the Tiangong Space Station, which might impact the future of the international space cooperation in this field. The agreement between China and Italy might also influence the future evolution of the ISS, strongly supported by Italy, possibly opening up new venues for the private space actors.

Furthermore, this framework agreement might enable other large space actors to warm up to the involvement of China on the global scene, since Italy is a long-term established partner in manned space activities, with key players such as ESA and NASA. However, it is important to figure out which kind of legal framework would be set up to guarantee a mutual exchange of science in exchange of the utilisation of the Chinese station by Italian astronauts.

3.4.2 ESA and CMSA: Which Framework for Europe-China Collaboration in Manned Space Activities?

It is important to recall that Italy has a relevant role within ESA. Here, Italy's agreement with China might help to influence and usher in the future evolution of the relations between ESA and the CMSA.

The ASI-CMSA agreement is in line with the ESA-CMSA cooperation agreement signed on 11 December 2014. This cooperation represents for both agencies an important milestone that helps to strengthen bilateral cooperation between both institutions and to offer to China a more friendly collaborative framework.

ESA member states did not raise reservations regarding this agreement that paves the way to a full collaboration between ESA and CMSA, which should facilitate bilateral relations between China and MS, as already achieved by France, Germany and Italy. It is important to remember that China, Russia and the USA represent important partners, so therefore bilateral and multilateral cooperation are opportunities for Europe to better position itself in manned space-related activities.[12]

In the global political context, it is too early to say if the evolution of a European-Chinese collaboration will depend on the positioning and evolution of the new US administration. It is therefore quite difficult to anticipate at the moment in which framework such collaboration could materialise.

3.5 Delimitation of Outer Space

3.5.1 Introduction

Over decades of devising a detailed regulatory framework for outer space activity, the global space community paradoxically has failed to define the clear delimitation of outer space. This seemingly simple question remains highly debated and has been a regular agenda item for the United Nations Committee on the Peaceful Uses of Outer Space for over half a century: should a clear border between airspace (nationally regulated) and outer space (governed by United Nations treaties) be defined and written into law?

Thus far, no consensus has been reached.[13] While some argue that a precise line should be drawn to avoid legal ambiguity, others deem it a negligible issue or consider delimitation of outer space a potential constraint to certain airspace- or spaceflight activity in years to come, among other reasons for objection. The first question the space community faces is whether such delimitation is in fact necessary to allow for proper functioning of global space activity now and in the future. More specifically for Europe, Denmark's 2016 declaration of national legislation concerning space[14]—which includes the delimitation of outer space at the

[12]European Space Agency (ESA) Ministerial Meeting (December 2nd 2014). Resolution on Europe's space exploration strategy. ESA/C-M/CCXLVII/Res. 2 (Final) of 2nd December 2014, Luxembourg. <http://esamultimedia.esa.int/docs/corporate/Final_resolutions_1_2_3_from_CM_2014_Releasable_to_the_public.pdf>.

[13]The USA, for instance, does not believe that a delimitation is necessary, stating that the absence of such a law has yet to cause problems, which stands in stark contrast to the views of Russia and China who are adamant about reaching a comprehensively developed international agreement on the matter. European nations like France or Germany, for example, do not currently support deviation from the status quo.

[14]<http://ufm.dk/en/legislation/prevailing-laws-and-regulations/outer-space>.

so-called Kármán line roughly 100 km above sea level—acts as a trigger for the discussion of the sensibility of a consolidated European legal framework that would include the definition of the boundaries of outer space.

3.5.2 Upside and Opportunities

The obvious primary benefit of delimiting space is that it leaves no question about where a sovereign government's airspace, and thus jurisdiction, ends and outer space, governed by UN treaties, begins. Having such an exact, universally accepted, quantitative measure in place to demarcate the end of a nation's airspace would help provide legal clarity, especially so in determining liability in certain cases. While a substantial amount of vehicles and objects already fill the skies above us, the looming presence of space tourism and commercial spaceflight along with a generally increasing reliance on space resources and applications can be expected to add to the traffic above Earth, thus giving more incentive to delimitate space in order to avoid a variety of potentially dangerous instances and legal disputes.

In line with the "Single European Sky" initiative that was put in place to allow for centralised coordination of European airspace by the European Union, it would seem logical to consider the possibility of a similar move toward a mutually agreed-upon vertical aerospace regulation in form of the delimitation of outer space. This would certainly allow for more logistical simplicity and if successfully implemented help facilitate the consolidation of forces between spacefaring European nations. However, reaching agreement between these nations that foster their various national space objectives may prove to be a challenging endeavour.

3.5.3 Challenges and Uncertainty

While delimitation on both a national and international level has plenty of upside, it certainly does not come without concerns. The rapid evolution of space technology—and thus human capability to travel through and interact with space, its resources and technology placed there—as well as the increasing prevalence of commercial spaceflight may render the law irrelevant or even an impediment to the future aspirations of spacefaring nations and perhaps even private entities. As it currently stands, the lack of manoeuvrability of most of space re-entering vehicles, among other challenges, is an exemplary cause for concern revolving around technology if in fact a delimitation law were to be passed and such vehicles were to unlawfully enter foreign airspace, for instance.

Furthermore, national security concerns are also of utmost relevance when deciding on an appropriate height for the border to outer space. As a matter of fact, it can be reasonably argued that the ceiling of airspace over national territory shall be kept as high as possible in order to allow for appropriate defensive reaction in case of potential military action aimed at a sovereign entity from above. In this respect, the avoidance of legislation setting a clear limit to airspace constitutes a radical preventative measure. This arguably justifies the strong resistance toward a delimitation as practised by many technologically advanced spacefaring nations.

As over 50 years of global debate has demonstrated, the delimitation of outer space—as simple of an idea as it seems to be—is a multifaceted issue, and it is unlikely that a widespread consensus can be reached among nations on this sensitive issue in the next few years. So far the absence of a clear delimitation of outer space has not hampered the development of space activities worldwide. However, the current multiplication of applications and services in the "grey zone" around 100 km above sea level might lead to additional pressure on the legislator to reduce such legal uncertainty.

3.6 Making Exploration Great Again

3.6.1 Introduction

The signature by Donald Trump on March 25 of NASA's Transition Authorization Act of 2017, the first comprehensive NASA authorization passed by Congress in more than 6 years, marks clearly the willingness of the US Congress and the executive to have NASA actively preparing for expanding "permanent human presence beyond low-Earth orbit,. . ., in a manner involving international, academic, and industry partners,. . . and to enable a capability to extend human presence, including potential human habitation on another celestial body and a thriving space economy in the twenty-first century".[15] It also highlights the unanimous bipartisan vision for a radically new approach on human space exploration, the final objective being the journey to Mars using cislunar orbits as an enabler and possibly the Moon as a terrain for experimentation (operations, infrastructure construction, science). Considering that human activity in low Earth orbit (LEO) has brought sufficient experience and knowledge to push the new frontier one step further, the USA wants to write a new page in human space exploration seizing the opportunity of learning how to work and explore far away from Earth, not having the immediate possibility to return to mother planet. By now, the American mind-set considers realistic to expand the technological and human spheres well beyond the Earth limes. Such a bold approach poses a series of questions, particularly to Europe: How robust in time is this new approach? Are we ready as Europeans to be part of this adventure and, if yes, to what extent since ESA has proposed its vision for a Moon Village? Or should we establish a cooperation with China to deepen our contribution to LEO space exploration and possibly to be part of a contemplated

[15]National Aeronautics and Space Administration Transition Authorization Act of 2017. <https://www.congress.gov/bill/115th-congress/senate-bill/442/text>.

permanent Chinese lunar base? Or be part of both endeavours? Because both of these, of a quite different nature, are moving ahead, it is difficult to conceive that Europe would not participate in at least one of them for a range of political, scientific, technical and industrial reasons.

3.6.2 The Architecture Proposed by NASA

The current plan consists in assembling a 40-ton space station called the Deep Space Gateway (DSG), between 2022 and 2026, manoeuvring on different lunar orbits with the support of a 40-kW solar electric propulsion (SEP), that receives a crew of four and docking with visiting Orion capsule and interplanetary spacecraft. The DSG would be used by crews departing from lunar orbits to deep space, e.g. to Mars, and as a waypoint on the journey back to Earth. These crews would use a Deep Space Transport (DST) assembled after 2027 via two Space Launch System (SLS) flights to the DSG and getting ready through additional cargo provisions shipped from Earth. The 150-kW SEP-equipped DST would then be used first for a long-term habitation checkout on lunar orbits starting in 2029 before journeying to Mars in the early 2030s, once refuelled and stocked with consumables delivered by subsequent SLS missions. This plan would involve a total of 12 SLS launches of different versions.

NASA has privileged a step-by-step approach for preparing the crewed journey to Mars as well as in assembling the DSG and the DSP. This is for safety reasons and also to provide sufficient flexibility for eventual reconfigurations resulting from lessons learned in the unwinding process. Other views may suggest a more economic approach[16] in terms of mass budget in LEO, crewed flights and expenses, but which may underestimate risk taking through the different steps.

William Gerstenmaier, NASA's associate administrator for human exploration and operations, indicated that the DSG and DST concepts were briefed to the agency's Advisory Council on March 28th as NASA prepares to submit a detailed budget request in May. In parallel, major space contractors, such as Boeing and Lockheed Martin, have already phase 2 contracts with NASA's NextSTEP public-private partnerships to develop ground prototypes of several modules[17].

Besides this NASA-led effort, one must also keep in mind the programme that SpaceX is putting together to fly-by to the Moon shortly and return, to land on Mars in the coming years and to orbit the Red Planet next decade, laying down the first elements of a massive undertaking destined to start colonisation of Mars in the 2040s. Blue Origin is also planning Mars travels in the same time frame.

---

[16]An alternative architecture for deep space exploration using SLS and Orion, Ari Allyn-Freuer, The Space Review, April 17, 2017. <http://www.thespacereview.com/article/3220/1>.

[17]Selling the Gateway, Frank Morring. Jr., Aviation Week & Space Technology, p. 22, 17–30 April 2017.

3.6.3 What Place for Europe?

In light of these developments in the USA, as well as those undergoing in China to pave the way for the construction of the Soviet-Russian Mir-class Tiangong space station in LEO, to be completed by 2022–2023, what should be the position of Europe which has no plans of its own of comparable breath for the moment? Indeed, at this stage, the proposed Moon Village by ESA's Director General is a vision which requires an implementation process to be developed and endorsed by ESA's member states.

Hence Europe has to address the strategic issue of human space exploration for the coming decades: how far are we ready to commit into this human endeavour considering that the ISS is approaching its last stretch of life? At this stage, the answer, based on December 2016 ESA's ministerial council decisions, and on the October 2016 EC communication on space strategy for Europe[18], calling for taking a much stronger role on the world stage but regrettably not addressing human space exploration, is that Europeans, in our view, are not (yet?) ready to change gears in space exploration. Yet, European space agencies and companies are involved through international cooperation in the robotic scientific exploration of the solar system and in the ISS, to the extent to be on the critical path of the development of the Orion spaceship via its service module. However, they are not part of these new endeavours unfolding in the USA and in China. The planned 2019 ESA Council at ministerial level in Spain would probably be the next opportunity to firm up a European strategy/policy on this matter. This would also be timely because we will have a much clearer view as to how the proposed deep space architectures are materialising and funded as well as how the Chinese programme is unwinding.

The European strategy may be drawn along the following lines:

- Considering NASA's plans, Europe could contribute to a second Orion capsule (Lockheed Martin project would take two Orions to Mars), manufacture one of the four DSG modules to be assembled on a Cislunar Near Rectilinear Halo Orbit in the first half of the next decade, prepare astronauts to crew the DSG and take advantage of the DSG for preparing concrete elements part of the Moon Village.

  It is reasonable to assume that once the 2018 NASA budget is approved, NASA will engage discussions with international partners for their contributions, both to alleviate the US budget burden and to benefit from recognised specific competences (for instance, Canadian excellence in robotic arms, European Columbus module contribution to the ISS). ESA, European national space agencies and industry should be prepared for such exploratory talks, and European governments, in a contemplated international agreement that could resemble to the ISS model, should secure shares of the future space exploration market for

[18]Communication on a Space Strategy for Europe; 26 Oct. 2016. <https://ec.europa.eu/transparency/regdoc/rep/1/.../COM-2016-705-F1-EN-MAIN.PD>.

European industries and ensure long-term stability of the programme through mutual binding commitments.

- If Europe does not feel ready to participate in the plans NASA is fleshing out, but still wants to pursue human space exploration in the post-ISS era, it should then consider how it could be part of LEO exploration on the Chinese Space Station (CSS) via experiments and astronaut participation, for instance. This could be seen as a natural continuation of the Columbus experience based on affordable budgets, learning to work in a different space environment, sharing European manned space knowledge with Chinese experts. If retained, it would contrast sharply with the policy conducted so far for human space exploration cooperation by the Europeans and would require therefore a political decision at the highest level.
- Further, one could also imagine Europe participating to both the American and Chinese ventures. But this then should imply that we would have sufficient budgetary and human resources to be a significant actor-otherwise it would be useless to participate- in each of these international cooperation, which is, under the current foreseeable resources, unlikely.

3.6.4 Conclusion

The USA, if their plans mature, will engage in space exploration on a broad scale, adding manned deep space exploration to their fleet of robotic missions, at least for the closest main celestial bodies to the Earth. It is a dramatic step in the extension of the human habitat in space, pushing a new frontier, opening new avenues preparing possibly to future commercial exploitation of space. Providing this is firmed up, the question for Europe is very simple: do Europeans want to be part of it or not? If yes, be ready to grab this one-time opportunity in preparing its technical, scientific and human contributions right now which may imply hard choices on existing or planned programmes, as well as facing the risk that the US endeavour be abandoned in a few years as this has happened in the past to other US-led initiatives. Yet, if as early as possible an international montage, robust enough, is crafted, we should have sufficient guaranties to build up the necessary resilience to face political, geopolitical and budgetary hurdles. After all, the ISS has proven successful in that regard. Quoting John F. Kennedy in his speech to a joint session of Congress in May 1961, where the Apollo programme was announced, but not applying his words to the space race, Europeans should reflect on "the impact of this [space] adventure on the minds of men everywhere, who are attempting to make a determination of which road they should take".

3.7 Are We on the Verge of Suborbital Flights?

3.7.1 Introduction

Recently, Richard Branson announced that Virgin Galactic will operate its first suborbital flight, which he will board, starting next year.[19] This latest development comes after a series of promising steps and announcements. However, in the last decade(s), technical drawbacks have been keeping this dream from becoming a reality for the about 650 people who have already paid up to 250,000 dollars.

This year in May, the European Space Agency announced the kick-off of the construction in the UK of a rocket engine test facility, in partnership with institutional and industrial actors. Within 3 years it will be possible to fire airbreathing engines with the potential to revolutionise space launches, powering vehicles that can take off and land like aircraft and allowing for flights up to five times the speed of sound, opening up the frontier of hypersonic air travel.[20]

3.7.2 What Are the Challenges and Opportunities for Suborbital Flights?

The risks associated with suborbital flights are certainly high, since they operate in the domain of experimental flight vehicles, employing new technologies and with a very limited number of flight hours. In addition, some critical failures occurred in the past years, causing the loss of the pilot's life as well as a substantial slowing of the development process. These elements clearly question the robustness of the initially foreseen business case and require a thorough and critical analysis of future strategies.

A game changer, which emerged in recent years, is given by industrial alliances and strategic partnerships with institutional actors. In this latter case, providing services in the microgravity sciences domain represents a new opportunity for both the suborbital flights industry and the scientific communities. These flight opportunities could become an integral part of space agencies' programmes and microgravity platforms portfolios, which nowadays encompass automatic capsules,

---

[19]Foust, Jeff. "All booked: Virgin Galactic says suborbital spaceflights are full until 2021." 19 May 2017. SpaceNews. Accessed 6 June 2017. <http://spacenews.com/all-booked-virgin-galactic-says-suborbital-spaceflights-are-full-until-2021/>.

[20]ESA. "Test site for ESA-backed airbreathing engine." 4 May 2017. Accessed 6 June 2017. <http://www.esa.int/Our_Activities/Space_Engineering_Technology/Test_site_for_ESA-backed_airbreathing_engine>.

sounding rockets, parabolic flights and drop towers, all currently used worldwide.[21] Many new actors entering the suborbital flights business are indeed considering strategic partnerships with space agencies, seeking technical support in the earlier development stages and pooling expertise in specific domains such as aerospace medicine and passengers' well-being.

However, the element which can finally make suborbital flights happen, boosting the development and deployment of large fleets, is the potential strategic alliance between the aeronautical and space sectors. This type of alliance can be more easily supported in the USA and Japan, where the space agencies clearly encompass both space and aeronautics in their mandates. A good example is also offered by European industrial actors that are already engaging in R&D activities with JAXA on hypersonic flight technology development.

On the European institutional side, despite the promising ESA initiative in the UK, actors are somehow limited by the mandate of the respective agencies and institutions and are not yet in a position to support a fully fledged programme that covers joint space and aeronautics developments, enabling the realisation of suborbital flights and hypersonic travel.

3.7.3 Way Forward

The key question is whether one can foresee a more substantial role for the European public sector, including the extension of the agencies' mandate and the appropriation of the necessary funding, or whether the development of the suborbital flights and hypersonic travel sectors should be left to private industry, significantly supported by the prompt definition and implementation of policy mechanisms and regulatory frameworks.

At this point in time, in the absence of even a clear definition of the air-space interface, the regulatory and legal aspects of suborbital flights and hypersonic travel are still far from being addressed in a comprehensive and exhaustive manner.[22] This aspect, though, is highly relevant to successfully support private ventures and national efforts, which are being spent toward the development of a promising sector, including both the flight and ground segments, as well as of other related sectors which can greatly benefit, such as tourism and transportation industries worldwide. While the USA is already tackling the issue since some time, in Europe it would be advisable to develop a framework that enables the full exploitation of this business opportunity. An initial step in this direction was taken by the Italian Space

[21]Foust, Jeff. "Blue Origin joins NASA's suborbital research flight program." 3 June 2016. SpaceNews. Accessed 6 June 2017. <http://spacenews.com/blue-origin-joins-nasas-suborbital-research-flight-program/>.

[22]United Nations Committee on Peaceful Uses of Outer Space (UNCOPUOS), Questions on suborbital flights for scientific missions and/or for human transportation. UN Doc A/AC.105/1039/Add.7 of 19 Dec 2016, Vienna, United Nations.

Agency and the Italian National Civil Aviation Agency (Ente Nazionale per l'Aviazione Civile—ENAC), who signed a Memorandum of Cooperation on "Commercial Space Transportation"[23] with the US Federal Aviation Administration, to support the nascent space tourism sector by developing and regulating Space Ports.[24]

3.7.4 What Could Be the Societal Impacts of Suborbital Flights?

The overview effect, i.e. the change in awareness that has been experienced by the few individuals observing the Earth from space, has been a key outreach element for space agencies around the world. The implications of this could clearly be largely extended if an increasing number of people were to see the Earth from space, eventually further developing in our societies feelings and behaviours associated with environmental awareness, democratisation of space, sense of oneness and belonging to our "common home". Considering the current status and the potential benefits, it is probably the right time to promote wider initiatives that encompass suborbital flights and hypersonic travel and that may create synergies among different sectors and actors on a global scale, ultimately making this dream a reality.

3.8 Satellite Data to Monitor International Agreements

3.8.1 Introduction

The use of data acquired through Earth observation satellites has become commonplace. Whether that data is used for weather predictions, for scientific research with respect to biodiversity or land use or for monitoring natural calamities, in the end satellite data has become more prominent and indispensable. The use of satellite data has even expanded as an extremely useful tool to implement international law since it provides factual, relevant and up-to-date information. Further technological developments will steadily increase the range of data which can be collected through Earth observation and further enhance its accuracy. Therefore, satellite data can be used to monitor compliance with obligations contained within international agreements or to resolve disputes before an international court. This brief will consider the current use of satellite data in assisting the implementation of international law and discuss the

[23]FAA. "Memorandum of Cooperation in the development of Commercial Space Transportation'." 30 June 2016. <https://www.faa.gov/about/office_org/headquarters_offices/ast/programs/international_affairs/media/Memorandum_of_Cooperation_FAA_Italy_ENAC.pdf>.

[24]ASI. "A spaceport in Italy?" 30 June 2016. Accessed 6 June 2017. <http://www.asi.it/en/news/a-spaceport-in-italy>.

possible future use in further implementing international law, with respect to current and future international agreements and to serve as evidence before international courts.

3.8.2 Current Use of Satellite Data in International Law

Satellite data is currently used both in monitoring international agreements and as evidence before the International Court of Justice (ICJ). For instance, the Globwetland projects, funded by ESA, support the Ramsar Convention on the conservation of wetlands and waterfowl through the use of satellite data. Globwetland I and Globwetland II demonstrated the potential of satellite data to monitor wetlands. The latest project, Globwetland Africa, aims to assist African states to fulfil their obligations and commitments toward the Ramsar Convention by providing maps which, inter alia, identify and delineate wetland areas and water cycle regime monitoring and map waterfowl habitats. Another example is the ESA-UNESCO Open Initiative project and the participation therein of DLR. Within the project, satellite data is used to support the World Heritage Convention, for example, by tracking changes in Mexico City with respect to terrain subsidence due to groundwater extraction or by assisting the governance of Machu Picchu through the mapping of activities and land use. Additionally, satellite data can be used to aid disaster management as is the case with UN-SPIDER which was established under UN General Assembly resolution 61/110. The goal of the programme is to provide space-based information and services, such as maps of the affected areas.

Further use of satellite data can be seen in the proceedings of the ICJ to aid the implementation of agreed rules of international law. In the Territorial and Maritime Dispute between Nicaragua and Honduras case, satellite imagery was used by Honduras to show that islands in the river mouth of the Rio Coco are formed due to sediment, which assisted in establishing the equidistance line to set the territorial sea of the two countries. In the Maritime Delimitation and Territorial Questions between Qatar and Bahrain case, both parties used satellite data in the proceedings to present support for their claims on disputed islands. Satellite data was used to get more accurate information on the sea level and thereby establish whether the "islands" should and could be considered as actual islands or sandbanks. Finally, satellite data was used in the certain activities carried out by Nicaragua in the border area case by Nicaragua to prove the existence of the "caño", or channel, by using satellite imagery from 1961 in which the caño was visible. Costa Rica contested the clarity of the satellite imagery because of thick vegetation, while producing its own satellite image from 2010 which would rule out the existence of a channel. Importantly, the ICJ concluded that the lack of clarity in the satellite imagery means that the satellite imagery could not be used as evidence to prove the existence of the caño. However, satellite imagery was further used as supporting evidence for the existence of river deltas prior to the construction of a road, while Nicaragua stated that these only came into existence after a road was created and caused significant harm to

Nicaragua. The aforementioned examples show the usefulness of satellite data, if it can be determined that the data is reliable and accurate.

3.8.3 Further Use of Satellite Data in International Law

The current use of satellite data, however, has not been laid down in the international agreements. Rather, it is dependent on initiatives by stakeholders or, in the matter of the ICJ cases, used as evidence. Nevertheless, the obvious benefit provided by satellite data calls for specific provisions detailing the use of space-based data to monitor international agreements. Although it has not yet entered into force, the Comprehensive Nuclear-Test-Ban Treaty (CTBT) explicitly states in Article IV paragraph 11 that satellite monitoring should be considered and examined as a means for verification. Such provisions could also be included in current or new international agreements which would benefit from the use of satellite data to verify whether the state parties comply with the obligations contained therein. Examples would be the monitoring of the emissions with respect to the UN Framework Convention on Climate Change (UNFCCC) Paris Agreement or the monitoring of compliance with provisions under the Treaty on the Non-Proliferation of Nuclear Weapons (NPT). However, such provisions will be politically contentious as states would be reluctant to accept international organisations or other states monitoring their compliance and would question the integrity of satellite data of other states. There is thus a question of confidence in the data, reliability of the processing of the data and neutrality of the interpretation of the data. This further leads to a twofold problem. Namely, how to determine the reliability, i.e. how can you verify the integrity of the data, and who will determine what can be considered to be reliable data, i.e. who has the authority on verifying the integrity of the data. A first step to resolve these issues would be to propose a process for the certification of space data in order to allow to further develop and encourage the use of satellite data in the monitoring of international agreements. Initiatives such as Treaty Enforcement Services using Earth Observation and projects such as Globwetland already assist in the development. Such initiatives can serve as the basis from where the use of satellite data can be further extended, processes to verify the data developed and standards to which satellite data has to adhere to be used to monitor international agreements set out.

Nonetheless, satellites can also be used in a less politically charged capacity, by assisting states in adhering to their obligations and commitments. For example, data provided by Earth observation could be used as the foundation in gaining more insight and taking more appropriate measures to adhere to international obligations. That data can be used to adhere to the nationally determined contribution under the UNFCCC Paris Agreement, to monitor greenhouse gases emissions over particular regions, to protect wetlands and waterfowl or to combat deforestation. However, before this data can be used, the necessary technology will have to be developed and validated, as of now the technology for the aforementioned examples is in different stages of development.

3.9 About the Enthusiastic Multiplication of Spaceports Worldwide

3.9.1 Defining Spaceports

A spaceport is defined as a ground base from which space-capable vehicles are launched either toward different orbits in outer space or to suborbital trajectories. Considered as key sovereignty assets, spaceports were historically funded by governments as part of their autonomous or independent access to space strategies. Still today, major spaceports are in general largely financially supported by public authorities (e.g. Europe's Spaceport in French Guyana; Kennedy Space Centre and Vandenberg Air Force Base in the USA, etc.), and only a few commercial spaceports are in operation.

Over the last few years though, spaceports have increasingly raised a new type of interest from private companies eager to rely on their own infrastructure to provide commercial launch services. As a consequence the number of spaceport projects has increased over the last years. This new generation of spaceports is not comparable to those that have been established decades ago in terms of size and operations. As a matter of fact, they mostly target specific subsegments of the commercial space transportation market and more specifically new markets such as space tourism and launch services offered by small launchers.

3.9.2 An Expected Booming Launch Market

The expected boost of the commercial space business—especially regarding small satellites—and the potential emergence of a commercial spaceflight market are game changers in the realm of access to space. These foreseen emerging markets have consequences on the ground segments and launch facilities. Indeed, from a business model perspective, an increased offer of low-cost space transportation services will require more cost-effective and flexible launch sites than the historical ones in order to be profitable.

Forecasted demand for such services certainly is the major driving force at work here, and the anticipation of various forms of return on investment (creation of new commercial opportunities in somewhat remote areas, development of land-based tourism, regional branding, etc.) motivates some governments or local authorities to provide legal and financial support to stimulate the development of such projects.

Beyond economic considerations, the appeal of an access to space—however limited—is probably a subsidiary motivation behind some governments' initiatives.

3.9.3 A Great Amount of New Projects

Currently, the major commercial spaceport is Spaceport America based in the Jornada del Muerto desert basin in New Mexico (USA). It is the first purpose-built commercial spaceport, whose principal tenants are Google for its SkyBender drone project, UP Aerospace targeting small satellites launches with the support of NASA, Virgin Galactic for its spaceship operations, the XPRIZE Foundation and—perhaps most importantly—SpaceX.

Several other spaceports, either recently built or leased, are currently used for commercial purposes in the USA: Mojave Air & Space Port in California; the Blue Origin launch facilities near Van Horn, Texas; the Launch Complex 36 in Cape Canaveral, Florida; and many more. Outside the USA, Canada, for example, is currently planning a new spaceport in Nova Scotia.

European continental soil also hosts a number of spaceport projects, mostly focusing on suborbital flights, including Kiruna (Esrange Space Center) in Sweden and Andoya in Norway. In recent years new initiatives have also been introduced in the UK and Italy who are also considering the possibility of setting up a spaceport.

3.9.4 Economic Perspectives

It can be expected that proven new launch services addressing existing space markets such as those proposed by SpaceX—which require full control of operations—will likely get the financial backing to carry out their spaceport projects. The success of their future exploitation will strongly depend on the commercial success of the services they offer.

A more speculative area of focus for many other initiatives is related to the expected growth of small satellites industry and space tourism. At this point in time, the increase of small satellites programmes observed worldwide over the last few years is likely to continue. However, it remains to be seen whether a market for individual launches of small satellites by small launchers will meet the expected success. As a matter of fact, at the moment, the launch strategy of smallsats operators is either to launch in batches or as piggyback of larger missions. Although (quasi-) individual launches by small launchers might come at a later stage, dictated by operational constraints, this segment of the market remains somewhat speculative. Similar doubts arise for the space tourism sector as companies engaged in this market still fail to meet the initial milestones planned a decade ago.

3.9.5 *Toward a New Paradigm for Launch Sites?*

Sovereignty and independent access to space has been the main—if not the sole—driver behind the development of the major launch bases currently in operation. Economic rationale did not play a major role in the motivation of governments to invest public money in such facilities, and their exploitation actually constitutes a financial burden accepted on pure political and strategic grounds.

The emergence of a new type of launch sites—based on the assumption of a profitable exploitation of new space transportation systems—implies a change of paradigm in which what used to be a costly public infrastructure could become a commercial source of profit. It is clear that private operators will necessarily implement radical optimisations to drive down exploitation costs of such a new generation of spaceports. However, stakes are quite high since the prospects for the targeted market segments remain rather uncertain raising questions about the potential profitability and sustainability of commercially driven spaceports with limited or no public support.

3.10 Spacetugs: Toward New Solutions for Satcom Operators

3.10.1 *Background*

The spacetug concept encompasses a variety of space systems and mission architectures, based on orbital rendezvous and docking capabilities, to deliver a range of in-orbit services to other spacecraft. In general, spacetug applications can be grouped into three categories: space object de-orbiting, satellite repositioning and tugging, satellite refuelling and maintenance. One aspect that will not be addressed in this brief is the dual-use nature of spacetug capabilities. The idea of satellite in-orbit servicing is not new, but recent announcements from the space industry suggest a renewed interest in the concept in particular for satellite lifetime extension purposes. Operators and governments have long waxed on the idea of cheaply extending the operational lifetime of their satellites, and many manufacturers are looking closely into how spacetugs can offer a cost-effective solution to prolong satellite operations. In this context, the spacetug concept can offer satellite life extension services in various ways: spacetug-based satellite disposal,[25] spacetug-based station-keeping[26] or satellite refuelling.[27]

[25]Spacetug-based disposal service: the spacetug docks with the satellite and moves it to a disposal orbit enabling the satellite to use the share of fuel that is usually saved for end-of-life operations.

[26]Spacetug-based station-keeping service: the spacetug docks with the satellite and takes over the attitude and orbit control of the combined vehicle, ensuring station-keeping for the satellite.

[27]Satellite refuelling: the spacetug docks with the satellite and refuels it.

Various spacetug initiatives have recently emerged on the budding GEO communication satellite lifetime extension service market. The most advanced initiatives are the ones of Orbital ATK and MDA/SSL which have signed customer agreements and have launches planned. Orbital ATK plans to launch its first Mission Extension Vehicle (MEV-1) toward the end of 2018,[28] to provide extensions services to Intelsat for an initial 5-year[29] period. And MDA/SSL has signed its first satellite life extension agreement with SES for an initial mission using SSL's on-orbit refuelling vehicle which is planned to launch in 2021.[30] In Europe, Airbus Defence and Space announced its intention to enter this market but has not yet communicated specific details on the project or agreements with potential customers.

3.10.2 *Business Perspectives*

The most notable business developments for spacetugs concern the delivery of life extension services to geostationary communication satellites. This actually already represents a rather large addressable market with more than 300 telecommunications satellites to be potentially served. Various factors will determine the market success of spacetug services, including cost-effectiveness and reliability of the concept, and also the actual interest from satcom operators to extend their satellites' operations. This last factor is certainly the most difficult to assess, and should not be underestimated, especially in a time when the satcom sector is experiencing substantial evolutions and uncertain prospects in its core markets. In regard to economic viability, several studies have already attempted to evaluate the potential profit from using spacetug-based services to extend satellite operations and consequently delay satellites replacement. In general, these studies compare the value of deferring satellite replacement, estimated by Analysys Mason in the order of $20–30 million per annum for a $300–350 million satellite,[31] with the possible price of spacetug services. In the case of station-keeping services, with spacetug leasing prices announced around $11–14 million per year,[32] some operators and manufacturers have warmed up to the idea of spacetug services as a cost-effective mean to delay the replacement of a satellite. In the case of satellite refuelling, the low technology

[28]MEV-1 will provide life extension services in the form of spacetug-based station-keeping.

[29]"Orbital ATK on Track to Launch Industry's First Commercial In-Space Satellite Servicing System in 2018." 24 Jan. 2017. Orbital ATK 1 Oct. 2017. <https://www.orbitalatk.com/news-room/feature-stories/MEV/default.aspx?prid=92>.

[30]"SES and MDA Announce First Satellite Life Extension Agreement." 28 June 2017. SES 1 Oct. 2017. <https://www.ses.com/press-release/ses-and-mda-announce-firstsatellite-life-extension-agreement>.

[31]Bates, Philip. "The Business case for satellite life extension: running on empty." 13 July 2016. Analysis Mason 1 Oct. 2017. <http://www.analysysmason.com/AboutUs/News/Newsletter/Satellite-life-extension-Jul2016/>.

[32]*Ibid.*

readiness level makes economic analyses of the concept even more uncertain, but the probable option to serve multiple satellites with a single spacetug could support a robust business case.

3.10.3 European Initiatives

In Europe, various ongoing initiatives are supporting the development of spacetug-related standards and technologies. ESA considers that grasping and refuelling are promising capabilities for future space missions and would require minimal add-ons on commercial or scientific satellites to enhance their serviceability. As a first step to prepare an international standard, ESA launched the ASSIST activity, in collaboration with the space industry and operators, to address the analysis, design and validation of provisions to support servicing/refuelling systems for GEO satellites with minimum impact on satellite architecture and no additional risks.[33] The European Commission also supports the development of in-orbit servicing related technologies in the frame of the Horizon 2020 R&D programme and more particularly within the orbital robotics track of the roadmap prepared by the PERASPERA consortium for H2020 Strategic Research Cluster on space robotics.[34]

3.10.4 Conclusion

The spacetug and related in-orbit services concepts have been considered for a long time; however, it has taken longer to reach an economic and technical model that would raise sufficient confidence to lead to the launch of actual missions. As for any highly innovative model, doubts will remain with the viability of the concept. Nevertheless, the commercial and technical landscape has changed and could grow more fertile for the spacetug. This is what the recent agreements between operators and industry appear to suggest. Spacetugs and in-orbit services would provide new capabilities and solutions for operators to deal with new business challenges. This new dynamic is likely what motivated the CTO of SES, Martin Halliwell, to declare that "in-orbit servicing is of upmost importance to next-generation [. . .] satellites (and) enables operators to have more flexibility in managing our fleet and meeting [their] customers' demands".[35] Should upcoming spacetug

[33]"Towards a standardized grasping and refuelling on-orbit servicing for geo spacecraft." 19 Jan. 2017. Acta Astronautica 1 Oct. 2017. <https://nereus.mech.ntua.gr/Documents/pdf_ps/AA17.pdf>

[34]"High Level SRC Roadmap." 21 May 2015. H2020 Peraspera 1 Oct. 2017. <http://www.h2020-peraspera.eu/?page_id=32>.

[35]Henry, Caleb. "MDA restarts satellite servicing business with SES as first customer." 29 June 2017. SpaceNews 1 Oct. 2017. <http://spacenews.com/mda-restarts-satelliteservice-business-with-ses-as-first-customer/>.

missions be successful from an operational and business perspective, the concept certainly holds the capacity to profoundly disrupt satellite operators, launch service providers and space system manufacturer businesses altogether.

3.11 Next Steps to the Moon: What Role for Europe?

3.11.1 Solidification and Internationalisation of the Deep Space Gateway

With an extension of operations beyond 2024 under consideration and promising opportunities of partnership with the private industry under development, the long-term future of the International Space Station (ISS) is not yet written. Notwithstanding, the preparation of the post-ISS era is a central topic for the partners who, despite multiple exchanges of ideas, declarations and precursor programmes, have been, so far, struggling to build a steady, robust and commonly shared vision for the future of human spaceflight and space exploration. Recent events and announcements suggest, however, that the state of affairs is now progressing as agencies seem to be converging toward a shared enthusiasm for the Moon and moving ahead with preliminary steps.

In ESPI Brief n°12 "Making Exploration Great Again", ESPI highlighted the important step forward made by the American administration in the field of space exploration with the signature of NASA's Transition Authorization Act by President Donald Trump in March 2017. The document underlined a strong willingness of the USA to engage more actively in human space exploration with the development of a Deep Space Gateway (DSG) in cislunar orbit as the next programmatic step to prepare the journey to Mars. Although building on the SLS and Orion capsule, US plans marked a turn from the Obama administration's "LEO-Asteroid-Mars" path and a come back to the former "LEO-Moon-Mars" Constellation programme supported by the Bush administration. Two major announcements recently highlighted a solidification of this renewed US posture and confirmed the emergence of an international cooperation dimension.

First, during the meeting of the reestablished National Space Council on 5 October 2017, Vice President Mike Pence delivered an engaging speech calling for a return to the Moon in cooperation with international and commercial partners under American leadership. The Council directed NASA to develop a plan within 45 days to carry out that revised policy. This presidential declaration and the development of a plan by NASA further consolidate both political support and programmatic implementation of the rehabilitated US strategy. Second, and perhaps more importantly, the DSG gained an official international dimension with the signature of a joint statement by Roscosmos and NASA on 27 September 2017 at the 68th International Astronautical Congress in Adelaide, Australia. Although NASA had already been discussing technical options for the DSG concept with

ISS partners through the ISS Exploration Capabilities Study Team (IECST) and International Spacecraft Working Group (ISWG) during meetings in Japan, Canada and Europe in 2017, the joint statement, focusing at the moment on preliminary studies, marks a noticeable milestone in the development of an international cooperation structure around the DSG project. Regardless of US considerable financial and technical resources, international cooperation will be critical to achieve ambitious objectives, to secure programme stability and to consolidate US leadership on the global scene. From this perspective the acceleration observed recently suggests that the DSG provides a more fertile environment for international partners to contemplate a financially and technically conceivable contribution to the programme than previous plans developed under the Obama administration did.

This being said, it is important to recall at this stage that, despite a strong political support, the programme still has to face major hurdles before becoming a reality. This includes the definition of an architecture meeting various objectives from different partners, the allocation of an appropriate budget (which may be challenged by the willingness to also expand ISS operations) and eventually an official endorsement by US and international partners' establishments. Nevertheless partners can now build on an existing and robust multinational cooperation framework, such as the one of the ISS, which will undoubtedly simplify future political and programmatic progress. Overall, in a context where drivers seem to outweigh barriers, the DSG seems particularly close to becoming the next stage of international cooperation in space exploration.

Another interesting factor that will certainly have to be closely looked into is the role that China may play in the future. As of today, with the Chinese programme ramping up toward the Moon and multiple legal and political constrains for NASA to engage in cooperation with China, the situation suggests that, following Cold War competitive era and ISS cooperation era, the third era of space exploration will very likely give way to a mix of cooperation and competition for leadership.

3.11.2 What Role for Europe?

Europe is already engaged, through ESA and at national level, in technical discussions with NASA and other partners and organised various consultations about the DSG. Maintaining such active dialogue will be essential to get an informed understanding of the evolution of the project and of European industrial and scientific communities' interest and to secure an active participation in the concept definition. Assuming that Europe intends to play a prominent role in the DSG as it did for the ISS, the acceleration of the project on both the American and international scene now requires Europe to bring the topic to a higher level and to reach a political momentum. As a partner, European decision-making process obviously depends on the progress of programme approval by US institutions; yet, reaching a shared European position will require a preliminary effort to build a political consensus, if not unanimity, among member states. Achieving this consensus shortly is a necessary condition to secure Europe capacity to react timely to future evolutions on the American and international scene and to already position Europe as a key partner.

When looking at the rehabilitated Moon objective and taking into account that current DSG plans foresee robotic and human Moon landings, one cannot overlook the renewed light that is shed on ESA Director General's Moon Village vision. Indeed, the current dynamic of the international space exploration scene could offer an interesting springboard for Europe to implement, at least partially, this ambitious vision. With the declared objective to prepare a journey to Mars, the deployment of a Moon Base as a potentially European-led component of an international DSG programme would certainly offer a relevant test bed for the development and validation of key capabilities for future Mars mission such as in situ resource utilisation, robotic-human cooperative operations or ground base assembly among many others. In general, the role that Europe will hold within the next international partnership framework will be first and foremost framed by the financial and technical resources it is ready to commit.

In this context, the upcoming International Space Exploration Forum in Tokyo in February 2018 could become an important milestone, on the occasion of which an ESA-NASA joint statement could be signed.

3.12 Super-Heavy-Lift Launch Vehicles: Global Status and European Perspectives

3.12.1 Overview

A super-heavy-lift launch vehicle (SHLLV) can be defined as a rocket with a capacity to low Earth orbit (LEO) substantially larger than the existing generation of heavy-lift vehicles—above 50 t—and whose main purpose is to fulfil human spaceflight missions Beyond Earth Orbit (HSF-BEO).

As of today, three space powers are engaged in the development of a SHLLV.

In the USA, NASA's Space Launch System (SLS) is at an advanced stage of development, following a block-upgrade approach allowing for cargo and crewed missions. The first version, which is expected to carry out a first launch in 2019, will have a capacity of 70 t to LEO, to be further upgraded in several variants ultimately reaching more than 130 t to LEO in 2030. In addition to NASA, two US private companies are also developing SHLLVs: SpaceX and Blue Origin. The former is pursuing the development of Falcon Heavy and Big Falcon Rocket (BFR), with respective capacities of 65 t and 250 t in their expendable versions, while the latter is developing New Glenn, which will be positioned in the lower end of the super-heavy segment with a capacity around 50 t. Interestingly, all these private vehicles include partial reusability in their design.

In Russia, plans to develop a SHLLV have been hampered by the country's difficult economic conditions, with priority given to modernising and consolidating the family of launchers in the medium-heavy segment. Yet, latest announcements indicate that the development of a super-heavy-lift vehicle has been accelerated, with the aim of achieving a first launch in 2028 and first manned lunar landing in 2030.

As for the Chinese SHLLV, the Long March 9 (LM-9 or CZ-9) is currently under study, with a planned capacity of 140 t to LEO, a first flight in 2025 and—similarly to the USA and Russia—a lunar landing in the 2030s.

3.12.2 *Objectives*

As noted, all institutional SHLLVs have HSF-BEO missions as their primary goal. The SLS in particular has, so far, a consolidated mission schedule of 11 launches between 2019 and 2033 primarily intended to deploy the Deep Space Gateway (DSG) in lunar orbit with modules and crew and ultimately to achieve a manned interplanetary flight to Martian orbit in 2033. Similar, though less detailed, objectives are contained in Russian and Chinese long-term exploration strategies, with SHLLVs seen in these three space powers as cornerstones of their space policies.

It appears therefore evident that the development of a SHLLV is still inherently interlinked with long-term strategies to reaffirm or establish a country's technological primacy, prestige and capacity to be autonomously at the forefront of future large-scale space endeavours. In this respect, it is not surprising that, while most current and future space exploration endeavours entail a certain degree of international cooperation, this does not seem to apply to access to space.

Additionally, it is worth noting that the emergence of privately developed super-heavy launchers is leading to a diversification of purposes in this segment, inasmuch they try to offer their very large orbital capacity not only for human spaceflight but also for satellite operators. Remarkably, New Glenn has already signed launch contracts with OneWeb and Eutelsat while Falcon Heavy with Viasat as well as the USAF, thereby paving the way for a role of private super-heavy launchers in the market of national security and defence missions.

3.12.3 *Considerations*

Europe has taken a policy decision not to develop a SHLLV in the foreseeable future. Given a forecasted launch cadence of less than one per year for the SLS (at a reported cost of over 500 M$ per launch at 2012 estimates), and amid a worldwide multiplication of SHLLVs efforts, concerns around these vehicles abound, ranging from potential global overcapacity to a risk of replicating the experience of Saturn V and the Space Shuttle, which were ultimately shut down due to unsustainable fixed costs.

Whether this will be the case or not also for the new generation of SHLLVs remains to be seen. In any case, without a strong set of long-term objectives with respect to human spaceflight and exploration, as well as lacking robust political and programmatic rationales for such a development, European stakeholders seem to be comfortable in relying on such foreign launch capacity, even for the highly ambitious human spaceflight missions of the post-ISS era.

In long-term scenarios of space exploration, Europe is thus poised not to compete, but rather to cooperate with international partners, most likely the USA, contributing with its own capacities (such as the Orion European Service Module) albeit not positioning itself—so far—on the highly critical space transportation segment for future global exploration endeavours.

However, the rise of commercial SHLLVs could affect the strict boundary between so-far clearly distinct purposes for this class of vehicles. Indeed, since in the space transportation's market the offer tends to shape the demand, the emergence of two launch service providers on the verge of offering commercial launches at the threshold of 50 t to LEO could influence satellite operators to envision larger satellites or establish different launch strategies to somehow exploit such an outsized capacity. This notwithstanding, it must be noted that since such plentiful capacity to orbit is being established before a strong demand is created, all possible uses are, as of now, highly speculative.

Today, Europe is strongly and successfully positioned in the small-to-heavy segment of space transportation. With most, if not all, applications linked to manned space exploration, no strong case to develop a European SHLLV can be reasonably made with regard to achieving autonomous institutional human spaceflight capabilities. This is likely to perdure, unless unforeseen adverse circumstances in the international cooperation mechanisms arise in the future or concrete space exploration objectives backed by political support are made in this regard. On the other hand, a future scenario in which manned space exploration services are purchased for European missions from private companies could also be envisioned, one advantage being represented by possible reduced investments and broader cooperation possibilities.

Moreover, on the purely commercial side, the already projected increase in competitive pressure from international launch service providers addressing the small-to-heavy market segment could be reinforced by a perhaps less anticipated competition from commercial SHLLVs on the upper side, as they enter into service.

3.13 Reigniting Europe's Leadership in Debris Mitigation Efforts

3.13.1 The Growing Concern of Orbital Debris

In the past decade, the count of trackable debris fragments has more than doubled, driven by collisions in space, several high-profile ASAT tests, the proliferation of cubesats and growing number of upper stages left in orbit. Today, around 23,000 objects reaching 10 cm or larger are in Earth orbit, along with 750,000 objects under 1 cm. While debris is a normal consequence of the exploitation of space, events such as the Iridium-33–Kosmos-2251 collision in 2009, which occurred near 800 km in altitude, will continue to aggravate matters since debris fragments will linger menacingly in orbit for decades before re-entering Earth's atmosphere, putting at risk manned spaceflight and operating satellites along the way. While 93.7% of the rocket bodies and other debris in orbit are attributed to the USA, Russia and China combined, the loss of control of ESA's bus-sized ENVISAT satellite in 2012 put Europe in a new position with a risk of adding to the maelstrom if it is shattered by debris collisions. Mitigating the growth of orbital debris is all the more pressing with the increasing commercial activity and impending smallsat mega-constellations from SpaceX, OneWeb and others which would sextuple the number of operating satellites in LEO from the 1071 in current operation as at 31 August 2017. If left unresolved, access to space and operations in space will continue to increase in complexity and to face growing threats, with the ultimate risk of losing capacity to explore and use space.

3.13.2 Europe's Effort in the Field of Orbital Debris

Europe has proactively undertaken a variety of actions in the field of orbital debris, including strategic, policy and diplomatic efforts at European and international levels, but also programmes through national institutions, ESA and EU, to develop European capabilities in this field. Examples of the European effort include:

- In the legal domain, France, acting as a trailblazer, introduced its Space Operations Act in 2008 which requires all components of any space system launched from the CSG to be eventually de-orbited in a controlled re-entry or to be put on a graveyard orbit by the "25-year" rule.
- In the operational domain, ESA formally launched its Space Situational Awareness (SSA) Programme in 2009 which addresses Space Weather, Near-Earth Objects and Space Surveillance and Tracking (SST). ESA also launched the Clean Space initiative aiming to promote, and to develop technical solutions for, a responsible behaviour throughout the entire life cycle of space activities, including removal of orbital debris. The European Union established an SST

Support Framework in 2014 with the participation of Europe's largest national space players (France, Germany, Italy, Spain and the UK).
- In the diplomacy domain, the European Union launched an initiative for an International Code of Conduct for Outer Space Activities (ICoC) in 2008 to achieve enhanced safety and security in outer space through the development and implementation of transparency and confidence-building measures (TCBMs). European actors have also played a strong role in the development of the UN COPUOS's Space Debris Mitigation Guidelines, endorsed by the UN General Assembly in 2007.
- In other domains, such as standardisation, for example, the European Cooperation on Space Standardization (ECSS) recently adopted a new branch of standards for space sustainability including standards for space debris and space situational awareness with, for example, the requirement to ensure passivation of components at the end of life to avoid risks of explosion.

Yet, additional European leadership will be needed to fully respond to the emerging space traffic challenges presented by various trends including the incoming LEO mega-constellations and increased risk of collisions. Currently, Europe's SSA capabilities are limited, and European agencies and operators must rely critically on data services supplied by the USA and other international partners to track smaller space debris. Dependence on others for SST should raise concern at the strategic level. While the EU's SST Support Framework is a good step toward filling in that potential gap, the current approach is viewed as too fragmented to provide the level of detail required to monitor smaller pieces in LEO and would require a much more substantial investment to develop a fully independent European SSA capability.

3.13.3 Conclusion: Breaking the International Deadlock in Debris Mitigation Efforts

While the international space community can agree that the mitigation of orbital debris should be made a priority, reaching the required global consensus on the steps to pursue seems unlikely in the near term. Because guidelines are unenforceable by their nature, orbital debris mitigation rests predominantly on the amount of goodwill that states are willing to extend in voluntarily restricting themselves and their national operators from creating debris. Here the major space powers in this debate will likely continue to privilege their freedom of action in their activities over submitting to binding restrictions from international organisations to ensure the security of their assets in orbit.

Since we are left with no choice but to accept this current status quo, where government activities are not bound by any kind of "hard law" or international binding agreement, we must continue to place our faith in the wisdom of institutional space players. Yet, the interests at stake would be different if the principals of the

ICoC applied only to commercial activity. Here, even the major space players could reach an accord, so long as the arrangement does not bias the international competition—i.e. the same rules would need to be applicable to all private operators worldwide and would require proper enforcement mechanisms to ensure their correct implementation. This could be a sound basis for a revised ICoC, limited in scope to commercial activities, but addressing the most critical and worrisome challenges anticipated with the increase in space traffic.

3.14 Uncontrolled Re-entries: Outstanding Issues and Considerations for the Future

As communicated to the UN Secretary-General in a note verbale released on 8 December 2017, China's Tiangong-1 space laboratory will re-enter the atmosphere between late February and late March 2018. The possibility that some parts of the spacecraft survive re-entry and impact on populated areas has inevitably generated media attention, raising a number of pressing questions on the issue of atmospheric re-entries.

3.14.1 Why Re-entry Is a Critical Safety Issue

In an increasingly congested space environment, atmospheric re-entries can be seen favourably as they contribute to reducing the debris population. Re-entries, however, can also pose a hazard to people, property or the environment, especially when the decayed space objects do not completely vaporise in the atmosphere and do not follow a controlled trajectory ending at a known and safe location—typically over an ocean. In the past, some atmospheric re-entries have had dramatic consequences. For instance, in 1978 the Soviet reconnaissance satellite Cosmos 954 crashed in the north-west part of Canada, causing the dispersion of numerous components contaminated with radiation from the nuclear reactor powering the satellite.

Since 1957, more than 24,000 orbiting objects have re-entered into the Earth's atmosphere, accounting for a total mass of ~32,000 metric tons. Every year, 200–400 trackable objects re-enter in the atmosphere. While the majority of these objects are small orbital debris that do not survive the extreme heat of re-entry, almost the entirety of the re-entered mass (~99%) corresponds to large objects (i.e. spacecraft, spent upper stages and platforms). Roughly 70% of these re-entries are uncontrolled, corresponding to about 100 metric tons per year. On average, there is one uncontrolled re-entry of a spacecraft or rocket body every week.

While the odds of being hit by incoming debris are very low—since 1957 only one person has ever claimed to be hit by a debris—hazards posed by uncontrolled re-entry cannot be overlooked. This primarily stems from the fact that it is extremely difficult to predict the exact time, location and impact of an uncontrolled re-entry.

Due to the great speed of re-entering objects, even predictions made few hours before the impact can be incorrect by hundreds to thousands of kilometres, making it impossible for civil protection authorities to take appropriate measures. In addition, recent studies presented at the 2017 European Space Debris Conference have highlighted the possibility that "uncontrolled re-entries of sizable space objects will become of growing concern in the coming years, due to a combined effect of the increasing use of space and population growth" as well as an increased utilisation of materials for spacecraft manufacturing (e.g. titanium) which survive re-entry.

3.14.2 International Measures for Re-entry: What Are the Gaps?

At national level, several organisations around the world have already adopted specific measures to minimise the risk to human life and property on the ground. These measures include the establishment of crisis units by national civil protection departments as well as contingency plans and programmes to dispatch information on the re-entry by means of issuing periodical reports to the potentially affected states and the general public.

At international level, the members of the Inter-Agency Space Debris Coordination Committee (IADC) have been conducting annual object re-entry prediction campaigns for performing risk and prediction analyses. In addition, the IADC has advised in its mitigation guidelines that "if a spacecraft or orbital stage is to be disposed of by re-entry into the atmosphere, debris that survive to reach the surface of the Earth should not pose an undue risk to people or property". To limit the risk of objects surviving atmospheric re-entry, the IADC recommends designing spacecraft in such a way that they will completely vaporise during re-entry. The International Organisation for Standardisation (ISO) has also been developing space systems disposal standards to support compliance with these guidelines. If such an option is not viable, the IADC underlines the necessity of performing a controlled re-entry that will confine the debris to uninhabited regions, such as broad ocean areas. Also, the guidelines affirm that "ground environmental pollution, caused by radioactive or toxic substances should be prevented or minimised in order to be accepted as permissible. In the case of a controlled re-entry of a spacecraft or orbital stage, the operator of the system should inform the relevant air traffic and maritime traffic authorities of the re-entry time and trajectory and the associated ground area".

Although international guidelines define rather clearly what should be avoided during re-entries, it does not go unnoticed that there are still many gaps as well as lack of transparency when it comes to the implementation of these re-entry standards and procedures. Different sets of problem arise. For one thing, it must be acknowledged that there is no international and legally binding definition of acceptable safety risk, which is left to national authorities. More broadly, the IADC guidelines—which are based on voluntary compliance—simply indicate what should be achieved, but not how to do it.

By the same token, it must be noted that implementing the recommended procedures might not always be possible. On the one hand, it was found that even in the case of objects designed to vaporise during re-entry, a fraction between 5% and 40% of the satellite's mass might reach the surface of the Earth, depending on its weight, size, shape and composition. On the other hand, it can be observed that even spacecraft designed for a controlled re-entry (e.g. ENVISAT) can suffer damages affecting this capability and forcing them to re-enter the atmosphere in an uncontrolled way.

Another set of problems stems from the current level of information sharing. Both the IADC Guidelines and the Transparency and Confidence-Building Measures (TCBMs) endorsed by the UN in 2013 invite states to notify all other potentially affected states, the Secretary-General of the UN and relevant international organisations of predicted high-risk re-entry events, i.e. by providing technical information on the estimated casualty area as well as indications on the material compositions of the spacecraft and the presence of dangerous material aboard. However, such information is not always shared for fear of revealing sensitive data. This inevitably impacts the accuracy of re-entry predictions. There are also several instances of classified military satellites' re-entries where notifications have not been even issued.

3.14.3 Considerations for the Future

Effectively mitigating the risks associated with atmospheric re-entries will require a comprehensive approach which addresses all of the above-identified shortfalls. Toward this, more complete regulatory procedures governing re-entries will have to be put in place along with appropriate technical solutions to implement these procedures and an effective international cooperation scheme to share information. This issue has been already identified within the long-term sustainability (LTS) guidelines, currently under finalisation at the UNCOPUOS. However, the content of the guidelines has been somehow weakened during recent negotiations, and questions remain on whether voluntary, nonbinding rules will suffice to mitigate effectively safety issues raised by atmospheric re-entries.

Among the procedures to be further elaborated in order to address the problem at its roots is the retrieval of large space objects, which is already acknowledged as a post-disposal option. In light of the upcoming Tiangong-1 re-entry or of the still-unsettled decommissioning of the ISS—due considerations should be paid to the definition of internationally agreed procedures for active removal of large space objects which may not be controlled during their atmospheric re-entry. Such procedures would inevitably raise a plethora of technical, legal, financial and operational issues that affects the practicality of the active removal option.

All in all, while no concrete breakthrough is on the horizon, should the re-entry of Tiangong-1 or of other space objects cause major accidents affecting human lives or critical infrastructures, the quest for effective solutions would certainly be brought higher in the international space agenda.

3.15 Commercial Activities from the Open Ocean to Outer Space

3.15.1 Regulation of Space Resources: Looking to the Past for Answers to Current Challenges

The foundational instruments of international space law do not provide answers to many increasingly relevant legal and regulatory questions raised by the current direction of space activities. Although the drafters of the Outer Space Treaty did foresee the involvement of commercial actors, the case of commercial space resource activities especially reflects the need for legal and regulatory development, and many across the space sector support the creation of an appropriate framework which includes space resources in its scope. With that background, this Brief (1) considers two historical companies whose activities and regulatory context share key similarities with those of emerging commercial space resources initiatives and (2) through making comparisons foresees the possible regulatory needs of such activity in the future.

3.15.2 Making Comparisons

Looking to the past, the activities and regulatory context of the British East India Company (EIC) and Dutch East India Company (VOC) share key similarities to those of the emerging commercial space resources initiatives. For example, many or all of these actors (using same tense for simplicity):

- Have close relationships with the states in which they were incorporated.
- Finance their initial operations with modern methods, including with state involvement.
- Employ modern methods regarding ownership and risk management.
- Are among the first to conduct their business activities at a large and profitable scale (or aim to).
- Conduct their business activities around the commercialisation of resources/ products derived from locations that are extremely distant from their countries of incorporation.
- Conduct their business activities in locations outside the national jurisdiction of their countries of incorporation and where the relevant international law and regulations are not yet highly defined and/or enforced (or defined at all).
- Their activities raise the need for common interpretations of existing law and the development of detailed regulatory regimes at national and international levels.

Further, the historical companies and early maritime trade in general had a major impact on international law and the law of the sea in particular, and this will be true of commercial space resources too. For example, the significant contention between the Dutch and Spanish in maritime activities led the VOC to request the jurist Hugo Grotius to prepare The Free Sea, which refutes claims of exclusive possession, navigation and trade in the sea and thereby defended its use for trade by all nations. Similarly, the first successful commercial space resource efforts will motivate states to develop a common understanding of space law in order to provide legal certainty where there is currently little. And just as Grotius' early ideas greatly shaped international law, so too will the solutions developed for space resources activities.

There are of course differences between the two cases. For example, the state of international relations today in space activities in which states are highly cooperative as opposed to fiercely competitive; space transport technology has not achieved the same readiness as the maritime shipping technology of the 1600s; the 1600s' volume of maritime trade was much greater than that of aspiring space resources companies, at least for the coming decade; and demand for space resources is not yet established and will depend on other initiatives. In particular, demand for seventeenth-century maritime trade products was a main industry driver, whereas today, profit as well as the intention to develop a space economy, which could support national scientific/exploration missions, is referenced throughout the space sector. A further difference is the regime of international space law, which, although not providing answers on commercial space resources, provides a clear framework around space activities; such a framework based on ratified treaties did not exist in the historical case.

Despite the differences, though, the similarities between the two cases converge around issues critical to the topic of regulation: Both involve (1) business models commercialising materials/products acquired from distant locations, and (2) in both cases, the actors are among the first to conduct their activities at a profitable scale while also (3) operating under regulatory frameworks that are not fully defined.

3.15.3 Looking Ahead

While still appreciating these differences, the broad development phases of commercial maritime trade and the regulatory regime which materialised around it are identified below. The position of space resources initiatives within this cycle is identified, and an attempt will be made to foresee the future conditions of this industry. This exercise neither promotes nor critiques the phases as described below:

(a) Commercial actors are established and some with significant state involvement.
(b) As knowledge/capabilities improve, more actors will participate and activities will expand.
(c) Leading to initial development of formal regulation.
(d) Eventually major evolution in international law/regulation.

Applying this cycle to the space resources case, phase (a) represents the situation today: Only a few companies are pursuing space resources activities, including Deep Space Industries, ispace, Kleos Space and Planetary Resources, which has received €25 million in funding from Luxembourg. Still in the mission design phases, with great technical challenges to overcome, and as it will be many years before envisioned commercial activities occur, the industry is clearly in its infancy. Considering phase (b), the lack of proven knowledge on the exact composition of target locations represents a major hurdle. If observations on target asteroids prove false after direct in situ measurements, current technology may not readily allow a mission to redirect to other targets. However, as knowledge and capabilities improve, space resource activities will eventually prove profitable, and more companies will participate while introducing new activities too. This will lead to phase (c), when initial regulation will develop. And it is in anticipation for a scenario with numerous companies conducting diverse commercial activities on a number of targets that initiatives such as the Hague Space Resources Governance Working Group have appeared. This initiative, for example, released in 2017 its Draft Building Blocks for the Development of an International Framework on Space Resource Activities.

3.15.4 Conclusions

To characterise the final phase (d) in the case of maritime trade is to describe the conditions today. The legal and regulatory developments in maritime activity seen during and even well after the EIC and VOC are significant: The international community has recognised in international law that the sea is open to all nations, and maritime trade has become a cornerstone of the global economy, operating under a robust regulatory system. If this cycle repeats in the space resources case, then in the last phase, their use will be seen in diverse industries, and the global community will eventually have to develop a common interpretation of the applicable international law, in particular Articles I & II of the Outer Space Treaty, and develop regulatory solutions. There are significant prospects for international cooperation, but a common approach will be necessary to enjoy its benefits. As there is no international authority to guide this process and likewise no established jurisprudence, a common understanding of the relevant international law is therefore essential. Complicating the matter is the speed of technological innovation: the cycle described in the historical case took centuries to play out; however in today's case, legal and regulatory developments must keep up with the pace of technological innovation. One approach to create an international framework on space resources activities may be for states to first develop regional agreements on the way forward, which is a process that Europe could begin now.

3.16 Brexit and Space

3.16.1 Introduction

On 23 June 2016, it was announced that the British public had decided through referendum that the UK would be initiating the procedures of leaving the European Union (EU), aptly titled Brexit. Little over 2 years into the withdrawal process and a great deal of uncertainty still remains over the final outcome—however it is the UK government's position that Brexit means leaving the single market, customs union and the European Court of Justice (ECJ). These three pillars of Brexit, and of course a non-EU membership status, will most likely have significant implications for the future of the UK space sector—and with no clear vision of the UK's future relationship with the EU and a final trade deal, the status of UK-European space activity is at question for both political institutions and the private space industry.

3.16.2 Access and Participation to EU-Funded Space Programmes: Galileo

A major component of UK-European collaboration in space ventures is facilitated through the European Space Agency (ESA). However, although ESA is an independent institution from the EU and membership of one is not a precondition of the other, and so would seem disconnected from the Brexit issue, the EU contributes around one third of ESA's overall budget and acts as ESA's primary procurer. This is of significance for the UK space sector in particular as EU-funded space programmes conducted through ESA will preferably be contracted to EU member states which are also members of ESA. For this reason, and because the UK will obtain a third-party status to the EU and no longer be placing funds in the EU space budget, concerns over access to and participation with EU-funded space programmes like Galileo, Copernicus or future potential Space Situational Awareness (SSA) and GovSatcom projects are in the limelight. While it is conceivable that an agreement could be formed so that the UK can contribute payments into such programmes, it is foremost a political issue regarding security and the status of third-party participants.

The matter of UK access to and participation with EU-funded space programmes is most apparent and encapsulated within current speculations regarding Galileo. The Galileo programme itself is a constellation of satellites to provide Europe with an independent navigation system equivalent to that of the US's GPS, costing about €10 Billion. To this point, the UK has played a large role in Galileo's development and operation, contributing 14% of its funding and 17% of the work. So far, Britain has been awarded the contracts to construct the "brains" of every Galileo satellite and hosts the Galileo Security Monitoring Centre, and a UK-based company CGI is responsible for the encryption service on the Public Regulated Service (PRS)—the security sensitive military grade component of Galileo. In terms of participation, it is highly unlikely that UK companies will benefit from any future development contracts. The UK has already been sidelined from the recently approved batch of ESA procurement contracts for Galileo as an agreement for Britain's continued participation is yet to be confirmed. In regard to access, the EC has stated that Britain will be excluded from one of the key aspects of Galileo, the PRS, once it becomes a third party and a foreign entity. While it is foreseeable that Britain can retain access to the services of Galileo, access to and operation of the restricted military elements will depend entirely on agreements for future UK–EU security relations. In response to discussions of removing the UK from these components, the UK has envisaged creating its own independent Galileo equivalent or even partnering with Australia—signifying the importance of Britain's past and current investments in a positioning, navigation and timing system for military capabilities in particular.

Ironically, after having been resistant against the Galileo programme in its initial phase, the UK is now making a clear demonstration of the high strategic value of this investment for Europe.

3.16.3 The UK's Industrial Position

As already outlined, non-participation in EU-funded space programmes poses risks to certain sections of the UK space sector. Though beyond this, leaving membership of the single market, customs union and the ECJ presents additional concerns in the absence of a defined future relationship and trade deal with the EU. It is important to note the UK space sector is export orientated, with 50% of satellites entering into the single market, and tariff-free trade with EU member states advantageously keeps construction costs down. Furthermore, leaving the customs union could lead to some serious disruptions to space component supply chains, with the possibility of time-consuming and bureaucratic procedures imposed at the borders. There is also a human aspect to these potential restrictions—the space sector is heavily dependent on international collaboration and the associated movement of expertise—and so limitations on not just the transfer of equipment but also access to talent can have a detrimental effect on the prosperity of the UK space sector. As well as this, and unless a dedicated agreement is quickly settled with the EU, the UK will no longer benefit from the availability of funding received from EU Research and Innovation Framework Programmes during the next MFF, which could have knock-on ramifications which will impact small- and medium-sized space companies greatest if the void is not independently filled by the UK.

The UK has successfully managed to play a prominent role and to position its industry at the forefront of the European space sector, thanks to an active and steady participation in European programmes through both ESA and the EC. However, except in the field of secured space telecommunications where it has deployed the Skynet series of satellites, it never seemed to raise high ambitions in this sector on a purely national basis. Following Brexit, the UK will need to further clarify its national space policy. As discussed above, continuation of R&D efforts and funding might be the most urgent issue, and UK continued contribution to ESA programmes is a major asset and can provide for this until further agreements are found with the EU. However, since EU is procuring most of European public space infrastructures, and in particular those implying the highest volumes of production, the UK space industry might be facing a difficult situation to keep the critical mass and secure its competitiveness if deprived from getting access to this market.

Part II
Rising Stakes in Security in Outer Space

Chapter 4
Outcome Report of the 12th ESPI Autumn Conference

The 12th ESPI Autumn Conference was held at the ESPI premises in Vienna on 27–28 September 2018 and had the theme of "Security in Outer Space: Rising Stakes for Civilian Space Programmes".

The conference was supported by the ESA; the Austrian Research Promotion Agency FFG; the Austrian Ministry for Transport, Innovation and Technology; and the George Washington University's Space Policy Institute.

4.1 Introduction

The democratisation of space-based services for governments, businesses and consumers is creating a growing dependence of society and economy on the space infrastructure. As a result, there is a recognition at national and global levels to uphold the substantial benefits drawn from space assets, pressing global awareness to better protect space assets from harm. Today, space-faring nations have to deal with a wide range of security challenges threatening the sustainable operation of space activities. In this sense, security in outer space is of growing importance not only for operators but also public and private entities using space-based data and services for their operations.

This increasingly challenging security situation has been acknowledged by the global community at large, recognising that space security can only be achieved as the outcome of a global effort. The European Union uphold "the protection and resilience of critical European space infrastructure" as a flagship objective of the

E. Burger, G. Bordacchini, *Yearbook on Space Policy 2017*, Yearbook on Space Policy, https://doi.org/10.1007/978-3-030-05417-5_4

Space Strategy for Europe and stress the importance of cooperation at European, but also at international, level, in particular with the USA. Across the Atlantic, space has always held a prominent defence and national security dimension, making space security of critical importance. Currently, in their respective policies, both the USA and Europe seek to contribute to a global effort and to reinforce international partnerships, in particular with one another.

Recent developments in space security in Europe and the USA have reshuffled the cards at strategic, policy and operational levels. A major development in the US approach to security in outer space was the adoption of Space Policy Directive-3 (SPD-3), which establishes the first "approach to Space Traffic Management (STM) that addresses current and future operational risks". The policy also announced that the USA would pursue and extend its relations with international partners for STM and SSA, edging toward a reinforced cooperation with Europe.

In this context, ESPI and the George Washington University's Space Policy Institute joined forces to conduct a research study entitled "Security in Outer Space: Perspectives on Transatlantic Relations" aimed at discussing European and US approaches to security in outer space, the role of transatlantic relations in this field and potential way forward toward a reinforced transatlantic partnership. The study served as groundwork for ESPI's 12th Autumn Conference that aimed to disentangle current and foreseen developments in space security as well as pre-requisites for a deepened dialogue between Europe and the USA in this respect.

The 12th Autumn Conference was a 2-day event and covered the following three sessions:

1. State of Play in Space Security Strategies: Space Assets in an Evolving Environment
2. Forging Ahead with European Space Security Efforts
3. Rethinking Transatlantic Cooperation: Toward STM

The event was intentionally designed to have a forward-looking vision on space security strategies to be endorsed on both sides of the Atlantic through a careful selection of speakers active in the field, both in the private and public sector. The present proceedings are aimed at providing the reader with a good understanding of the main ideas brought forward during that event.

4.2 Conference Sessions

4.2.1 Introduction and Presentation of ESPI Studies in Space Security

The welcome address was delivered by Jean-Jacques Tortora, Director of ESPI, in which he introduced the session themes and provided perspectives and background information on the topic of the conference. Mr. Tortora put the emphasis on the changing nature of policy approaches in security in outer space on both sides of

Atlantic and their associated challenges and opportunities, notably with the announcement of the US new STM policy. He also restated his enthusiasm for ESPI's fruitful partnership with the George Washington University's Space Policy Institute.

The opening remarks were followed by a presentation of Sebastien Moranta, ESPI Coordinator of Studies, who presented the preliminary findings of an ESPI study on "Security in Outer Space: Perspectives on Transatlantic Relations". This study, scheduled for publication in November 2018, was conducted in close collaboration with the George Washington University's Space Policy Institute with the objective to provide a comprehensive overview of current approaches to space security in Europe and the USA in view of revisited and reinforced cooperation in this field. The study is a follow-up of an ESPI report published in August 2018 on "Security in Outer Space: Rising Stakes for Europe" and will be followed by an upcoming ESPI study on "European Space Traffic Management Policy".

During the presentation, it was underlined that challenges to space infrastructure security, which include intentional (e.g. ASAT, malicious interferences, cyberattacks, etc.) and unintentional threats (e.g. space debris, unintentional interferences, etc.) but also space weather hazards (geomagnetic storms, solar storms, etc.) are:

- **Multiple and diverse**, and therefore require different mitigation and protection measures
- **Interrelated and interdependent**, creating a dependence between actors involved and requiring a holistic approach
- **Ubiquitous and inclusive**, although space systems are unequally exposed or vulnerable to threats
- **Intensifying**, because of various trends (e.g. increasing space activity, new concepts, connected space, strategic target, "space control" capabilities)

In giving a brief overview of the "security in outer space" field, which is a subset of the "space security" domain, Mr. Moranta introduced the ESPI Matrix used for mapping activities (see Fig. 4.1), recalling that three main areas should be taken into account in a holistic approach to the issue:

- **Space Situational Awareness (SSA):** Current and predictive knowledge and understanding of the outer space environment including space weather and location of natural and manmade objects in orbit around the Earth
- **Space Environment Protection and Preservation (SEPP):** Preventive and curative mitigation of negative effects of human activity in outer space on the safety and sustainability of the outer space environment
- **Space Infrastructure Security (SIS):** Assurance of the infrastructure ability to deliver a service that can justifiably be trusted despite a hazardous environment

The presentation then continued with the description of European and American approaches to security in outer space.

Policy areas and activities: ESPI Matrix

	Capacity-building programmes	Legal and regulatory regimes	Diplomacy and cooperation
Space Situational Awareness	• SST capabilities development • Space weather models development • SSA services delivery	• Space objects registration obligations and procedures • SSA data policy	• SSA data sharing agreements • TCBMs
Space Environment Protection and Preservation	• CleanSpace technologies development (e.g. active debris removal solutions)	• Space law (e.g. end-of-life obligations) • Standards for space environment-friendly design (e.g. passivation devices)	• Space Debris Mitigation Guidelines • Long-term sustainability guidelines • International Code of Conduct proposal
Space Infrastructure Security	• Security enhancing technologies development (e.g. secure links, materials) • Resilient system architectures (e.g. fragmented systems)	• Space programme security rules and procedures • Security and safety standards • Supply chain control processes (e.g. import, qualification)	• Collision avoidance procedures and coordination • Deterrence through hosted payloads on allies' satellites

Fig. 4.1 Matrix for the European approach to space security (ESPI)

ESPI identified that, in Europe, the main space security policy drivers are to:

- **Secure the results of the continuous and substantial investment** made by public and private actors
- **Protect the European economy and society** against risks related to its pervasive and sizeable dependence on the space infrastructure
- **Contribute to a service-oriented policy** by assuring the ability of the infrastructure to deliver a service that can be justifiably be trusted, in particular for users in defence and security
- **Guarantee European autonomy and freedom of action** in the field of security in outer space with implications on the space domain at large (nondependence).

In line with the strategic objective to "promote Europe's position as a leader in space, increase its share on the world space markets, and seize the benefits and opportunities offered by space", additional long-term stakes can be identified:

- Since space security now holds a central position in space diplomacy, Europe needs to play a prominent role on the international scene, as a promoter of a clear, united and consistent "European way".
- Taking into account that security will play an increasing role in commercial space markets, Europe should support the European industry competitiveness and stimulate the emergence of commercial markets for SSA-related data and services.

A few key elements of the European approach were then described:

- National governments remain the core actors of space security in Europe. Space security policy is primarily addressed at national governmental level and driven by domestic security and defence strategies with some reluctance to transfer sovereignty. Although European cooperation is progressing, it remains mainly structured through bilateral and intergovernmental frameworks. A pan-European cooperation in this field still faces hurdles. Recent declarations by high-ranking officials in some countries (France, UK, Germany) suggest a growing importance of space security and defence in policy agendas.
- ESA launched a number of initiatives including an SSA programme (limited Space Surveillance and Tracking component on member states' request), a cybersecurity excellence centre and the CleanSpace initiative, which positioned the Agency as a key player for capability building. Additionally, ESA made efforts to improve its capacity/legitimacy to handle security-related activities.
- The EU is consolidating its role in space and security and defence policies. In this context, Space security is of special importance for the EU, at the crossroad of space and defence and security policies, two strategic areas of development of the EU. EU awards an increasingly great importance to the security of its space programme (e.g. security accreditation, security architecture, data security policy, synergies with defence, etc.). A number of activities are supported by/embedded in EU programmes (e.g. EU SST support framework, H2020 R&D projects, and international diplomacy initiatives). The EU proposal for the Multiannual Financial Framework 2021–2027 suggests an increase of efforts in various fields including in particular SSA.

Comparatively, the US approach to space security is building on the legacy of successive administrations reflecting that space became an important component of US national security and military but also economy and society. Since 2017 the Donald Trump administration ("Peace through strength") still follows the main core objectives of its predecessors in terms of space superiority. What has changed is the emergence of space as a warfighting domain leading to political will to prioritise investments in resilience, reconstitution and operations to assure [the USA] space capabilities and the promotion of a "whole-of-government" approach to US leadership in space, in close partnership with the private sector and allies. In line with this new context, the 2018 space strategy builds on four essential pillars, all related to security and defence: (1) mission assurance, (2) deterrence and warfighting, (3) organizational support and (4) conducive domestic and international environment. ESPI identified that, in the USA, the main space security policy drivers are to:

- **Assure military superiority in space and on the ground:** Space infrastructure is a critical component of warfare; strengthening of deterrence in space and on the ground; re-emergence of a space warfare doctrine and need for tactical response options in space.
- **Safeguard national security against space vulnerability:** Growing perception of vulnerability in space (i.e. concept of Space Pearl Harbor); need to protect

critical space assets against threats (e.g. ASAT, cybersecurity, jamming and spoofing).
- **Foster commercial space:** Give way to a potentially promising commercial market; foster competitiveness from cost-effective economic agents; avoid diverting valuable public assets from their strategic missions.
- **Reinforce global leadership in space:** Position the USA as leader in a domain of increasing strategic significance on the global scene; promote space safety standards and norms across the international community.

It was noted that American SSA capabilities are unmatched as a result of massive investments from defence budgets and that they are expected to be even further improved with the deployment of the Space Fence programme. The promotion of deterrence and resilience-oriented architectures consider fragmented systems (e.g. constellations, miniaturised systems) and hosted payloads (e.g. hosted DoD payload on Norwegian satellite) as valuable enhancers. The signing of Space Policy Directive-3 announced the establishment of a national-led approach to Space Traffic Management (STM) and a reorganisation of responsibilities across military and civil/commercial branches with the objective to:

- Encourage and facilitate US commercial leadership
- Improve SSA data interoperability and enable greater SSA data sharing
- Develop STM standards and best practices

In addition to these efforts, the announcement of the possible creation of a Space Force symbolises a shift in the US posture and may lead to possible reactions on the international scene. Mr. Moranta recalled that, to be effective, the decision to establish a Space Force requires an approval by the Congress which is not secured at the moment.

With regard to transatlantic relations in space security, Mr. Moranta explained that, today, they involve a complex mix of arrangements and different channels:

- **Bilateral government-to-government channels:** SSA data sharing agreements/ operational liaison and exercises (military field)
- **Europe-wide to US channels:** Regular EU-US Space Dialogues; case-by-case cooperation between US and European organisations (e.g. NOAA and EUMETSAT)
- **Multilateral channels:** NATO, UN COPUOS, Conference on Disarmament, IADC, ITU, etc. (different stakeholders represented)
- **Industry-to-industry channels:** Satellite operators relying on governmental and commercial data and services, cooperation through Space Data Association

This somewhat scattered cooperation framework builds on strong roots of value-based partnership in various sectors (economy, security, space, etc.). Notwithstanding, and although cooperation in space often remains unaffected by ups and downs at political level, the recent deterioration of political relations between Europe and the USA may have implications, still unclear at this stage.

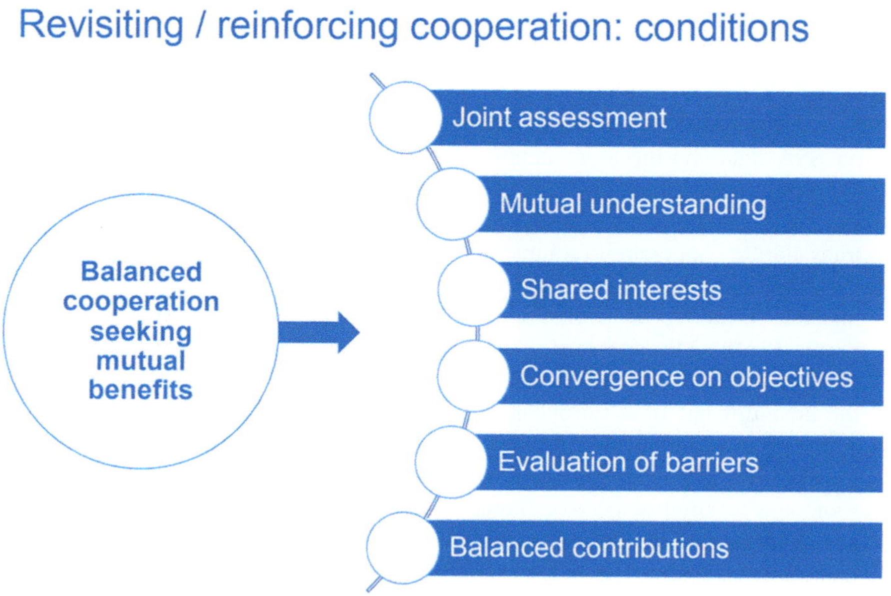

Fig. 4.2 Conditions for revisiting/reinforcing cooperation (ESPI)

To conclude, Mr. Moranta highlighted that space security is now a priority of space and security and defence strategies on each side of the Atlantic. In this context, both Europe and the USA share a comparable assessment of security challenges, however based on different policy drivers. In the USA, there is a growing perception of a national security vulnerability in the space domain (i.e. Space Pearl Harbor) driving the protection of space assets against threats and justifying groundwork for the preparation of tactical response options. In Europe, stakeholders are ready to consider multiple challenges with different priorities but still face difficulties to translate it into operational cooperative decisions. Both conduct similar activities (SSA, SEPP, SIS) but with a lopsided state of affairs (policy, resources, capabilities). Although active, transatlantic relations do not build, yet, on an integrated framework at political level which gives way to cooperation on a case-by-case basis. Recent and foreseen developments in the USA and in Europe create a fertile ground to revisit and reinforce transatlantic relations. In the USA the introduction of new initiatives and the reorganisation across military and civil/commercial branches with the SPD-3 may have consequences for international partners but also offer new opportunities (e.g. cooperation in STM). In Europe, the expected increase of efforts in the space security domain and the consolidation of the European approach may imply considerations for cooperation with third countries (in particular the USA).

To revisit transatlantic relations toward a reinforced cooperation seeking mutual benefits, a number of conditions have to be met (Fig. 4.2).

As a preliminary conclusion (the study was not finalised at the time of the conference), ESPI identified three key areas with great potential to reinforce transatlantic relations:

- **Space Situational Awareness:** Changes in US organisation will have an impact on SSA data sharing agreements (channels, data fusion, confidentiality, etc.). More ambitious efforts on the European side may also create a lever to revisit data sharing granted that the development of European capabilities takes into account transatlantic cooperation (complementarity/redundancy of capabilities, data policy). Streamlining cooperation in civil space security aspects may benefit from a mandate of the EU in specific areas such as international relations.
- **Space Traffic Management:** To weigh in the accelerating international discussions on STM, the nationally led initiative announced by the USA should trigger the development of a European approach to STM. The development of such approach would require, before anything, setting up a forum gathering European stakeholders to discuss this issue. Transatlantic cooperation in this field should be promoted to align respective frameworks through a convergence of objectives and development of compatible requirements.
- **Commercial policy:** There are expected benefits and prerequisite conditions for an open market, with the USA and Europe sharing the same vision on commercial space at large but are at different level of maturity in the integration of such involvement in space security policy. Additionally, more discussions on the conditions for an open market (shared delineation of commercial areas, common standards and compatible security requirements) will be necessary.

Mr. Tortora and Mr. Hertzfeld concluded the presentation with a few complementary remarks. Mr. Tortora highlighted that the European approach is clearly changing; the state of affairs is still lopsided with both supranational and intergovernmental factors in play. Mr. Hertzfeld stated that Europe should not overreact to recent announcement. He argued that the US space policy did not change substantially (the Obama administration's Space Policy of 2010 was not revoked) with a great emphasis put on cooperation. He mentioned the role of Congress for the creation of the Space Force with upcoming elections that might reshuffle the cards. Most of the new measures are designed for internal reorganisation within DoD. In terms of resources, more investments are made in the USA; however, more capabilities are available around the globe announcing plenty of opportunities.

4.2.2 Keynote Speech by Kai-Uwe Schrogl, ESA

Mr. Schrogl, Chief Strategy Officer of ESA, opened the conference with a keynote speech presenting ESA perspectives on the topic of security in outer space. He welcomed new ESPI initiatives in space security studies as a positive development.

Mr. Schrogl started his presentation by highlighting the existence of a seamless beam of citizens' protection needs ranging from safety to security and defence needs.

Traditionally, while the safety side is in practice more of a civilian issue, the defence side is hold by the military. However, as technology progresses, the area of security is getting broader, encompassing new issues such as food security, environmental security, etc., resulting in blurred lines between the traditional segmentation of safety and defence. The responsibility to promote and safeguard security is therefore not always easy to attribute and sometime falls to both the military and the civilian realms.

Mr. Schrogl recalled that, in Europe, a number of initiatives were launched in the field of security, including space security from both ESA and the EU. In fact, an ESA-EU joint statement in space policy was signed in November 2016, highlighting the key goals to set for Europe at the political level in order to ensure a cohesive European identity in space. The joint statement also sets socioeconomic objectives and strategic goals notably, the question of European autonomy in space, which involves the development of launchers and a reinforced cooperation between EU-ESA and with other space actors, and calls for the need for Europe to operate "in a safe and secure environment".

With this last objective, ESA is actively pursuing the establishment of a strategy concerning ESA space security activities, which will be proposed and decided at ESA Ministerial Council in 2019. This strategy will be part of an overarching ESA strategy that structures ESA activities in four main programmatic categories including (1) science and exploration, (2) applications, (3) enabling and support and the lastly added (4) safety and security. The safety and security pillar itself includes three components:

- Space safety (in space), which is grouped under a single programme, will oversee initiatives in the field of space debris, space weather, collision avoidance autonomy and planetary defence.
- Safety and security applications (from space), which englobes several programmes in different domains such as disaster management, maritime safety, Quantum Key encryption, etc.
- Cybersecurity, which is part of all programmes and basic activities conducted by ESA.

With regard to cybersecurity, which is an essential element across security applications, Mr. Schrogl underlined that the vulnerability of space systems is largely underestimated and that there is a growing need to catch up in this field through partnering with international organisations, both in the space and ground segment.

Mr. Schrogl then addressed the topic of STM recalling the publication of the International Academy of Astronautics (IAA) studies, first published in 2006 and reedited in 2018, which tried to think of a comprehensive approach to STM, and on how STM could be set up and implemented. He explained that the STM debate is likely to get traction and that there is a need to establish a common roadmap in Europe along common positions and joint ideas between the different stakeholders (ESA, EEAS, EU, etc.). He mentioned two possible approaches to STM:

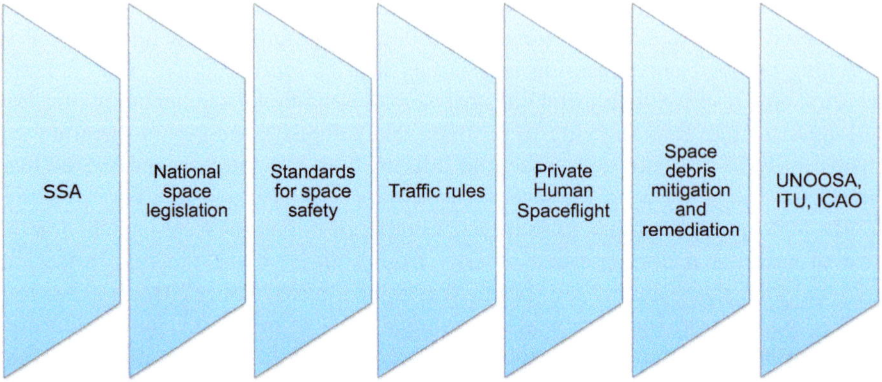

Fig. 4.3 Incremental bottom-up approach for STM (ESA)

Fig. 4.4 A comprehensive top-down approach to STM (ESA)

- An incremental bottom-up approach (see Fig. 4.3), involving work on SSA capabilities, national space legislation, standards for space safety, traffic rules, space debris mitigation and remediation, UNOOSA, ITU and ICAO.
- A comprehensive top-down approach following the ITU and ICAO models (see Fig. 4.4), involving work on outer space convention, outer space traffic rules and outer space traffic technical standards.

In a nutshell, Mr. Schrogl concluded that the various threats to space activities from military, non-military and natural origins are growing and are more properly and precisely assessed. As a result, the willingness to take on these multidimensional challenges is perceptibly rising, but political solutions to achieve to an efficient organisation are still required.

4.2.3 Session 1: "State of Play in Space Security Strategies: Space Assets in an Evolving Environment"

The first thematic session of the conference gathered following speakers:

Jana Robinson	*Space Security Program Director, Prague Security Studies Institute*
Philippe Brunet	*Director for Space Policy and Research, Copernicus Programme*
Richard Buenneke	*Senior Space Policy Advisor, US Department of State*
Regina Peldszus	*Co-Chair, EU SST Consortium*
Carlo des Dorides	*Executive Director, European GNSS Agency*
Andrew Monham	*Spacecraft Operations Manager, EUMETSAT*

The session was moderated by **Ms. Jana Robinson**, Space Security Program Manager at the Prague Security Studies Institute, who contextualised the discussion with backdrop information highlighting the topicality of security in outer space. Ms. Robinson noted that a holistic approach to space security is required, taking into consideration all its dimensions, notably when it comes to space exploration, regulations for the economy, STM and possible defence strategies.

4.2.3.1 Presentations and Panel Discussion

Mr. Philippe Brunet, from the European Commission, started his speech by highlighting the changing role of the Commission in space security and putting forward the new proposal under review by the European Council and European Parliament (EP). To add to this moving environment, he also recalled that, following European elections in 2019, the Commission can be expected to go through changes. Mr. Brunet structured his speech around three main points: (1) state of play, (2) challenges ahead and (3) possible approach.

First, regarding the state of play, he explained that space security is increasingly important for the EU since, beyond the immediate value of space assets, the potential impact of a loss of space services for the European economy which benefits greatly from Copernicus and Galileo data would be disastrous. For this reason, it is essential to collaborate to protect the integrity of these infrastructures, address their vulnerabilities and ensure delivery of services. The EU launched an EU SST Consortium Support Framework in 2014 that started delivering SST services since mid-July 2016.

The EU/European Commission considers four main challenges ahead:

- The first challenge for Europe is to develop adequate capabilities to be able to cooperate with the USA on a balanced basis, bringing forward the concept of "nondependence".
- The second challenge is to overcome fragmentation and achieve the pulling, sharing and optimisation of all European space surveillance capabilities. This

will create a healthy basis to develop additional capabilities in a coherent and coordinated manner according to operational needs.
- The third challenge is to look at all the threats to space assets and develop a holistic response addressing both intentional and nonintentional risks.
- The fourth challenge is to achieve synergies between civil and military needs and capacities (90% of current military operations supported by satellite services). This last objective is in line with the goals of the EU Space Strategy, the European Defence Action Plan and the Global Strategy.

Finally, Mr. Brunet concluded with an overview of the possible approach, which would include three main areas of activities:

- *Space Situational Awareness* to benefit from a comprehensive view and understanding of what is happening in space through surveillance and tracking activities. That segment will primarily address unintentional risks (space debris, NEOs, space weather, etc.). In addition to the current EU SST Support Framework, these activities will allow the development of a full SSA programme. The SSA component is meant to be dual source and dual use, which indicates potential for relevant synergies with the European Defence Fund and some PESCO initiatives, for instance, in the field of ballistic missile detection (BMD) systems.
- *Space Protection and Deterrence*: this is aimed at reducing the vulnerability of space assets against unintentional or intentional threats, through the development of specific "counter-space capabilities". A coherent and integrated approach at EU level is still to be developed, using tools and policies already in place for:

  - Improving the resilience of the space infrastructure through the use of innovative materials, equipment and engineering solutions to build the satellites, to improve their manoeuvring capability and to deploy on-board collision detection systems; the research programmes and work on critical technologies could deliver some solutions;
  - Improving the resilience and continuity of the services through the development of "disruptive" space architectures, such as multi-constellations of small satellites, to reduce the impact of the destruction or unavailability of a satellite; this can also be achieved through the development of more agile launching facilities, to be able to easily and rapidly replace a damaged satellite.
  - Improving the cybersecurity of the space systems, including the physical security of the ground segments and the links to guarantee the integrity, continuity and reliability of the services provided; in this regard, GOVSATCOM will play an important role in the EU Space Programmes to ensure that the necessary satellite communication links are secure and do not make the EU space programmes dependent on non-EU actors.
  - Limiting the risk of intentional threats through new business models in which the ownership of space assets is shared among several states or national interests to lower the geopolitical risks; another way to limit the intentional threats is to work at global level for the adoption of multilateral norms or, ultimately, to prepare and apply economic or military retaliation.

- *Space Active/Sustainable Management*: this section aims at preventing issues resulting from the congested and contested nature of the space domain, with a view to guarantee free, safe, efficient and sustainable access to space for all. It may involve reflecting on a series of areas for action such as:
  - Reviewing the mechanisms for the allocation of frequencies/spectrum and orbital positions; expanding the obligation to de-orbit and manage satellites end-of-life; in parallel, discuss and study possible active removal solutions
  - Working at international level to ensure a level playing field and fair competition for commercial space transportation through certification/licencing of vehicles and pilots, protection of passengers and protection of third parties in the air, in space and on the ground
  - Preparing a "space traffic management" system at global level in the longer term, inspired by air traffic management and based on operational rules adopted at multilateral level and implemented by an international body

Ms. Regina Peldszus, Co-Chair of the EU SST Consortium, made a presentation of activities performed within the EU SST Consortium involving France, Italy, Germany, Spain and the UK as the main contributors. Poland, Portugal and Romania are scheduled to join the consortium in the near future, and eight other EU member states are aspiring to join the cooperation scheme at later point.

Ms. Peldszus noted the 2014 Decision of the European Parliament and the Council establishing the EU SST Support Framework, underlining the prevailing objective of establishing an autonomous European SST capability, with three functions (sensor function, processing function and services provision) to be developed to meet this goal. Recalling the specific governance structure of EU SST Consortium, Ms. Peldszus pointed out dual nature of systems, services and actors involved within the overall framework. Significant military legacy, accompanied today by rather civilian orientation of the EU approach to SST generate delicate issues, which, for now result in national security considerations affecting the cooperation scheme and national control, operation and ownership retained over SST sensors (Fig. 4.5).

Continuing interest of other European stakeholders in provision of SST services resulting from autonomous European SST capabilities creates a rationale for establishment of a stronger European dimension of EU SST capabilities. Mr. Peldszus stated that the current network encompasses 34 European sensors (11 radars, 4 laser stations, 19 optical telescopes) and relies also significantly on US data. The model on which the EU SST Consortium is based is evolving since its inception, and the aim is to move toward the specialisation of the Consortium according to the European Commission's needs and internal member states' requests. Three services are delivered since July 2016: collision avoidance, re-entry and fragmentation. The Consortium provides data to 16 states, 43 organisations and 80 user accounts. On collision avoidance (CA) services, 34 of the satellites covered are in LEO, 30 in MEO and 53 in GEO. The role of SatCen in the EU SST framework will evolve in the near term.

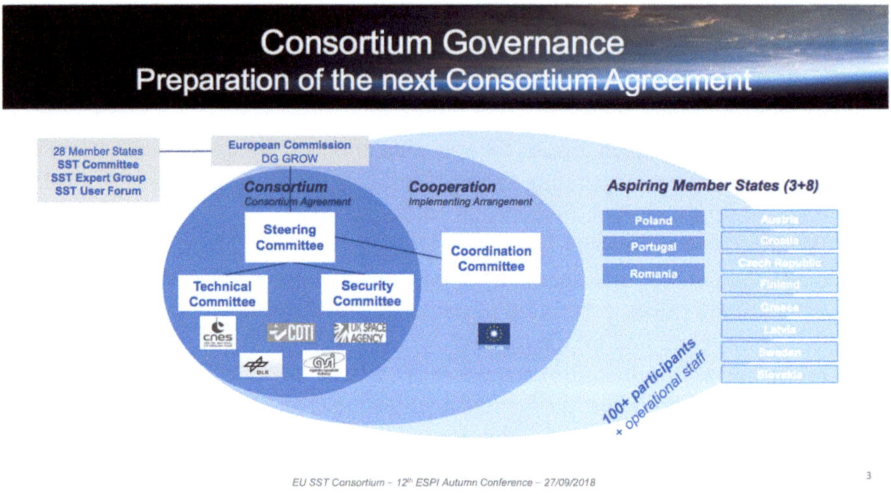

Fig. 4.5 Current EU SST Consortium governance scheme (EU SST Consortium)

Mr. Carlo des Dorides, Executive Director of the European GSA Agency, noted the progress made on the Galileo programme and highlighted the complexity of managing the dual-use nature of the programme within the same house. Significant growth was achieved since December 2016 (declaration of initial services) with four hundred million devices now compatible with Galileo. Mr. des Dorides presented that the full operational capability of Galileo system is expected to be achieved by 2020.

Keeping the presentation within the lines of the topic of the conference, Mr. des Dorides then discussed mostly all the different security aspects related to the development, operation and maintenance of Galileo programme. He noted harmonised procedural aspects of signal in space provision through "Galileo Service Operator". Concerning the Public Regulated Service (PRS), Mr. des Dorides acknowledged PRS user segment being under development, pointing out to first preoperational PRS receivers 3 parallel PRS pilot projects for PRS field testing involving 18 member states with a coordination model led by the GSA and definition of PRS-related technologies. Additionally, he underlined the importance of the Galileo Security Monitoring Centre, an integral part of Galileo infrastructure that monitors and takes action regarding security threats, security alerts and the operational status of systems components and enable users access PRS service.

With regard to the European augmentation system EGNOS, Mr. des Dorides explained that the system is planned to evolve from version 2 to version 3 seamlessly, ensuring continuous service provision. For EGNOS V2, the service provider (ESSP) was certified by EASA based on the security provisions of the Single European Sky (SES) regulation. This will also be the case for EGNOS V3, currently under development, in addition to the security accreditation process by the EU GNSS Security Accreditation Board (SAB).

With regard to this last point, Mr. des Dorides stresses that security accreditation is an enabler of GNSS Programmes. The Security Accreditation Board (SAB) for EU GNSS is an independent body vested within the GSA, comprising the representatives and experts from the member states. The SAB responsibilities include the approval of satellite launches; the authorisation to operate the systems in their different configurations and for the various services; and the authorisation to operate the ground stations.

In the concluding part of his presentation, Mr. des Dorides addresses the potential policy developments affecting the role of GSA resulting from the recent EC proposal including a new regulation for the whole EU space programme (Galileo and EGNOS, Copernicus, GOVSATCOM, SSA) with EUR 16 billion of budget over the next Multiannual Financial Framework. At its current stage, it includes evolution of GSA into EU Agency for the Space Programme with new tasks for market development, security accreditation and possibly other tasks.

Mr. Andrew Monham, Spacecraft Operations Manager at EUMETSAT, provided the audience with EUMETSAT perspectives on the topic of space security, dealing primarily with impact of evolving space security environment on space assets, EUMETSAT measures to tackle growing security risks and the importance of international cooperation in this regard. Concerning the evolving space environment, two issues were brought to spotlight in this respect—space debris environment and space weather environment.

Mr. Monham recalled that EUMETSAT services and products are critical for weather forecasts and specialised forecasts, allowing to save hundreds of lives each year and bringing substantial benefits in the fields of defence and security (infrastructure, property, transport), air quality, marine forecasts or dispersion of pollution in the context of Copernicus (see Fig. 4.6).

Mr. Monham explained that the criticality of EUMETSAT services and products requires to guarantee a level of quality (accuracy, latency, etc.) and availability in delivering services through a robust constellation (see Fig. 4.7).

BENEFIT AREA	LIKELY BENEFIT
Protection of property and infrastructure	€5.4 billion/year
Added value to the European economy	€41.0 billion/year
Private use by European citizens	€15.0 billion/year
TOTAL	**€61.5 billion/year**

Fig. 4.6 Estimated benefits for protection of the space infrastructure (ESPI)

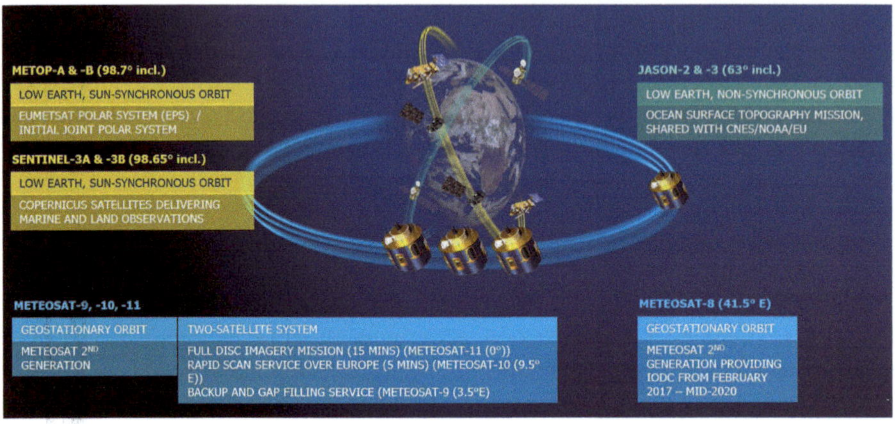

Fig. 4.7 Current EUMETSAT constellation (EUMETSAT)

Having in mind this important role of EUMETSAT space infrastructure, Mr. Monham stated that EUMETSAT is aware of stakes and challenges resulting from current developments in the space environment. Following this line of thought, Mr. Monham argued that Earth observation community, particularly in cases, where it relates to provision of services in weather and climate forecast thus needs to take measures to safeguard its space assets from potential hazards.

In this context, one example could be the long-lasting cooperation of EUMETSAT with the NOAA that augments resilience aspects of services that these two organisations provide. To elaborate more on the changes in the space environment, Mr. Monham informed the conference participants that EUMETSAT is both user and provider of space weather data and that EUMETSAT is also involved in an international cooperative framework, targeting primarily SSA data sharing, to improve its space debris mitigation capabilities. EUMETSAT uses a European public database to assess its needs for collision avoidance manoeuvers and draws from various sources including the EU SST Consortium and SSA sharing agreement with the USSSTRATCOM.

As a final point, Mr. Monham elaborated in detail on the EUMETSAT Strategy 2025 aiming at protecting operational space assets, meeting needs through cooperation, planning for future programmes, consolidating EUMETSAT and partner capabilities and assessing how they can be used together with a focus on operational space weather needs.

During the session a few comments were made on the US approach to space security. It was noted that the Trump administration's national space strategy takes a whole-of-government approach to US leadership in space, in close partnership with allies and the commercial sector and based upon shared values. Implementation of the US space strategy is overseen by a revived National Space Council under the chairmanship of Vice-President Mike Pence. The council's regular public meetings review the implementation of President Trump's space policy directives and discuss emerging space policy issues.

One key catalyst for US space strategy has been the growing realisation of emerging threats to space security, with both Russia and China moving forward with destructive ASAT capabilities which are expected to be operational in the next few years. Of particular concern, Russia and China continue to launch "experimental" satellites that conduct sophisticated on-orbit activities, at least some of which are intended to advance ASAT capabilities. The US Intelligence Community assesses that some technologies with peaceful applications—such as satellite inspection, refuelling and repair—can also be used against US and allied spacecraft. At the same time, Russian and Chinese diplomats continue to promote an international treaty on "prevention of placement of weapons in outer space", which Washington has argued is hypocritical in light of the Russian and Chinese militaries' development of orbital ASAT systems.

It was also noted that the US national space strategy is putting emphasis on dynamic and cooperative interplay between the national security, commercial and civil space sectors, to ensure American companies lead in global markets. There is room for cooperation for weather satellites and PNT satellites allowing for higher resiliency.

With regard to deterrence and warfighting options, it was noted that bilateral agreements are still relevant to cooperate, while the role of NATO is evolving with new efforts to develop an overarching NATO Space Policy. In this sense, it is important to consider what issues and policy discussions are required to move forward, with Europeans being encouraged to participate along with North American allies. The US space strategy also seeks to improve foundational capabilities for Space Situational Awareness (SSA), which also supports the Trump administration's Space Policy Directive-3 (SPD-3) on Space Traffic Management (STM). One key point of both the space strategy and SPD-3 is that the US military will take a proactive role in protecting and defending space assets. The US policies also seek to leverage new commercial sensor capabilities as well as new technologies such as cloud computing. It was also highlighted that the US department will lead efforts to promote two-way SSA cooperation, which was not possible with the DoD's current organisation and missions. The promotion of a domestic conducive environment has stimulated efforts to streamline US regulation of satellite servicing missions and improve US domestic competitiveness. Other reforms include President Trump's proposal to establish a US Space Force and other fundamental reforms for organisation of space activities because of emerging threats.

At international level, multilateral engagements made substantive progress, notably on UN Committee on the Uses of Outer Space sustainability guidelines—a process that included newer space-faring nations. It was mentioned that last June, Russia single-handedly stopped the long-term sustainability process. In Moscow, a senior official justified this blockage because of non-proliferation and arms control concerns. The USA is still involved in multilateral engagements such as COPUOS but remains cautious about top-down process such as a Code of Conduct. However, there are still potential opportunities for US bilateral transparency and confidence building measures with Russia and China, although progress is impeded by larger geopolitical issues.

Following the last presentation, a Q&A session was opened. A thought-provocative discussion started on the nature of space security and the collaboration with the USA in this field. It was mentioned that there is a need to look at each of the current cooperation schemes from a different light: in the past, relations involved mostly ad hoc bilateral cooperation, but a change can be expected in the future, including in multilateral bodies (the recent SPD-3 notably states that the USA will work with the UN COPUOS). As well, it was mention that NATO has probably a role to play in deterrence issues. On the European side, it was stressed that the EU SST Consortium is currently developing SST capabilities and integrating new member states and that, in the future, additional work will be done in Europe to address space weather, NEO and cybersecurity. In all these areas, it was agreed that there is a need to integrate more partnerships. About the role of private industry, Ms. Peldzus mentioned that the industry is already involved in SST as manufacturers of the sensors but that different levels of involvement should be considered now.

4.2.3.2 Highlights

All speakers of the first session acknowledged increasing security challenges to space infrastructures. Space debris environment, space weather events, radiofrequency interference or both intentional and unintentional irresponsible space behaviour were introduced as relevant examples in this sense. It was made clear that ensuring an appropriate quality of service increasingly implies to work on the security component of space programmes. Indeed, for space-based services to reach an adequate level of availability, reliability, safety, integrity and confidentiality, satellites must be better protected.

In order to fulfil this objective, two main lines of action were consequently discussed. The first relates to the necessity of significant improvement of actual capabilities required to protect one's own space infrastructure. It is essential for all engaged stakeholders to recognize individual responsibility and increase efforts to tackle emerging risks stemming evolving space environment. The second line of action voiced by the panel underscored that overall only a coordinated international approach could be an effective tool to tackle security challenges to space infrastructures, taking into account the global impact of space activities and security challenges that pose risks to space operations.

In achieving greater coordination, it was noted that both the European and US approaches to space security are changing to better address institutional and private security needs. In the USA, internal reorganisation is materialising across military and commercial branches (following the initiation of the National Space Traffic Management Policy in recent Space Policy Directive-3), with the Department of Commerce taking a more proactive role in distributing SSA data both to commercial and public actors. This is likely to have an impact on the way cooperation with international partners is organised. In Europe coordination efforts are gaining momentum, with the EU SST Consortium moving forward and reaching milestones in its deployment strategy but also with the European Commission's proposal implying a more proactive role of the institution in SSA capabilities or other related undertakings in space security.

4.2.4 *Keynote Speech by Daniel Oltrogge, CSSI*

Daniel Oltrogge, Director of the Centre for Space Standards and Innovation at Analytical Graphics, started the second day of conference with a keynote speech.

His presentation titled "Marshalling Space Traffic Management requirements and expectations in the international context" provided detailed technical insight into a spectrum of issues associated with SSA and STM and illustrated different approaches of governments, international organisations and private industry.

Mr. Oltrogge began his speech by discussing various existing definitions of Space Situational Awareness and Space Traffic Management and providing an insight into current situation of space traffic in LEO and GEO orbits. "As a consequence of limitations of current global SSA capabilities", he argued "small LEO space population is largely unknown and what is more, small GEO space population especially unknown". This is further reflected by data in currently available public catalogue, providing information about objects around 10 cm in size in LEO region and 1 m in size in GEO region. Accordingly, Mr. Oltrogge assessed current SSA capabilities as insufficient. Recognizing this state-of-play contributed to growing international interest (private as well as governmental) in development of new tools to deal with risks stemming from growing space traffic. In this regard, he underlined recent initiation of the National Space Traffic Management policy in the USA or emerging private SSA capabilities (Fig. 4.8).

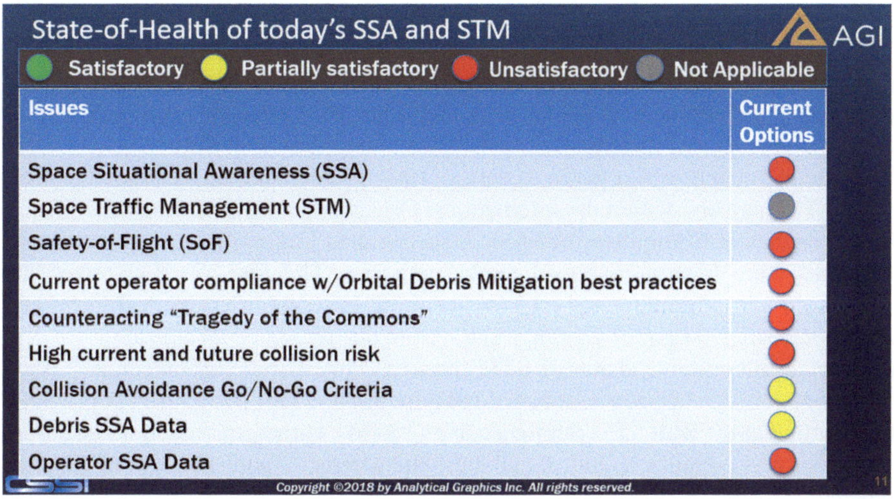

Fig. 4.8 State of health of today's SSA and STM (AGI)

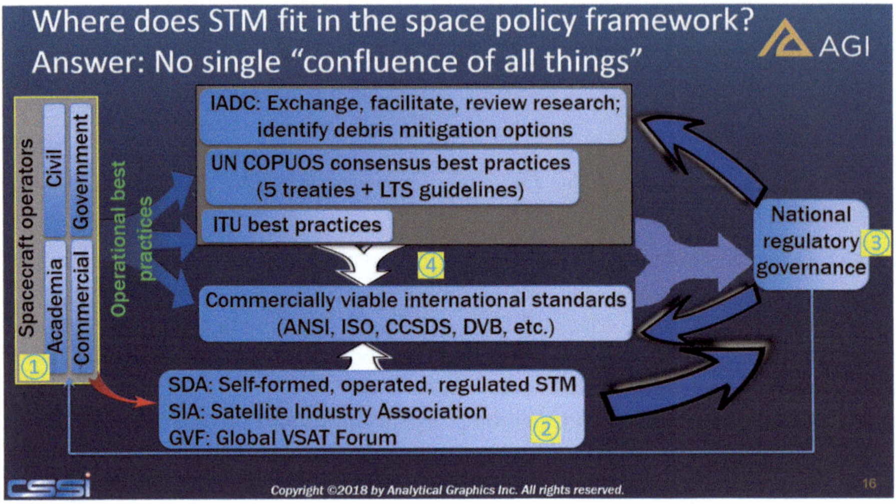

Fig. 4.9 Stakeholders engaged in STM (AGI)

After illustrating former orbital collisions with models of generated debris and exploring forensics of LEO and GEO conjunctions, he claimed that LEO and GEO operators do not know which conjunctions are "too close". Furthermore, LEO and GEO operator covariance are largely unavailable, and not all orbit regimes are created equal. Through clarifying these and some additional challenging aspects of technical nature required for effective international STM regime, he demonstrated that sound SSA capabilities are the backbone of any STM efforts. Due to vulnerable nature of different components of SSA assets, he then added, SSA value chain has become significantly complex and fragile.

Next part of Mr. Oltrogge's speech explored various actors engaged in development of STM, including spacecraft operators, industry associations, national governments and international organisations. In an effort to synthesise this complex mix of engaged stakeholders and their roles in STM, Mr. Oltrogge developed a chart depicting channels of different relationships among them (Fig. 4.9).

Particular attention was given to the existence of international standards applicable to STM, namely, within the International Organization for Standardization (ISO) and Consultative Committee for Space Data Systems frameworks. Afterwards, two messages were further conveyed to the audience. The first one recognized the significance of radiofrequency interference, both in terms of space operations and of relevance in STM policy-making. The second one presented an appealing concept to develop a long-term vision of STM through "crowd-sourcing" method. Taking into account all the aspects associated today with the STM, Mr. Oltrogge provided the participants with a comprehensive "STM Paradigm" recommendation, recognising its complex technical and legal characteristics and spectrum of engaged stakeholders (Fig. 4.10).

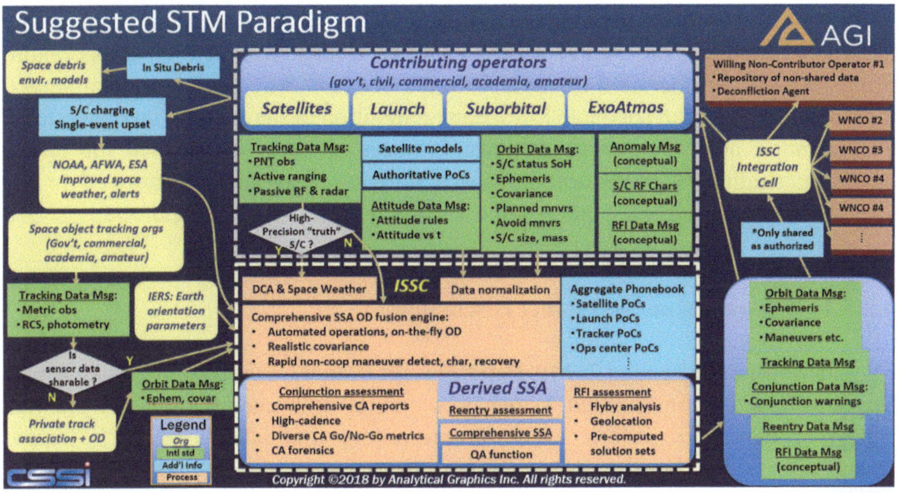

Fig. 4.10 Comprehensive STM Paradigm (AGI)

In concluding part of the keynote speech, Mr. Oltrogge once again reaffirmed the main themes of the approach he presented, in particular the notion of inadequate current SSA capabilities (making collision risk in both LEO and GEO real but also manageable—by witnessing that both government and private interest in SSA and STM is recognisable), the fragility of the SSA value chain and the suggested crowd-sourcing approach toward future STM efforts in international space sector.

4.2.5 *Session 2: "Forging Ahead with European Space Security Efforts"*

Session 2, first thematic session of the 2nd day of the Conference gathered five panellists:

Gustav Lindstrom	*Director, EU Institute for Security Studies*
Petr Havlik	*Senior Space Policy Advisor, European External Action Service*
Nicolas Bobrinsky	*SSA Programme Manager, ESA*
Pascal Legai	*Director, EU SatCen*
Mark Dickinson	*Chairman, Space Data Association*

This session was moderated by **Mr. Gustav Lindstrom**, director of the European Union Institute for Security Studies, who opened the session with a brief speech stressing three main points with respect to an increasing relevance of space security for European stakeholders. Firstly, he acknowledged the forward-looking nature of this panel and invited the speakers to reflect the ongoing relevance of strategic autonomy objectives associated with European space activities and set this course of action into broader contemporary political developments in Europe, such as the EU Global Strategy, European Defence Fund or Permanent Structured Cooperation. Following the consideration of strategic value of space, Mr. Lindstrom then argued that there is still not an adequate European recognition of the need to have a strategic consideration about European presence in space. Finally, he recalled the newly emerging challenges to security of space infrastructures and transitioned to the speakers highlighting that "things are not going to get easier" (for Europe).

4.2.5.1 Presentations and Panel Discussion

The first speaker of Session 2, **Mr. Peter Havlik** from the European External Action Service (EEAS), concentrated in his presentation on contemporary and planned European diplomatic efforts aimed at promoting multilateral approach toward space security. Recognising the unsuccessful outcome of the International Code of Conduct for Outer Space Activities (ICoC), he informed the audience that EU diplomatic corps is not leaving the idea of a multilateral voluntary instrument but rather aims to refurbish this idea and initiate new multilateral discussions about a similar proposal, with a number of adjustments already incorporated, as a result of previous EU engagement in these issues.

Mr. Havlik recalled that the European Union has been engaged in space security, sustainability and safety for multiple years and has a solid experience with leading multilateral negotiations. The ICoC was initiated in 2007 with formal talks launched few years afterwards in 2012. Building on this experience, he presented to the audience a new EU approach, which has been introduced for the first time earlier in 2018 at the UNISPACE + 50 conference in Vienna and subsequently also at UN platforms in New York in Geneva. The EU statement on the occasion of the UNISPACE + 50 was subsequently distributed in the conference room in physical copies.

Mr. Havlik reflected on criticism, which was voiced toward the ICoC, explaining that both substance and process were criticised by some UN member states. Even though the ICoC process has not produced the desirable outcome, the European Union is still confident that an international voluntary instrument for space security and sustainability, setting basic principles of space activities with political commitments, is needed. The "new" proposal is not yet on the table, but the key idea behind it will follow the objective of the original ICoC effort. Taking into account the experience with the ICoC, Mr. Havlik stated that the idea to launch discussions should be supported and mandated by the UN General Assembly. He concluded that the EU will also not aim to be the only proponent of this initiative.

While introducing the next speaker, Mr. Lindstrom provided a consideration that EU objectives and processes described by Mr. Havlik were strikingly similar to efforts in cyber realm.

Mr. Nicolas Bobrinsky, SSA Programme Manager at the European Space Agency (ESA), provided the audience with remarks on genealogy and current and projected activities of ESA SSA Programme. He began his speech by recognizing

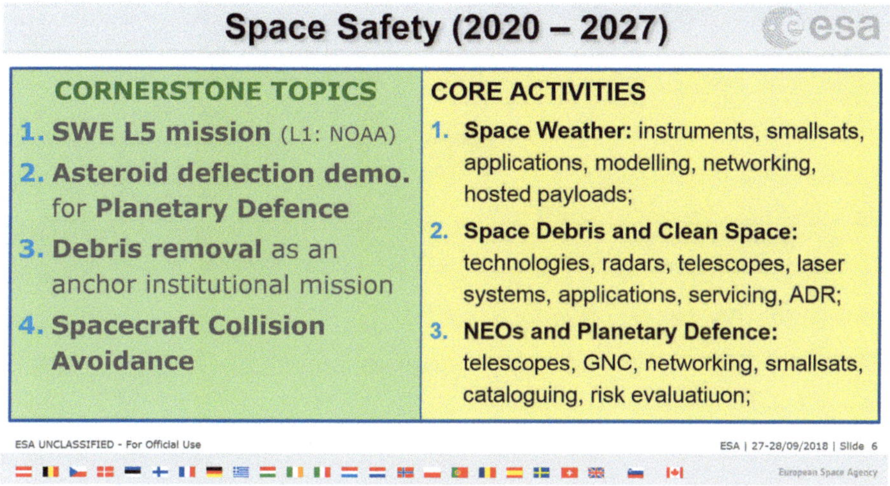

Fig. 4.11 Key themes of ESA's SSA Programme (ESA)

that (not just European) society is heavily dependent on space, which in turn creates a need for protection of space infrastructures. In November 2008, this assessment was recognized in the programme declaration at ESA level, which gave the Agency the legal mandate to execute the SSA Programme. Currently, the SSA Programme has a threefold structure: Space Surveillance and Tracking (SST), near-Earth objects (NEOs) and Space Weather (SWE). Mr. Bobrinsky pointed out that the primary focus is on activities within NEOs and SWE segments and that the SST segment is mostly oriented toward R&D.

After introducing ESA views and definitions of several terms associated with space security, Mr. Bobrinsky introduced in more detail some examples of missions in each of aforementioned fields of the SSA Programme: enhanced SST capabilities, surveillance of space debris for collision avoidance, laser ranging to non-cooperative targets, debris mitigation and removal, enhanced SWE observation system and NEO detection and threat analysis (Fig. 4.11).

In concluding his presentation, Mr. Bobrinsky acknowledged the good progress achieved in the development of an SSA system in Europe, as a result of the cooperation of many actors and discussed the potential way forward in Europe with a couple of ideas addressing policy challenges for the upcoming years. Considering the high number of stakeholders in European space sector, he argued for a consolidation of the distribution of roles. Taking into account the promising perspectives but at the same time sensitive nature of exploitation of radars, he emphasised the need to agree on a suitable governance scheme for this utilisation. He stressed that the development of high-performance radar can be achieved within 3 years, and SWE and NEO systems will reach preoperational status by 2020.

Third speaker, **Mr. Pascal Legai,** informed the audience on views and role of EU Satellite Centre in the domain of space security. Before delving deeper into the EU

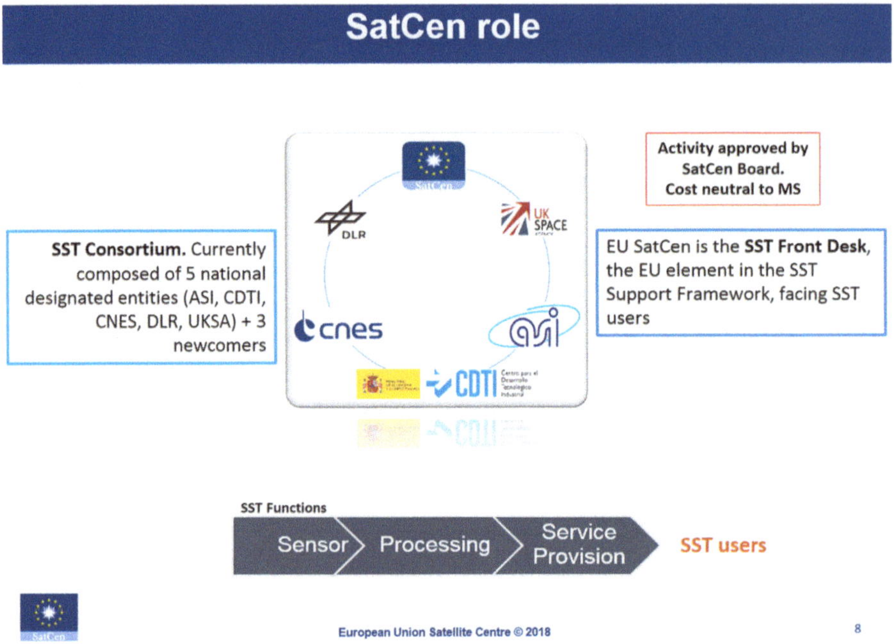

Fig. 4.12 EU SatCen role in SST Consortium (SatCen)

SatCen perspectives, Mr. Legai reaffirmed the notion voiced by previous speaker regarding critical dependence of European economy on space assets. The role of EU SatCen in this regard has been progressively increasing, which could be demonstrated by inclusion of EU SatCen in the EU SST Support Framework, where it serves as a front desk for dissemination of SST services resulting from cooperation within the EU SST Consortium (Fig. 4.12).

Mr. Legai noted that, although the EU SatCen is a purely civilian entity, it is covering some dual-use activities as well. He highlighted that an emerging area of focus was artificial intelligence and its use in satellite data analysis. Mr. Legai highlighted that roles of EU SatCen are not limited to Earth observation but cover also secured communications, positioning or meteorology. He then noted the underlying objective of EU SatCen to support the execution of EU's Common Foreign and Security Policy (CFSP) and Common Security and Defence Policy (CSDP).

Following the elaboration on recent developments, Mr. Legai welcomed the ongoing cooperation with other stakeholders in the EU SST Support Framework and discussed in more details key space security challenges that require immediate policy discussion:

- Explosion of the satellite population, with particular growth in the number of small satellites
- Increasing role of private companies, being the driving force behind planned large constellations and having prominent position in the smallsat market

- Maturing propulsion concepts for microsatellites, expanding the application opportunities of smallsat missions

Finally, **Mr. Mark Dickinson**, representing the Space Data Association (SDA), took the floor as the last of scheduled speakers. Mr. Dickinson brought a different angle to the discussion, introducing the views, rationales and goals of the private industry, satellite owners and operators, who engaged in information sharing concerning satellite trajectories and other related situational awareness in Earth orbits. After presenting some general information about the Space Data Association and benefits that this organisation brings to its members explaining that the SDA is an international non-profit association of satellite owners and operators supporting the controlled, reliable and efficient sharing of data critical to the safety and integrity of the space environment and the radio frequency (RF) spectrum.

Mr. Dickinson pointed out to existing limitations of public catalogue and operator provided data. In order to protect one's own space systems and safeguard long-term utilisation of satellite services, the rationale to cooperate and share SSA data (precise satellite orbits, manoeuvres, etc.) among private operators has been recognised. He also stressed that advancements in capabilities (e.g. EU SST) with additional sensors is highly welcomed.

Mr. Dickinson argued that SDA contributes to exploring a next generation of STM through demonstrating effective private cooperation in data sharing. In addition to basic model of cooperation, an advanced form of cooperation among SDA members was illustrated by Mr. Dickinson—the Space Data Centre—a platform within the SDA cooperative framework that ingests flight dynamics information from member companies as well as other available sources of space object information to provide conjunction assessment and warning services.

Following the last presentation, a Q&A session was opened. A thought-provoking debate emerged as a result of few questions and comments tackling the risk management perspective of SSA value chain, namely, dealing with the question, who is pressured most urgently to "do something" concerning potentially negative developments in on-orbit collision, space debris and major space weather events. In this sense, the notion of vulnerability of SSA value chain, voiced already in the keynote sessions, was reaffirmed by additional conference participants. Eventually, taking into account the complexity of the issue, it was noted that both public and private actors' interests are deeply linked to the SSA, which might stimulate new cooperative instruments between industry and governments.

Given the European nature of this panel, several remarks from conference participants in the audience targeted the performance of EU SST consortium and ESA SSA Programme. The progress achieved over the last 4 years was highlighted, together with new aspiring EU member states aiming to join the Consortium. On the other hand, the "European" dimension of SST services is seemingly still inferior to national security considerations and prominent roles of military stakeholders at national level.

Additionally, the role of Europe (collectively as well as individually on national levels) in larger international thematic frameworks, such as the Inter-Agency Space

Debris Coordination Committee, the International Asteroid Warning Network or the Space Mission Planning Advisory Group, were discussed, with prevailingly positive assessment of European presence there. Ultimately, sharing of responsibilities, governance issues or involvement of private sector in future European or broader international measures in space security domain were brought to the discussion among conference participants.

4.2.5.2 Highlights

The views presented by panellists and the open discussion with the audience underlined the rising importance of development of SSA capabilities and the necessity of coordinating approaches of different stakeholders (countries, public/private, etc.) toward enhanced SSA capabilities and shared rules of behaviour in space.

It has been generally agreed that European stakes in space security are increasing. The growing European presence in space and importance of space assets in European economy urges European stakeholders to develop the means to safeguard space assets against new threats.

Growing private capabilities, particularly in the SSA domain, are creating new opportunities for more ambitious public-private cooperation, which is seemingly becoming of mutual interest to both private sector and public sector in Europe.

Finally, reinvigorated Institutional interest in developing SSA capabilities accompanied by emerging private capabilities in this domain and continuing diplomatic efforts remain, for now, the key themes of European response toward increasing relevance of space security.

4.2.6 Keynote Speech by Henry Hertzfeld, GWU SPI

Henry Hertzfeld, Director of the George Washington University's Space Policy Institute, opened the afternoon with a keynote speech on Outer Space and Economic Security.

Mr. Hertzfeld gave a presentation on the concept of Economic Security, noting that this policy area is addressed differently across the world as all nations have different policies, structures and priorities to handle economy and security and defence. As a result, Mr. Hertzfeld informed the audience that Space Economic Security could be understood as the guaranteed use and stability of valuable assets in space supporting non-military users or as the economic stability, growth and better standards of living through the use of space applications on Earth.

Mr. Hertzfeld explained that the dual use of space implies that security in space may become indistinguishable between the military and civilian responsibility which, by extension, will make Space Economic Security more challenging to address as it will be dealing with national security. Mr. Hertzfeld recalled that, in the US space policy, defence and national security are the highest priorities, while

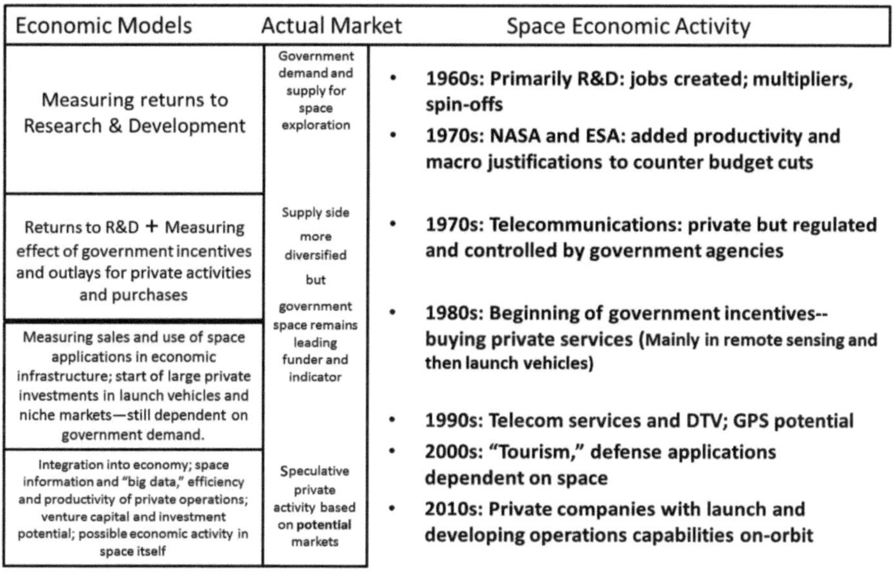

Fig. 4.13 Evolution of thinking about space economics (GWU SPI)

economy tends to be a stronger driver of space policy in Europe. From this perspective, economic policies in the United States must necessarily account for national security aspects, but the opposite is not always true and national security policies sometimes overlook their potentially negative economic impacts. Commercial space policy will therefore not be the main driver of the US space security policy. Nevertheless, space security policy can offer some commercial opportunities.

Subsequently, Mr. Hertzfeld explained that the space infrastructure cannot be considered as a critical infrastructure per se but rather as a critical element of critical infrastructures. Mr. Hertzfeld invited the audience to reflect on the shift in US space policy with the rise of the commercial space sector which led to an evolution of space economy and its measurement that he summarised as presented in Fig. 4.13.

Mr. Hertzfeld pointed out that space economics evolved from being measured through jobs and spin-offs (imagery, medicine, etc.) to now talks of profit, strengthening its importance within national security. With the emergence of new private actors in space, Mr. Hertzfeld acknowledged the need to accommodate regulations as the changing space environment creates new vulnerabilities, not only in the USA but internationally, with a globalisation of networks (industrial, financial, information) that affects commercial space in both supply and demand.

Mr. Hertzfeld explained that these regulations are instrumental to structure global markets in the long term, but that are not easy to put in place. To illustrate this, Mr. Hertzfeld took the example of the US commercial space policy, explaining that space economic policy cannot be separated from non-space economic policy, and that contradictory policies should be avoided. In some situations, the results of policies encouraging the development of commercial space can be hindered by export regulations, for example.

Mr. Hertzfeld formulated two key questions: "do national objectives require space business investment?" and "are foreign commercial space assets essential to a nation's domestic security?" To these questions, Mr. Hertzfeld answered yes in both cases but reminded conference participants that if so, then a question remains: "What factors will determine whether private space assets will be protected/defended by a government?"

Continuing on the interdependence of international actors, Mr. Hertzfeld explained that, since space is global, it requires international coordination and cooperation. He presented some options for the future of the US commercial space:

- Treating commercial space as "just another commodity", which ignores the dual-use nature of most space applications
- Ensuring the dominance and control through military actions, which will encourage counter measures by others with uncertain outcomes and increase commercial risk factors
- Stimulating renewed economic competitiveness in the USA, which may not be consistent with export restrictions and other US policies related to free trade and competitiveness

Mr. Hertzfeld concluded that if there is a security solution to be found through commercial strength, the USA should encourage R&D in areas likely to advance commercial space, which would constitute an "offence" rather than a "defence" for future commercial products, meaning that we may want to build more resilient and resistant assets, not build more weapons to defend them. As well, Mr. Hertzfeld suggested that commercial space should produce the best products to encourage worldwide purchase of goods and services, which would promote leadership through best practices and market dominance, and that finally, regulatory disincentives should be eliminated without violating treaty provisions and without jeopardising security or public safety.

4.2.7 Session 3: "Rethinking Transatlantic Collaboration: Toward STM"

The third and last session of the ESPI 12th Autumn Conference gathered following speakers

Xavier Pasco	*Director, Fondation pour la Recherche Strategique*
Joe Pelton	*Executive Board Member, IAAS*
Victoria Samson	*Head of Washington Office, Secure World Foundation*
Didier Alary	*Head of Advanced Space Systems, Airbus Defence and Space*
Rob Heron	*Senior Sales Manager, SES TechCom*

State	Constellation	# of Sats	Radio Frequency Bands
Canada	CANPOL-2	72	LEO and highly elliptical Earth orbit in VHF-, UHF-, X-, and Ka-bands
Canada	Telesat Constellation	117 satellites plus spares	LEO in Ka-band
Canada	COMSTELLATION	Nearly 800 Satellites	LEO in Ka-band
France	Thales Group's MCSat	between 800 and 4000	LEO, MEO, and highly elliptical Earth orbit in Ku- and Ka-bands
Liechtenstein	3ECOM-1	264	Ku- and Ka-bands
Norway	ASK-1	10	Highly elliptical Earth orbit in X-, Ku-, and Ka-bands
Norway	STEAM	4257	Ku- and Ka-bands
U.K.	L5 (OneWeb)	650-750	Ku- and Ka-bands
US	Boeing	1396-2956	V-band in 1200 km orbit
US	SpaceX	Up to 4000	Ku-Ka band
US	SpaceX	7500 plus	V-band
US	LeoSat	Initially about 80	Ka-band

Fig. 4.14 Problem of LEO constellation (Joe Pelton)

This session was moderated by **Mr. Xavier Pasco**, Director of the Fondation pour la Recherche Strategique, who introduced the issue and the speakers. Mr. Pasco reflected on who the USA and Europe are as partners, and on what they want to achieve. While recalling elements of previous presentations and in particular the divergence between the US and European approaches, Mr. Pasco underscored that a discussion on STM is timely to identify converging interests between the USA and Europe.

4.2.7.1 Presentations and Panel Discussion

The first panellist, **Mr. Joe Pelton**, Executive Board Member of the International Association for the Advancement of Space Safety, presented "A Path Forward to Better Space Security: Finding New Solutions to Space Debris, Space Situational Awareness and Space Traffic Management". Mr. Pelton started his presentation by recalling past events that generated significant amount of space debris and their effect on space policy. Mr. Pelton mentioned that current studies on space debris carried out by NASA, the European Space Agency and other agencies have concluded that, even without new launches, a new major collision will likely happen every 5–10 years on average due to the already high number of satellites in orbit, especially in LEO (see Fig. 4.14).

As a result, Mr. Pelton suggested that five to ten major debris elements—i.e. those with the greatest potential for collisions—need to be actively removed from orbit in order to cope with the rising space debris threat. Mr. Pelton explained that "achieving enhanced space infrastructure security requires consideration of a global system of space traffic management as well as active and passive space debris mitigation, collision avoidance and active debris removal". Mr. Pelton then mentioned that the current failure requires at least to develop an agreement on international processes that should push national entities and associated regulatory bodies to undertake these actions individually and then coordinate such actions on a global basis.

For Mr. Pelton, the ability to provide timely alerts of possible collisions is increasingly critical. This is not only a matter of enhancing SSA capabilities but also to have reliable means to share data with all satellite operators—particularly those with spacecraft in polar orbits and those operating networks in the 300–1500 km range. Dr Pelton insisted that a more extensive and real-time global sharing of information about potential collisions is a high priority and that governmental systems, private SSA capabilities and international commercial systems for sharing potential conjunctions of space objects need to continue to create better ways to share information and avert collisions since this creates liabilities for all.

Finally, Mr. Pelton stressed that the recent US initiatives outlined in Space Policy Directive-3 offers opportunity for sharing of information that can be helpful. Dr Pelton told the audience that the outcome of this Directive should be widely shared and considered at global level. Dr Pelton pressed the audience to be forward-thinking, saying "What is clear is that time to act in the matters of space safety and security is now. Certainly the new U.S. initiatives with Space Policy Directive-3 in particular, must be considered an opportunity to make progress in this area. Progress is essential, because failing to address the space debris problem in particular is no longer a viable option."

The second panellist, **Ms. Victoria Samson**, Head of the Secure World Foundation's Washington Office, addressed "Major Issues Affecting Space Traffic Management Policy and A Look at Future Challenges". Ms. Samson spoke about changes in the space domain and in particular challenges related to the increasing role of private actors in space that the international community, used to dealing between nation-states, have difficulties to tackle. She also went through issues facing SSA and STM that need to be dealt with in the near-term.

Ms. Samson reminded the audience that space is an existential component of national security but that space security challenges cannot be addressed alone. For example, enhancing SSA capabilities would require a large network with well-thought placement of sensors which cannot be achieved by a single country. She also recalled that the space landscape is evolving. To illustrate this point, Ms. Samson pointed out the increase in the launch of satellites by private actors, which means that space is no longer going to be dominated by governments but by commercial actors.

As a result, Ms. Samson stressed the need to reflect on this change and thing about how policy should develop. She shared the view that regulation is important and a necessity for a government but that the time for more treaties and legally binding regulations is gone. Today the industry is already establishing best practices and will have, soon, sufficient bargaining power to establish them as norms.

Ms. Samson concluded that there is a growing need at policy-making level to understand why, where and when legislation is needed, and how policy should take on new emerging issues. Then, Ms. Samson invited the audience to critically reflect on recent STM policy developments in the USA. Finally, she shared the results of scenarios analysed in recent Secure World Foundation research project showing that SSA would most probably remain government-led and that the idealistic scenario of a globally governed SSA is not realistic.

The third panellist, **Mr. Didier Alary**, Head of Advanced Space Systems at Airbus Defence and Space, presented Airbus views on STM. Mr. Alary explained to the audience how the industry is challenging the status quo in space with advanced technologies (Ariane 6, OneWeb system, etc.) and is now "preparing the future" with in orbit services and test options for debris removal and thinking about ADR ready infrastructure.

As seen from a space system manufacturer or a space operator point of view, the devil is in the variety of guidelines. Indeed, the industry operates in a competitive international environment but regulations can be more or less restrictive according to the country which creates competitive bias. Mr. Alary told the audience that in fact, the current set of regulations are lacking many points to ensure an effective and sustainable management of the future space traffic. Mr. Alary claimed that guidelines should be applied strictly by the industry as it is the only way to make space activities sustainable in the long term, minimise the risk on population and ecosystem and preserve the competitiveness of European Industries. He also touched upon the important topic of active debris removal and elaborated in more detail that is needed to maintain good practices within the industry.

Mr. Alary then shared few ideas that could trigger a cycle of Space Sustainability and Safety conferences that could lead to a decision to pursue global STM initiative. He also addressed the possibility to have an international organisation between treaties and national laws, acknowledging the difficulty to set up such organisation and the fact that it could be considered as a barrier by new entrants and a burden by some players. Mr. Alary presented a paper setting the foundations of such organisation, stating that this idea is not new, but that the time to actually develop it is coming.

Mr. Alary concluded that the private industry may lead the change, pointing out to business interests linked to safe and secure satellite operations. Finally, Mr. Alary gave the audience food for thoughts expressing that if agreeing on a new treaty is a utopia (and having an international space agency is too), why not to explore the idea of a civilian space organisation (Fig. 4.15).

The fourth panellist, **Mr. Rob Heron**, Director of SES TechCom, presented an operator's take on security for outer space. Mr. Heron stressed the importance of STM for the commercial side of space operations, and that there is a great interest in Europe to provide secure and reliable satellite communications at much lower price than for military satellites. For Mr. Heron, one of the key areas of concern is resilience and reliability, which all call for the development of STM polices.

Mr. Heron then illustrated future developments of satellite fleets to be put in orbit, in an already congested space environment. Mr. Heron stressed that the security of the entire network is essential, from the ground segment to the in-orbit one. Mr. Heron mentioned that, if the value of satellite fleets is worth billions of US

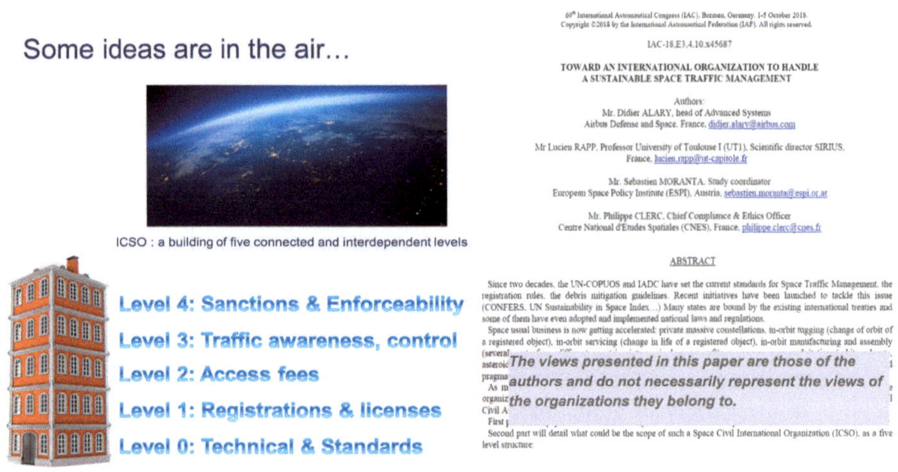

Fig. 4.15 Idea for a civilian space organisation (Didier Alary)

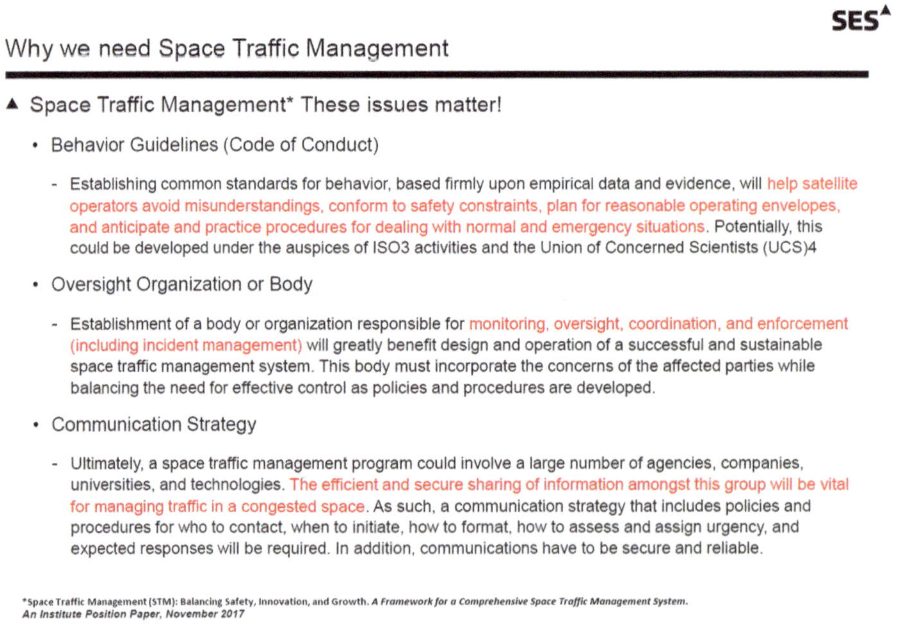

Fig. 4.16 Why we need Space Traffic Management (SES)

dollars, the generated societal value is even more significant, pushing operators to build-in technologies and promoting research like Quantum Key distribution.

Mr. Heron concluded by referring to the AIAA position paper on STM from November 2017 that echoes his views on STM (see Fig. 4.16).

After the session ended, time was allocated for a Q&A session. The debate first focused on the issue of interoperability. Panellists argued that it cannot be forced (i.e. obligated through hard law) but rather pursued through diplomatic ways as it was done for the GPS/Galileo systems. On that topic, the establishment of standards also seem required from the industry. The panellists see the recent SPD-3 as a framework that could contribute to an international approach and noted that SDA is already a useful model. The debate then moved on to the topic of European ambitions in space security. The panel answered that Europe needs to be ambitious, but also recognised that achieving anything at European level is a protracted process, although it can eventually lead to great success. It was also noted that, in defence policy, Europe is moving forward with PESCO and shows a growing appetite to work together on sensitive topics. The final part of the debate focused on commercial space in Europe. It was said that the industry itself ought to take more initiatives and that the European Commission should systematically favour further development and evolution of commercial space. It was noted that the space sector in Europe is still very segregated. This assessment points out to the fact that space ventures in Europe are traditionally done by specialised actors, whereas in the USA stakeholders from other industries often engage in space-related projects.

4.2.7.2 Highlights

Due to the invaluable socioeconomic value that space is bringing down to Earth, the need to better protect space assets is a pressing issue. Panellists stressed the need to progress on suitable space policy and legislation (when needed) to take this into account. Space is global, and Europe and the USA are facing shared challenges and, as such, need to find a way to deepen their cooperation. In this regard, the panel consensually agreed that STM is an urgent issue that would greatly benefit from a convergence between US and European policies. STM should contribute to:

- Mitigating the risk of collision from orbital debris and other spacecraft
- Fostering the development of commercial space, which requires to develop new rules and regulations
- Ensuring that governments oversee activities in space and develop cooperative mechanisms

An enhanced space infrastructure security will require consideration of a global approach to Space Traffic Management coherent with frameworks for space debris mitigation, collision avoidance and active debris removal.

Certainly, the new US Space Policy Directive-3 is an opportunity to make progress in this area. The panel also remarked that the US and Europe need to seek better cooperation mechanisms to further promote mutual understanding of emerging space security issues and strengthen the dialogue toward establishing STM and global standards. Finally, the panel offered some recommendations for the development of STM:

- Need to address technical challenges of sharing data
- Need for guidance at the international level
- Establishment of common standards for responsible behaviour
- Update and enforcement of regulations at the national level
- Establishment of a body or organisation responsible for monitoring, oversight, coordination and enforcement of agreed rules

4.3 Conclusion

Ms. Genevieve Fioraso of the ESPI Advisory Council offered a wrap-up of the conference, which stressed that there is a "growing awareness around the need for an integrated space defence policy that coordinates European efforts towards securing space assets".

Ms. Fioraso reminded the recent evolution in terms of security policy developments, both at national and EU level and explained that these actions are not sufficient: there is now a need to build a coherent, clear and visible European policy for space security. She acknowledged that bilateral or multilateral partnerships are still prevailing, whereas a supranational European vision for defence (and especially for space defence), including cybersecurity, "appears urgently necessary" to face threats to space systems.

Ms. Fioraso recalled previous presentations and the ongoing democratisation of space that generate an ever-increasing flux and traffic coming from public and private actors. In this context, Ms. Fioraso stressed the need to anticipate the consequences of the changing space landscape, with setting up clear rules respected at an international level. Most importantly, Ms. Fioraso expressed that Europe will not act alone and that the relationship with the USA is crucial. Ms. Fioraso expressed that it is critical to "further promote mutual understanding of emerging space security issues and strengthen the dialogue toward establishing STM and global standards".

Ms. Fioraso also pointed to the increasing role of commercial space and the role of the industry in related policy debates. For Ms. Fioraso, the institutions and agencies should take advantage of the partnership with the private sector, favouring innovation and competitiveness on the global market.

To conclude, Ms. Fioraso mentioned that such a venue where partners can exchange their ideas and intentions is essential in opening new frontiers and opportunities for collaboration.

Mr. Pierre Delsaux, Deputy Director General in European Commission and Directorate General for Internal Market, Industry, Entrepreneurship and SMEs, offered the closing remarks and stressed that interactions between both sides of the Atlantic support a mutual understanding of developments taking place. Mr. Delsaux referred to the conference topic, *Security in Outer Space: Rising Stakes or Civilian Space Programmes*, by explaining that the theme is at the crossroad of two interrelated issues: the use of civilian programmes for security purposes (Space FOR Security) and the protection of space assets (Security IN Space).

On the first issue, Mr. Delsaux mentioned that more and more programmes are used for security interests. In the case of Galileo, which is a civil programme, the military dimension is essential which creates a number of challenges. The Copernicus programme is also concerned. These two examples illustrate the blurred lines between civilian and security dimensions of space programmes. On the second issue, Mr. Delsaux stressed that if Europe wants to build an economy based on space assets, this infrastructure needs to be protected. He highlighted, among other security challenges, the importance of the cybersecurity dimension, explaining that the next attack will probably be cyber and that member states, the European Commission and the industry at large are very well aware of this emerging threat and are investing in cybersecurity.

Mr. Delsaux shared with the audience the progress made in defence and security policies at EU level with the creation of the Defence Fund among other major developments. The Defence Fund represents a major milestone for the EU which would have been impossible to talk about just a few years ago. In the space domain, Mr. Delsaux assured that there is a wide support in member states and at EU level to enhance SSA capabilities and that, with its recent proposal for a new regulation, the EU aims to help member states to work better together, acknowledging that cooperation in defence and security areas will take time as it is a new area for the EU.

Mr. Delsaux then reminded the audience that the proposed EUR 16 billion for space activities in the next Multiannual Financial Framework 2021–2027 still needs to be approved, and it is not yet a done deal, so that all engaged stakeholders need to continue promoting the importance of space for society and security.

Mr. Delsaux concluded that an emphasis should be put on the notion of international cooperation. Europe and the USA are facing the same challenges, so they need to find a way to deepen their cooperation. From this standpoint, ongoing developments in the USA, in particular in the area of STM, need to be looked at so that international partners can move forward together.

4.4 Conference Programme

Day 1 (14:00 - 17:30) - ESPI 12th Autumn Conference

13:00 - 14:00	*Registration and Welcome Coffee*

Time	*Speaker*	*Overview*
14:00 - 14:10	Jean - Jacques Tortora *Director, ESPI*	Welcome
14:10 - 14:40	Sébastien Moranta *Coordinator of Studies, ESPI*	Introduction
14:40 - 15:00	Kai - Uwe Schrogl *Chief Strategy Officer, ESA*	Keynote speech

15:00 - 15:30	*Coffee Break*

Session 1:
State of Play in Space Security Strategies, Space Assets in an Evolving Environment
Moderator: Jana Robinson, *Space Security Program Director, PSSI*
Rapporteur: Martin Sarret, *Resident Fellow, ESPI*

Time	*Speaker*	*Overview*
15:30 - 15:45	Jana Robinson *Space Security Program Director, PSSI*	Introduction
15:45 - 16:00	Philippe Brunet *Director for Space Policy and Research, Copernicus Programme*	Presentation (1)
16:00 - 16:15	Richard Buenneke *Senior Space Policy Advisor, US Department of State*	Presentation (2)
16:15 - 16:30	Regina Peldszus *Co-Chair, EU SST Consortium*	Presentation (3)
16:30 - 16:45	Carlo des Dorides *Executive Director, EU GSA Agency*	Presentation (4)
16:45 - 17:00	Andrew Monham *Spacecraft Operations Manager, EUMETSAT*	Presentation (5)
17:00 - 17:30	Jana Robinson *Space Security Program Director, PSSI*	Q&A, Wrap-up

Day 2 (09:30 - 16:15) - ESPI 12th Autumn Conference

Time	*Speaker*	*Overview*
09:30 - 10:00	Daniel Oltrogge *Director, CSSI*	Keynote speech

10:00 - 10:30	*Coffee Break*

Session 2:
Forging Ahead with European Space Security Efforts

Moderator: Gustav Lindstrom, *Director, EUISS*
Rapporteur: Tomas Hrozensky, *Resident Fellow, ESPI*

Time	*Speaker*	*Overview*
10:30 - 10:45	Gustav Lindstrom *Director, EUISS*	Introduction
10:45 - 11:00	Petr Havlik *Senior Space Policy Advisor, EEAS*	Presentation (6)
11:00 - 11:15	Nicolas Bobrinsky *SSA Programme Manager, ESA*	Presentation (7)
11:15 - 11:30	Pascal Legai *Director, SatCen*	Presentation (8)
11:30 - 11:45	Mark Dickinson *Chairman, SDA*	Presentation (9)
11:45 - 12:15	Gustav Lindstrom *Director, EUISS*	Q&A, Wrap-up

12:15 - 13:45	*Lunch Break*

Time	*Speaker*	*Overview*
13:45 - 14:15	Henry Hertzfeld *Director, SPI*	Keynote speech

Session 3:
Rethinking Transatlantic Collaboration: Towards STM
Moderator: Xavier Pasco, *Director, FRS*
Rapporteur: Lisa Perrichon, *Research Intern, ESPI*

Time	*Speaker*	*Overview*
14:15 - 14:30	Xavier Pasco *Director, FRS*	Introduction
14:30 - 14:45	Joe Pelton *Executive Board Member, IAAS*	Presentation (10)
14:45 - 15:00	Victoria Samson *Head of Washington Office, Secure World Foundation*	Presentation (11)
15:00 - 15:15	Didier Alary *Head of Advanced Space Systems, Airbus Defence and Space*	Presentation (12)
15:15 - 15:30	Rob Heron *Senior Sales Manager, SES TechCom*	Presentation (13)
15:30 - 15:45	Geneviève Fioraso *Advisory Council, ESPI*	Q&A, Wrap-up
15:45 - 16:15	Pierre Delsaux *Deputy-Director General, DG GROW*	Closing remarks

ESPI Newsletter:
Sign up for free and receive the latest from ESPI directly into your mailbox
https://www.espi.or.at/newsletter or

4.5 Speakers' Biographies

Jean-Jacques Tortora, Director, ESPI	
	Jean-Jacques Tortora is the Director of the European Space Policy Institute (ESPI), in Vienna, Austria. He served as the Secretary General of ASD-Eurospace from 2007 to 2016. He was the Head of the French Space Agency (CNES) office in North America and Attaché for Space and Aeronautics at the Embassy of France in Washington, D.C. from 2004 to 2007 and served as Deputy Director for Strategy and Programs responsible for the Industrial Strategy of CNES from 2001 to 2004. He advised the French Ministry of Research for Industrial Policy Funds management aiming at competitiveness support and new space applications and services development and promotion Mr. Tortora was based in Kourou, French Guiana, from 1990 to 1996 as Head of Arianespace Operations Quality Department where he led the Quality Management of the Ariane 4 launchers' final integration and of the launch pad operations and maintenance. He supervised the industrialisation and the implementation of the associated control plans. He started his career in 1984 at the French military procurement agency, DGA, as a naval weapons integration and test engineer, expert in signal processing and warships acoustic discretion
Henry R. Hertzfeld, Director, Space Policy Institute, George Washington University	
	Dr. Henry R. Hertzfeld is the Director of the Space Policy Institute and Research Professor of Space Policy and International Affairs at the Elliott School of International Affairs, George Washington University. He is also an Adjunct Professor of Law at GW. He is an expert in the legal, economic and policy issues of space and advanced technological development. Dr. Hertzfeld has served as a Senior Economist and Policy Analyst at both NASA and the National Science Foundation and is a consultant to both US and international agencies and organisations. He is the author of many articles on the legal and economic issues concerning space and technology and teaches courses in Space Law and Space Economics. Professor Hertzfeld is a member of the Bar in Pennsylvania and the District of Columbia and holds a B.A. from the University of Pennsylvania, an M.A. from Washington University, a Ph. D. from Temple University and a J.D. from George Washington University. He can be contacted at: hrh@gwu.edu

(continued)

Sebastien Moranta, Coordinator of Studies, European Space Policy Institute	
	Sebastien Moranta is the Coordinator of Studies at the European Space Policy Institute in Vienna. Prior to supervising the research activities of the Institute, he was a Senior Associate at PricewaterhouseCoopers Advisory where he managed space-related studies for the European Commission, ESA, GSA and other space sector stakeholders. Former analyst at Eurospace, the association of the European space industry, he has worked on a variety of space-related topics including strategy and policy, programme management, R&D&I, standardisation, security, business and economics across different space domains. Mr. Moranta has an aerospace engineer degree from the Institut Polytechnique des Sciences Avancées (IPSA) in Paris
Kai-Uwe Schrogl, Chief Strategy Officer, ESA	
	Prof. Dr. Kai-Uwe Schrogl is the Chief Strategy Officer of the European Space Agency (ESA, Headquarters in Paris, France). From 2007 to 2011, he was the Director of the European Space Policy Institute (ESPI) in Vienna, Austria, the leading European think tank for space policy. Prior to this, he was the Head of the Corporate Development and External Relations Department in the German Aerospace Center (DLR) in Cologne, Germany. Previously he also worked with the German Ministry for Post and Telecommunications and the German Space Agency (DARA) in Bonn, Germany. He has been a delegate to numerous international forums and has served from 2014 to 2016 as the Chairman of the Legal Subcommittee of the UNCOPUOS. He also was the Chairman of various European and global committees (ESA International Relations Committee and two plenary working groups of the UNCOPUOS Legal Subcommittee, the one on the launching State and the other on the registration practice, both leading to UN General Assembly Resolutions). He presented, respectively testified, at hearings of the European Parliament and the US House of Representatives Kai-Uwe Schrogl is the President of the International Institute of Space Law. He holds a doctorate degree in political science and lectures international relations as an Honorary Professor at Tübingen University, Germany

(continued)

Jana Robinson, Space Security Program Manager, Prague Security Studies Institute	
	Dr. Jana Robinson is the Space Security Program Director at the Prague Security Studies Institute (PSSI). She previously served as a Space Policy Officer at the European External Action Service (EEAS) in Brussels as well as a Space Security Advisor to the Czech Foreign Ministry. From 2009 to 2013, Ms. Robinson worked at the European Space Policy Institute (ESPI), seconded from the European Space Agency (ESA), leading the Institute's Space Security Research Programme. Prior to joining ESPI, Mr. Robinson served as Development Director at PSSI from 2005 to 2009 and administered its affiliate organisation in Washington DC, PSSI Washington. Ms. Robinson is an elected member of the International Institute of Space Law (IISL) and the International Academy of Astronautics (IAA). She is also a member of the Advisory Board of the George C. Marshall Missile Defense Project of the Center for Strategic and International Studies (CSIS) in Washington, D.C. Ms. Robinson holds a PhD from the Charles University's Faculty of Social Sciences, Institute of Political Studies, in the field of space security. She also holds two Master's Degrees, from George Washington University's Elliott School of International Affairs and Palacky University in Olomouc, respectively
Joe Pelton, Executive Board Member, International Association for the Advancement of Space Safety	
	Dr. Joseph N. Pelton is on the Executive Board of the International Association for the Advancement of the Space Safety (IAASS) and Associate Editor of the Journal of Space Safety Engineering. He is Emeritus Director of the Space and Advanced Communications Research Institute (SACRI) at George Washington University. He played a key role in the establishment of the International Space University of Strasbourg, France, where he served as the Chairman of the Board (1992–1995) and Vice-President of Academic Programs and Dean (1996–1997). From 1989 to 1996, Dr. Pelton served as the Director of the Interdisciplinary Telecommunications Program at the University of Colorado where he also headed the Center for Research in Telecommunications. From 1969 to 1989, Mr. Pelton held a number of management positions with the COMSAT Corporation and the INTELSAT global satellite organisation, including the position of Director of Strategic Planning. Dr. Pelton is the author of over 50 books or research studies in the field that include writings on satellites, advanced telecommunications technology and regulation and especially about the long range impact of technology on society. His book Global Talk received a Pulitzer Prize nomination and won the literature award of the American Astronomical Society. He was the founding President of the Society of Satellite Professionals International and has been elected to this organisation's Hall of Fame. He has

(continued)

	also received the 2001 Arthur C. Clarke Prize, and in 2017 both the IAASS da Vinci award and the Guardian Award of the Lifeboat Foundation
Victoria Samson, Head of Washington Office, Secure World Foundation	
	Victoria Samson is the Washington Office Director for Secure World Foundation and has 20 years of experience in military space and security issues Before joining SWF, Ms. Samson served as a Senior Analyst for the Center for Defense Information (CDI), where she leveraged her expertise in missile defence, nuclear reductions and space security issues to conduct in-depth analysis and media commentary. Prior to her time at CDI, Ms. Samson was the Senior Policy Associate at the Coalition to Reduce Nuclear Dangers, a consortium of arms control groups in the Washington, D.C. area, where she worked with Congressional staffers, members of the media, embassy officials, citizens and think-tanks on issues surrounding dealing with national missile defence and nuclear weapons reductions. Before that, she was a researcher at the Riverside Research Institute, where she worked on war-gaming scenarios for the Missile Defense Agency's Directorate of Intelligence Ms. Samson holds a Bachelor of Arts (B.A.) degree in political science with a specialisation in international relations from UCLA and a Master of Arts (M.A.) in international relations from the Johns Hopkins School of Advanced International Studies
Mark Dickinson, Chairman, Space Data Association	
	Mark Dickinson is the Inmarsat's Vice-President of Space Segment and Deputy CTO. In March 2017, Mark was elected as the new Chairman of the Space Data Association (SDA), an international organisation that brings together satellite operators to support the controlled, reliable and efficient sharing of data critical to the safety and integrity of the space environment. Mark, who has a PhD in High Energy Astrophysics from the Durham University, joined Inmarsat in 2000. Between 2009 and 2017, he was the Head of Satellite Operations, responsible for Inmarsat's geostationary telecommunication fleet. He is now responsible for the design, procurement and launch of Inmarsat's future space segment assets

(continued)

Daniel Oltrogge, Director of Center for Space Standards and Innovation, Analytical Graphics	
	Daniel Oltrogge is the Director of AGI's Center for Space Standards and Innovation and Program Manager of the Space Data Center, providing space situational awareness and analysis support to 33 government, civil and commercial space operators in the Space Data Association. Dan is also a modelling, simulation and analysis analyst for AGI's Commercial Space Operations Center. Dan's specialties include launch and early orbit operations, nanosatellites, collision avoidance, RF interference mitigation, space situational awareness, astrodynamics and international standards development for space operations and orbital debris mitigation. Dan has authored over 100 technical papers and professional journal articles including the consensus international standard on estimating orbit lifetime. He received a Bachelor's Degree in Aerospace Engineering from Iowa State University and a Master's Degree in Aerospace Engineering from the University of Southern California
Andrew Monham, Spacecraft Operations Manager, EUMETSAT	
	Andrew Monham has 32 years' experience in the space industry, having graduated with a Master of Engineering in Aeronautics and Astronautics from Southampton University in England. Andrew led the system preparation and transition into operations of the ground station and near real-time product processing for ESA's first European Remote Sensing satellite (ERS-1) and subsequently led the flight operations system preparation for ENVISAT. He performed consultancy to EUMETSAT as System Manager on the European Polar System/Metop mission, following which he became Director of Operations at Rhea Group. Andrew then moved back to EUMETSAT where he is responsible for Low Earth Orbit Spacecraft Operations. Andrew is also deeply involved in debris and space weather management, as well as EUMETSAT's technical, management and financial interfaces with external partners and stakeholders.

(continued)

Xavier Pasco, Director of the Foundation for Strategic Research	
	As Director of the Foundation for Strategic Research since 4 October 2016, Xavier Pasco coordinates research on space, high-tech industries and security. In particular, he has a specific interest for US space policy both on civilian and military issues, as well as its approach toward international affairs in high-tech industries. He is also responsible for studies on future national and European programmes in the field of space—civilian and military. He is the author of numerous study reports on civil and military space activities carried out on behalf of national and European public entities. Xavier Pasco has been a member of numerous ministerial and interdepartmental working groups. He has been rapporteur for the report "A Space Ambition for Europe" published by the Center for Strategic Analysis He has participated in several projects related to "preparatory activities for security research" launched by the European Commission in 2004 and in the framework of the 6th and 7th Framework Program for European Research and Development (2007–2013). In this context, he coordinated for the period 2010–2011 the European activities in transversal research on "Conditions for Space Policy and Related Action Plan Consolidation in Europe" supported by the Executive Research Agency of the European Union
Gustav Lindstrom, Director, EU Institute for Security Studies	
	Gustav Lindstrom is the Director of the EU Institute for Security Studies (EUISS)—an agency of the European Union dealing with the analysis of foreign, security and defence policy issues. Previously, Dr. Lindstrom served as the Head of the Emerging Security Challenges Programme at the Geneva Centre for Security Policy (GCSP). While at the GCSP, he also served as the Head of the Euro-Atlantic Security Programme, Director of the European Training Course, co-chair of the Partnership for Peace Consortium (PfP-C) Emerging Security Challenges Working Group and as a member of the Executive Academic Board of the European Security and Defence College. Prior to his tenure at the GCSP, he was a Research Fellow and later a Senior Research Fellow at the EUISS. He has also worked with the RAND Corporation and the World Bank Dr. Lindstrom holds a doctorate in Policy Analysis from the RAND Graduate School and MA in International Policy Studies from Stanford University

(continued)

Didier Alary, Head of Advanced Space Systems, Airbus Defence and Space	
	Didier is presently the Head of Advanced Space Systems at Airbus Defence and Space. Prior to that position, he was the Head of Innovation, head of Earth Observation advanced concepts and several positions in the design and development of spacecraft and associated ground segments. Thirty years of space experience. Before 1988 he worked in the Manufacturing Engineering at Schlumberger for 7 years. Didier Alary is graduated in Electronics at the Ecole Centrale de Paris
Nicolas Bobrinsky, SSA Programme Manager, ESA	
	Mr. Nicolas Bobrinsky has been managing ESA's Space Situational Awareness (SSA) Programme since November 2008, following its approval at the ESA Ministerial Conferences in The Hague (2008) and in Naples and Lucerne (2012 and 2016). This Programme encompasses activities relating to Space Weather (SWE), Near Earth Objects (NEO) as well as Space Surveillance and Tracking (SST) domains. 19 ESA member states have to date committed ca. 200 M€ to the SSA activities covering the time frame 2009–2019. Nicolas Bobrinsky has been with ESA since 1987, mostly at the European Space Operations Centre (Darmstadt, Germany), and held engineering and managerial positions in the areas of Engineering and Operations in support of various Space Missions ranging from Near Earth to Deep Space, as well as ESA's network of Tracking Stations (ESTRACK). He holds an engineering degree in aerospace engineering from the Ecole Nationale Supérieure de l'Aéronautique et de l'Espace (France)
Regina Peldszus, Co-Chair, EU SST Consortium	
	Dr. Regina Peldszus is a policy officer with DLR Space Administration, Department of Space Situational Awareness. She leads the German delegation to the Consortium implementing the European Space Surveillance and Tracking (SST) Support Framework for the European Commission, serves as Co-Chair of its Steering Committee and handles studies at the intersection of space policy, foresight, security and infrastructure. From 2013 to 2015, Regina was an Internal Research Fellow with ESA's European Space Operations Centre (ESOC), Studies and Special Projects Division, focusing on resilience of flight control teams during critical mission phases. Prior to this, she contributed to various future systems and safety projects, including at ESA's Astronaut Training Division, Special Skills and Exploration Unit. Regina holds a Ph.D. from Kingston University, London, on human systems integration in exploration class missions. Her interests

(continued)

	include transfer of practice on issues such as governance and dual use from high reliability domains, including polar and nuclear. She is a member of the AIAA, the IFIP WG 13.5 on Resilience, Reliability, Safety and Human Error in System Development and the ESA/CNES/DLR committee for Human Dependability
Petr Havlik, Senior Space Policy officer, European External Action Service	
	Petr Havlik serves as a Senior Space Policy Officer at the European External Action Service in Brussels. He was the First Counsellor of the European Union Delegation in Vienna from 2011 to 2016 focusing on space policy, international security, international trade law, and industrial development. He was the Counsellor at the Ministry of Foreign Affairs of the Czech Republic for 16 years posted in Prague, Geneva, the Hague, and Vienna. He worked in the fields of space policy, international trade, security policy, international transport, and industrial development. Petr Havlik worked as Senior Economist and International Trade Expert at Motokov International from 1986 to 1995, dealing with export of agricultural tractors and agricultural machinery. Petr Havlik holds an Engineer's degree in Information Technology from the University of Economics, Prague, and a MBA from Sheffield Hallam University
Carlo des Dodrides, Executive Director, European GNSS Agency (GSA)	
	Carlo des Dorides is the Executive Director of the European GNSS Agency (GSA). In his role prior to Executive Director of the GSA, he held a management role at the European Commission and was responsible for the definition of the Galileo/EGNOS exploitation phases Previously, he led the Concession Department at the European GNSS Supervisory Authority and served as the Chief Negotiator of the Galileo Public-Private-Partnership/ Concession contract at the Galileo Joint Undertaking. This in-depth understanding of Galileo and EGNOS, as well as a career in management, helped him excel in his first term as Executive Director After obtaining a degree in engineering from the University of Rome, he went on to achieve an M.B.A. from CUOA, Vicenza, Italy, and obtained additional education at the International Space University in Toulouse, France. From there, he worked in management in the aerospace sector, including a position as Head of Advanced Telecommunication Programmes and Programme Manager for major satellite telecommunication projects at Alenia Spazio. Later, he served as Director of Programmes and Engineering at ENAV, the Italian air navigation service provider

(continued)

Pascal Legai, Director, European Union Satellite Centre	
	Pascal Legai has been a Brigadier General in the French Air Force since 1988. He acquired experience mainly in the fields of geography, imagery, international relations, space and security issues. He had been appointed as a geographer in the French Air Force staff in Paris for 5 years (1994–1999). He was also the Head of the French Imagery Intelligence Centre (2004–2006), Human Resources adviser of the French Air Force (2006–2008) and the Commanding Officer of the French Air Force Base in Grenoble (2008–2010). He has extensive international experience in geospatial and intelligence domains. He has a PhD in International Relations; a law degree; several master's degrees in Imagery Processing, in History, in British Civilization and in Mathematics; and an Engineer Diploma in the field of geographic sciences. After 4 years as Deputy Director, Pascal Legai was elected by the member states as the Director of the EU Satellite Centre, taking up his duties on 1 January 2015.
Philippe Brunet, EC Director for Space Policy and Research, Copernicus Programme and Defence	
	Doctor in Medicine (MD) and Iuris Doctor in Community Law (LLD), he entered the Commission in 1988 (DG V – Social Affairs). He joined the DG III (Industry) in 1993 in order to prepare the establishment of the EMEA (European Medicines Evaluation Agency, now the EMA) and complete the legal framework of the EU authorisation scheme for medicinal products. Philippe Brunet was appointed Deputy Head of Unit "Pharmaceuticals and Cosmetics Unit" in 1998 and subsequently Head of the Unit "Pharmaceuticals, legislative framework and market authorisations" in April 2000. In October 2007 he was appointed Head of Cabinet by Commissioner Kyprianou The Commission appointed on 19 December 2012 him as Director in DG Directorate General for Internal Market, Industry, Entrepreneurship and SMEs (GROW) for Directorate "Aerospace, Maritime and Defence Industries" as of January 2013. Mr. Brunet's main institutional files comprise, inter alia, the implementation of COPERNICUS, a constellation of EU-satellites offering Earth Observation services on a full, open and free of charge basis and the development of civil and military synergies in security and defence matters to render the defence industry competitive and underpin a credible and effective Common Security and Defence Policy. Since 2016, his directorate has in charge the design and development of the new European defence fund Philippe Brunet has two sons. He has been knighted in the French Order of the Legion of Honour in 2011

(continued)

Rob Heron, Managing Director, SES Defence UK	
	Rob has over 35 years of experience in the satellite communications industry. His activities include developing business and products in international environments. He has extensive experience of people and project management and has previously managed other business in the UK. Rob has worked on the ground segment ranging from product development to system engineering. Rob has also worked on the space segment ranging from autocompatibility analysis to antenna array performance on military satellites Within SES, Rob has occupied several roles including market development and the creation of the GovSatCom business. Previously also responsible for sales on a global basis and before that project management of innovative projects such as the implementation of a global geolocation system and the design and implementation of an advanced spread spectrum ranging and IOT system. Previously Rob managed a UK R&D company of 86 professional engineers specialised in satellite communications and related technologies. Rob has worked in several countries including ESA in the Netherlands and Dornier in Germany. He is fluent in several languages and has wide experience of working in international environments
Richard Buenneke, Senior Space Policy Advisor, US Department of State	
	Richard H. Buenneke is a senior advisor for national security space policy in the Bureau of Arms Control, Verification and Compliance (AVC) at the United States Department of State in Washington, D.C. In his current position, Mr. Buenneke advises senior State Department officials on the planning and implementation of diplomatic and public diplomacy activities relating to US national security space policy. He served from June 2011 to June 2018 as the co-chair of an international expert group on space debris, space operations and tools for collaborative space situational awareness, which was established as part of a United Nations Committee on the Peaceful Uses of Outer Space's working group on the long-term sustainability of outer space activities. He also serves as the AVC bureau's lead for US-European Union exchanges on space security cooperation Before joining the State Department in March 2007, Mr. Buenneke was a senior policy analyst at The Aerospace Corporation, where he served as lead analyst for the US Department of Defense's efforts relating to mission assurance and resiliency of commercial space capabilities. Mr. Buenneke has also led research tasks in support of US government assessments of policies and architectures for national security and commercial space systems as an analyst with Booz Allen Hamilton and the RAND Corporation. He holds bachelor's degrees in economics and systems engineering from the Wharton and Engineering

(continued)

	schools of the University of Pennsylvania. Mr. Buenneke also holds masters' degrees in policy analysis from George Washington University's Elliott School of International Affairs and the Pardee RAND Graduate School
Genevieve Fioraso, Former French Minister for Higher Education, Research and Space, and Chair of the ESPI Advisory Council	
	From 1979 to 1986, collaborator of Hubert Dubedout, Mayor of Grenoble and member of the French Parliament, in charge of press, documentation and Parliament relations In 1989, joined the private sector, as a member of the management board of CORYS, a spin-off from the CEA (French Nuclear Energy Commission), specialised in the design of simulators for training operators in normal and accidental situations in nuclear and thermal power plants in France and abroad (Eastern Europe, USA, India, China, Indonesia, Pakistan, Spain, etc.) as well as train and tram drivers: in charge of Corus institutional relations and European R and T projects (Tacis, Phare and Eurêka programmes). From 1995 to 1999, director of cabinet of the Mayor of Grenoble. In 1999, creation and general management of the Rhône-Alpes Agency for the diffusion of digital in regional SMEs In 2001 and until 2012, election as deputy mayor and first Vice-President of the metropolitan area of Grenoble, in charge of economic development, innovation, university and research. Part-time marketing counsellor for France Telecom (now Orange) in the field of health and social care. From 2003 to 2012, CEO of a new semiprivate-public company, Minatec enterprises, welcoming within an innovative technological pole start-ups and cooperative projects between public and private research in the field of micro-, nano- and biotechnologies Elected as a member of the French Parliament in 2012, member of the Parliament Office in charge of evaluating public choices in the field of science and technology (OPECST) From 2012 to 2015, French Minister of University and Research, also in charge of space. Much committed for the European decision for Ariane 6 2015–2017: member of the French Parliament, Commission of defence and Parliament group for space. Report for French Prime Minister in July 2016 entitled "Open space, an answer to the challenges of the space sector" 2017 : Chair of ESPI (European Space Policy Institute) advisory board, Vienna, member of several scientific foundation boards in Paris, Marseille, Grenoble in the field of space, health and scientific diffusion

(continued)

Pierre Delsaux, Deputy DG, DG GROW, European Commission	
	Pierre Delsaux is Deputy Director General at the European Commission Directorate General for the Internal Market, Industry, Entrepreneurship and SME's. After studying Law at the University of Liège, he obtained his Master of Law at the Northwestern University, Chicago, in 1983. He was Legal Secretary at the European Court of Justice from 1984 to 1987. He worked in the private sector before joining the European Commission in 1991. He started his career within the European Commission in the Directorate General for Competition. He was appointed Director responsible for regulating the financial services in 2007. Following this, in 2011, he was appointed Deputy Director General with responsibilities for the Single Market in the EU. Since December 2015, he is in charge of Space Policy and Defence

Chapter 5
Security in Outer Space: Rising Stakes for Europe

5.1 Introduction

5.1.1 Rationale

The ecosystem of the space sector has shifted drastically over the past decades with new technical concepts and business endeavours building on a changing institutional and economic environment. Now a pillar of the modern economy and society, the global space infrastructure enables key services across vital sectors and directly supports public actions to address economic, societal, environmental and security issues at a national and global level. This ever-growing use of space-based data and services by a variety of public and private actors/users has created a virtually invisible dependence on space technologies, which closely relates to Information and Communication Technologies (ICT). As the use of space applications becomes more pervasive, brings more benefits and becomes part of the business-as-usual routine, dependence on space infrastructure intensifies, which creates new vulnerabilities for the economy and society at large.

In the European Union only, at least 10% of the GDP depends to some extent on space assets.[1] Thus, even a partial incapacitation of these assets could lead to a substantial economic loss of up to EUR 50 billion per year and would put up to one million jobs at risk. The growing importance of space infrastructure raises new stakes concerning its protection from harm. Experts routinely caution governments and operators about the rising threats to space infrastructure security, underlining that space is increasingly congested and contested, which poses intensifying challenges to safely deploying, operating and exploiting space assets. Driven by a variety of trends from inside and outside the space sector, challenges to the security of space

[1]PwC (2017). *Dependence of the European Economy on Space Infrastructures*. Brussels: EU Publications. Retrieved from: http://www.copernicus.eu/sites/default/files/library/Copernicus_SocioEconomic_Impact_October_2016.pdf

E. Burger, G. Bordacchini, *Yearbook on Space Policy 2017*, Yearbook on Space Policy, https://doi.org/10.1007/978-3-030-05417-5_5

infrastructure include the proliferation of space debris, accidental or malicious radio interferences, cyberattacks, antisatellite technologies (ASAT) and natural space hazards such as geomagnetic and solar radiation storms.

This deteriorating situation has already been widely acknowledged by European stakeholders including member states from the European Space Agency (ESA) and the European Union (EU). Recently, the EU identified the reinforcement of "Europe's autonomy in accessing and using space in a secure and safe environment" as a top priority of its Space Strategy for Europe.[2] Accordingly, European stakeholders have taken steps in this field. Organised at national, intergovernmental and European level, efforts encompass capacity building in the field of Space Situational Awareness (SSA), development of technologies and standards, reinforcement of space programmes security architecture, establishment of legal and regulatory regimes but also diplomatic initiatives and cooperation frameworks.

Despite multiple initiatives, European efforts, in particular in the field of capacity building, remain limited in comparison with other spacefaring nations and especially with the USA. Across the Atlantic, securing space assets has long been a strategic priority as underlined by the development of an ambitious SSA programme by the US Department of Defence. Recent developments, including the launch of the $1.6 billion US Space Fence programme aiming to further increase domestic SSA capabilities and the recent publication of a Space Traffic Management (STM) policy, demonstrate the desire of the US government to accelerate its effort in this domain. As the space security dialogue gains momentum, Europe sits at the crossroads of important decisions with far reaching consequences. Among these decisions, the priorities and budget envelopes of the post-2020 Multiannual Financial Framework (MFF) will be decisive to shape the role that Europe will play in this field in the coming years.

5.1.2 Research Scope

"Space security" is understood here primarily as "security in outer space" referring to the protection of space infrastructure from threats so that this infrastructure can fulfil its specific functions as expected. From a general standpoint, ensuring "security in outer space" is a multifold challenge encompassing three main areas (Fig. 5.1).

The space infrastructure can be described as a network of space-based and ground-based systems interconnected by communication channels and enabled by access to space capabilities. It includes a space segment, a ground segment, a user segment and downlinks and uplinks to interface between these segments. As the present research focuses on "security in outer space", the analysis addresses predominantly security challenges to the "space segment" and, whenever relevant, to the downlink and uplink channels. The research does not address security challenges specific to ground and user segments, such as Earth natural hazards, physical attacks on facilities, or eavesdropping. This research also focuses on security challenges to space infrastructure in operation, including system disposal. Security challenges at earlier stages of space programmes (e.g. security-by-design, supply chain security)

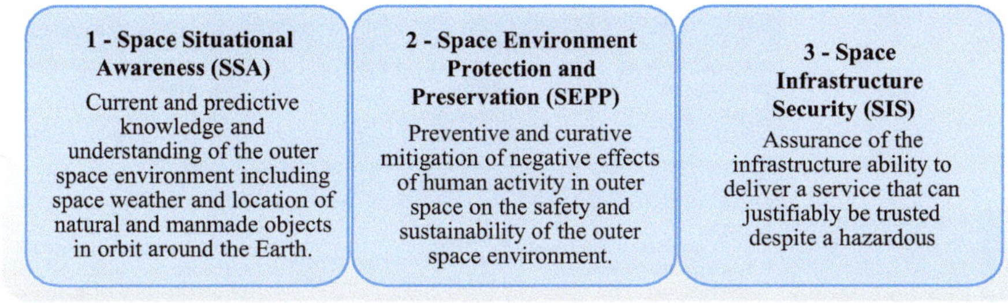

Fig. 5.1 Core components of the "security in outer space" challenge

are mentioned whenever relevant but have not been investigated specifically. By extension, the security architecture of space programmes (e.g. security accreditation, threat response architecture), which requires to be adapted to the specificities of each infrastructure, is not addressed in the present work, which focuses on common and external dimensions of security in outer space.

5.1.3 Security in Outer Space: Action Matrix

Addressing security in outer space challenges requires a variety of actions across multiple domains. The initial and overarching action is the elaboration and endorsement of a comprehensive space security strategy and policy framework including clear objectives and associated actions supported by a coherent governance framework and the allocation of appropriate resources.

Key space security actions include:

Capacity-building programmes: Development of technical capabilities required for the implementation of space security measures. Capacity building covers the entire life cycle from research and development to operations and delivery of service, through acquisition, deployment and upgrade of assets.

Legal and regulatory regimes: Preparation, ratification and application of a set of laws, regulations, rules, procedures and standards.

Diplomacy and cooperation frameworks: Setting up of cooperative political and operational frameworks supporting a collective approach to space security.

The following Fig. 5.2, based on a categorization matrix applied to security in outer space, provides a list of examples of actions and measures related to "security in outer space" organised by domain and field of action. The list is intended to be illustrative rather than comprehensive. The matrix of action provided the key methodological tool to organise the mapping of European activities.

Security in Outer Space subdomain	Field of action		
	Capacity-building programmes *Develop and deploy operational capacities to ensure security in outer space*	**Legal and regulatory regimes** *Establish a reference framework to conduct space activities in compliance with space security requirements*	**Diplomacy and cooperation frameworks** *Harmonise and coordinate space security efforts among stakeholders*
Space Situational Awareness (SSA) *Monitor space environment threats*	Examples: • SST capabilities development • Space weather models development • SSA services delivery	Examples: • Space objects registration obligations and procedures • SSA data policy	Examples: • SSA data sharing agreements • Transparency and Confidence-Building Measures
Space Environment Protection and Preservation (SEPP) *Keep the space environment safe to operate in*	Examples: • Curative technologies development (e.g. active debris removal solutions)	Examples: • Space law (e.g. end-of-life obligations) • Standards for space environment-friendly satellite design (e.g. passivation devices)	Examples: • Space Debris Mitigation Guidelines • Long-term sustainability guidelines • International Code of Conduct proposal
Space Infrastructure Security (SIS) *Protect the space infrastructure from threats*	Examples: • Security enhancing technologies development (e.g. secure links) • Resilient system architectures (e.g. fragmented systems)	Examples: • Space programme security rules and procedures • Security and safety standards • Supply chain control processes (e.g. export/import rules, testing procedures)	Examples: • Collision avoidance procedures and coordination • Deterrence through hosted payloads on allies' satellites

Fig. 5.2 Examples of "security in outer space" measured by field and domain category

5.2 Increasing Need for Space Security in Europe

5.2.1 Rising Threats to the European Space Infrastructure Security

In the context of an increasingly congested, contested and competitive space environment, the European space infrastructure is exposed to a number of security challenges, including:

- *Passive man-made threats*, including unintentional man-made hazards to space infrastructures such as space debris or unintentional interferences to signals
- *Active man-made threats*, including intentional attacks on space infrastructures such as antisatellite capabilities, malicious interferences and cyberattacks
- *Natural threats*, including space weather hazards such as geomagnetic storms, solar radiation storms or disturbance of the ionosphere

Several key takeaways were identified concerning security challenges and threats faced today by the European space infrastructure. These challenges and threats are:

- *Multiple and diverse* in nature and origin and as a consequence require a set of different mitigation and protection measures. Although the focus here is on "security in outer space" and therefore on threats in orbit, it can be established that a share of space infrastructure vulnerabilities during operations in space can result, at least partially, from earlier stages (e.g. system development and production).

- *Interrelated and interdependent*, with space infrastructure being vulnerable to some hazards that have the potential to create new threats. For example, vulnerability to cyberattacks aiming to take control of a satellite to weaponise it (e.g. by commanding it to collide with another satellite) can create new ASAT threats. Similarly, challenges and threats to space security depend on the approach of all space stakeholders, underlining the relevance of the OSCE's underlying premise that, comparable to other security domains, space security is indivisible.
- *Ubiquitous and inclusive*, although some systems are less exposed or vulnerable to specific threats.
- *Intensifying*, driven by endogenous and exogenous trends including:
  - Increasing space activity in terms of the number of launches and objects in orbit but also in the number of governmental and commercial actors owning and operating space systems
  - New concepts, technologies and capabilities
  - An ever more connected space infrastructure, including with other ground networks and systems
  - The increasing importance of space infrastructure, which makes it a key target for a variety of actors pursuing different objectives
  - The rehabilitation of a "space warfare" doctrine encompassing activities to develop "space control" capabilities

This situation, expected to further deteriorate in the future, results from a variety of factors consistent with the growing strategic and socio-economic significance of space infrastructure. From this standpoint, the pervasive dependence on space systems suggests that rising threats to the space infrastructure mean, ultimately, potential risks for the modern economy, society, security and, more generally, modification of the geopolitical scene.

5.2.2 Strategy and Policy Rationales

The growing importance awarded to security in European space policies is driven by four key policy rationales:

- *Secure the results of the continuous and substantial investment* made by public and private actors.
- *Protect the European economy and society* against risks related to its pervasive and sizeable dependence on the space infrastructure.
- *Contribute to a service-oriented policy* by assuring the ability of the infrastructure to deliver a service that can justifiably be trusted, in particular for users in defence and security.
- *Guarantee European autonomy and freedom of action* in the field of security in outer space and in the space domain at large.

5.2.2.1 Protect the Value of the European Space Infrastructure

The value of the European space infrastructure, which is the result of a continuous and substantial investment by public and private actors, lies first and foremost in the substantial socio-economic benefits it enables across a multitude of sectors for Europe. More and more turnkey services are available to end-users—households, companies and public actors. As a result, the exploitation of space services stimulates growth and generates considerable benefits for the European economy. More than 10% of the EU GDP depends on space infrastructure.[2] The Galileo system now has more than 200 million users[3], and its full potential has yet to be reached. Financial assessments investigating the downstream sector and economic benefits to end-user sectors conclude that the total economic benefit is around € 53.5 billion per year in gross value added, supporting one million workers directly or indirectly. The Copernicus programme is expected to generate more than € 13.5 billion of cumulated economic benefits in gross value added by 2020. The European space infrastructure is also a critical asset for the implementation of governmental policies benefiting society and the environment at large (EU Digital Agenda dealing, the Common Fisheries Policy (CFP), and the Common Agricultural Policy (CAP)).[4] Space systems are also critical for defence and security policies dealing with traffic monitoring, border surveillance and humanitarian aid. Additionally, the uptake of space technology plays a huge part in innovation policy, laying the groundwork for the development of new services in Europe. The dependence of governments, businesses and individuals on space infrastructure grows, creating new risks if space assets were to be incapacitated, even partially. With the intensification of challenges to the security of space systems, the need to protect space assets to safeguard the benefits they enable can be considered, alone, as a reasonable argument for Europe to position security at the top of the space policy agenda.

5.2.2.2 Contribute to a "Service-Oriented" Policy

The rising need for enhanced space security in Europe also lies in the significant progress of EU programmes (i.e. Galileo, EGNOS, Copernicus) and in the potential introduction of new initiatives such as GOVSATCOM. As a matter of fact, the

[2]PwC (2016). Study to examine the socioeconomic impact of Copernicus in the EU. Report on The socio-economic impact of the Copernicus programme. Brussels: European Commission. Retrieved from European Commission: http://www.copernicus.eu/sites/default/files/library/Copernicus_SocioEconomic_Impact_October_2016.pdf

[3]https://www.numerama.com/politique/386721-galileo-le-gps-europeen-compte-200-millions-dutilisateurs-dans-le-monde.html

[4]PwC (2016). Study to examine the socioeconomic impact of Copernicus in the EU. Report on The socio-economic impact of the Copernicus programme. Brussels: European Commission. Retrieved from European Commission: http://www.copernicus.eu/sites/default/files/library/Copernicus_SocioEconomic_Impact_October_2016.pdf

security needs of the EU are growing, inasmuch as common foreign, security and defence policy objectives are increasingly being incorporated into the agenda and DNA of the EU. The EU's role in such matters was reinforced and institutionalised by a number of policy documents including the Common Security and Defence Policy (CSDP), the European Agenda on Security, or the most recent European Defence Action Plan[5] (followed by launching of the European Defence Fund in June 2017). These recent developments indicate the growing importance of security in EU Policy.[6] The European GNSS System Galileo will play a pivotal role in the implementation of national and European policies for localisation and navigation for troops and vehicles, mission planning, delivery of cargos and search and rescue.[7] GNSS-based services are also key in case of conflict for guidance, strikes and other related operations.[8] The Copernicus programme will provide critical data for policies related to the atmosphere, marine environment, land, climate change, emergency management and security. The new GOVSATCOM initiative aimed at ensuring reliable, secure and cost-effective satellite communication services for EU and national public authorities will also elaborate on security aspects.

5.2.2.3 Reinforce European Autonomy and Freedom of Action

From a strategic perspective, Europe also seeks to guarantee the security of its space infrastructure autonomously through independent capabilities (i.e. systems, data and technologies). Here, and although cooperation with third countries is essential in the field of space security, Europe must ensure a capacity to control the level of reliance on its partners and to maintain it within acceptable boundaries. That policy rationale is acknowledged in the Space Strategy for Europe stating that "reinforcing Europe's autonomy in accessing and using space in a secure and safe environment" is a pillar objective.[9] Elaborating further, the EC seeks to "ensure [Europe's] freedom of action

[5]European Commission. (2015). The European Agenda on Security. Brussels. Retrieved from European Commission: https://ec.europa.eu/home-affairs/sites/homeaffairs/files/e-library/documents/basic-documents/docs/eu_agenda_on_security_en.pdf

[6]European Commission (2016). Communication From The Commission To The European Parliament, The Council, The European Economic And Social Committee And The Committee Of The Regions Space Strategy For Europe. COM (2016) 705 final. Brussels: European Commission. Retrieved from https://ec.europa.eu/transparency/regdoc/rep/1/2016/EN/COM-2016-705-F1-EN-MAIN.PDF

[7]Mutschler, M. M. (2010). Keeping Space Safe. Towards A Long-Term Strategy to Arms Control in Space. Frankfurt: Peace Research Institute Frankfurt.

[8]Sitruk, A., & Plattard, S. (2017). The Governance of Galileo. Vienna: European Space Policy Institute. Retrieved from ESPI: https://www.espi.or.at/images/Rep62_online_170203_1142.pdf

[9]European Commission (2016). Communication From The Commission To The European Parliament, The Council, The European Economic And Social Committee And The Committee Of The Regions Space Strategy For Europe. COM (2016) 705 final. Brussels: European Commission. Retrieved from https://ec.europa.eu/transparency/regdoc/rep/1/2016/EN/COM-2016-705-F1-EN-MAIN.PDF

and autonomy".[10] Such strategic goals are not being pursued at the expense of cooperation, however, as cooperation in this field is a necessity (i.e. data sharing, efficiency of measures, transparency and coordination). The level of reliance European seeks to achieve remains an open question: at the moment, Europe relies on US data for its operational SSA needs. This position underscores a number of rising issues. First are the SSA data and service restrictions that mean that the USA does not grant full access to its database because of its intrinsic military nature.[11] Additionally, the US system has its own technical limitations (i.e. unverified warnings), and cooperative agreements do not accommodate accountability mechanisms.[12] And, last but not least, the US government has the right to terminate the user account, to limit both access duration and data amount, to deny access to SSA data and information and to change or modify the terms and conditions at any time for any reason and without prior notification.[13]

5.2.2.4 Long-Term Stakes for Europe's Position on the International Space Scene

Beyond short-term policy stakes, the need for a reinforced approach to space security in Europe is also driven by longer-term considerations stemming from the strategic ambition of Europe to "promote its position as a leader in space, increase its share on the world space markets, and seize the benefits and opportunities offered by space." As space security now holds a central position in space diplomacy, playing a prominent role in international dialogues and negotiations in this field, and promotes a clear, united and consistent "European way", it will be essential to build European leadership in the global space sector at large. Taking this into account, equipping Europe with a comprehensive and independent capability to protect its space infrastructure is a priority to position Europe as a credible interlocutor on the international scene.

Last but not least, space security will play an increasing role in commercial space markets. First, the implementation of new practices (e.g. laws, regulations, standards, procedures) promoting space security and impacting the way space activities are conducted will inevitably create a competitive bias for the industry. Second, space security may open up a business opportunity for private industry seeking to enhance public capabilities with commercial data and value-added services. For

[10]Ibid.

[11]Specifically, DoD resistance to open SSA data sets, algorithms, and processes to external review and scrutiny, results in the uncertainty of the data and in possible false positive rates. See: https://www.ida.org/idamedia/Corporate/Files/Publications/STPIPubs/2016/P-8038.ashx

[12]Froeliger, J. L. (2017). Greater Industry Cooperation Needed to Avoid Space Collisions. INTELSAT http://www.intelsat.com/news/blog/greater-industry-cooperation-needed-to-avoid-space-collisions/

[13]Space-Track. User agreement. Retrieved from https://www.space-track.org/documentation#/user_agree

these two reasons, Europe must contribute actively to future developments in the field of space security to support the competitiveness of its industry and reap the benefits of new potential markets.

5.3 Overview of European Activities and Capabilities

5.3.1 Space Situational Awareness

The field of Space Situational Awareness (SSA), which encompasses all activities aimed at monitoring space environment threats, constitutes the bulk of activities in the field of security in outer space.

In Europe, just as in many parts of the world, interest in SSA data is growing to meet the challenges of the space environment. Several countries are implementing various national SSA programmes, although the main stakeholders—France, Germany, Italy, Spain and the UK—still rely on data sharing agreements for their operational needs. This subsection provides an overview of European SSA capabilities. However, assessing the operational needs of European end-users on the basis of open source documents is a challenging task. In investigating national SSA capabilities, the research team focused mainly on the five most involved countries in the field. Organisational culture is diverse, although one can see that in most cases, the military plays a predominant role in SSA activities, highlighting the strategic stakes for the military. Indeed, European states pursue SSA development strategies that involve updating existing systems, funding complementary software able to double check data and others. However, that very data is still required for manoeuvres and making well-informed decisions.

In the field of Space Security, the development of national sensors was the first step when securing space assets first emerged as a national security priority. However, in an attempt to create synergies between European countries and avoid duplication of efforts, the European Space Agency was considered as the most competent body to manage consolidation efforts. The agency today is especially active in the field of space debris, in which it develops software to incorporate raw SSA data into databases, and collision assessment algorithms.[14] As part of its engagement in Space Security, in 2009 ESA launched an SSA programme aimed at supporting "the European independent utilisation of, and access to, space for research or services, through the provision of timely and quality data, information, services and knowledge regarding the space environment, the threats and the

[14]ESA. (2012). ESA deploys first orbital debris test radar in Spain. Retrieved from http://www.esa.int/Our_Activities/Operations/Space_Situational_Awareness/ESA_deploys_first_orbital_debris_test_radar_in_Spain

sustainable exploitation of the outer space surrounding our planet Earth".[15] The objectives of the programme were to produce an independent database, support R&D in industry and strengthen the accuracy of available data through greater focus on European space assets.[16] For the period 2009–2020, € 200 million was budgeted, including the contributions of 19 member states. The programme covers R&D, the creation of data and coordination centres, systems development and procurement. The programme has awarded over 100 industrial contacts. ESA is building on the solid base of existing capabilities, relying on the expertise of European and international actors (e.g. the USA), and seeks to actively contribute with additional data to strengthen them. The SSA Programme is coordinated from ESOC in Darmstadt (Germany) but also other centres such as ESRIN and ESTEC.

ESA has developed its SSA programme around three major areas, which are:

- *The Space Surveillance and Tracking (SST) Segment*: which aims at maintaining a data catalogue up to date. Funding allocated to that segment has been reduced, and part of ESA activities have been transferred to the EU SST Consortium.
- *The Space Weather (SWE) Segment*: which aims at developing the European ability to monitor solar activity in a timely matter and inform the main stakeholders—public and private—by enhancing existing capabilities.
- *The Near Earth Objects (NEO) Segment*: which aims at generating data on NEOs—asteroids or comets, making impact likelihood assessments and developing deflection methods. There are estimated to be around 17,000 threatening NEOs that require monitoring.

Part of ESA's mandate in space security deals not only with the management of centres and data integrity processes but also the conduct of studies on hosted payload and sensor development deployments and asteroid impact mitigation studies. Funding is also allocated to developing new infrastructure such as databases, software tools, applications and hardware (optical telescopes and radars). Several achievements illustrate ESA's SSA Programme objectives: funder of preliminary studies of a future space weather satellite, two new radars built in Spain and France to support SST for civilian uses, ESA's Proba-2 solar observatory satellite, the establishment of the ESA Space Security and Education Centre (ESEC) in Redu (Belgium), the Space Surveillance and Tracking Data Centre at ESAC in Madrid, three coordination centres both at Space Pole in Brussels for space weather and at ESRIN in Frascati (Italy) for NEOs and one at ESOC in Darmstadt. Test-bed optical telescopes were installed in Cebreros (Spain) and are being installed at la Silla (Chile). The project "FlyEye", which is an initial automated telescope with high-tech European optical hardware, will be deployed by 2018 and is expected to contribute to a global asteroid survey system to be updated every day. ESA's

[15]ESA. Space Situational Awareness. https://www.esa.int/Our_Activities/Operations/Space_Situational_Awareness

[16]Suzuki, K. Space Security: Is Europe a Credible Diplomatic Actor? Hokkaido University/ Princeton University. Retrieved from https://swfound.org/media/91262/suzuki_kazuto.pdf

involvement in Space Security also requires constant interactions with the operational entities dealing with SSA at national levels, where ESA actively participates in the coordination of efforts. In the field of security in outer space, ESA fuses data from various sources, primarily from the US SSA System, to provide them to European stakeholders. In this sense, ESA acts as an SSA-providing platform for European users.

ESA is also actively involved in the field of Space Weather (SWE). It coordinates a European network SWE Expert Service Centres (ESCs). Additionally, ESA is conducting a mission at the Lagrangian point L5 (LGR) in cooperation with the USA expected to take place after 2019.

In the field of NEOs, ESA is maintaining its work in developing the field-of-view "Fly-Eye" telescope, conducting daily sky surveys for objects from 15 m up to 30 m. The Consortium in charge is led by OHB. The recently proposed Hera mission will aim at testing planetary defence mechanisms and will be complementary to the NASA DART mission whose objective is to destroy an asteroid.[17] The Hera mission will be instrumental in investigating the kinetic effects of such explosion.[18]

The emergence of the European Union as a major player in Space Security is one of the most important developments of European Space Policy as a whole. As the European Union does not have to deal with institutional constraints, such as ESA in several areas, especially those with strong security dimensions, the EU has more agency to implement programmes in this respect. The European Union funds projects related to "security in outer space" under the umbrella of large R&D programmes. Since 2010, the EU has almost continuously supported R&D projects directly linked to "security in outer space" as part of FP7-SPACE and H2020-SPACE,[19] but also through other non-space instruments. In total, the EU has supported at least 35 R&D projects in this domain for a total of € 66 Million over 7 years (i.e. excluding funding to the EU SST support framework). These projects have covered a wide scope of activities including topics such as space debris mitigation, in-orbit collision avoidance, and space weather. A spike in EU contributions can be observed in 2010 corresponding to financial support for a number of

[17]ESA. (2018) Earth's first mission to a binary asteroid for planetary defence. Retrieved from: https://www.esa.int/Our_Activities/Space_Engineering_Technology/Hera/Earth_s_first_mission_to_a_binary_asteroid_for_planetary_defence

[18]Jeff Foust. (2018) ESA plans second attempt at planetary defense mission. Retrieved from: https://spacenews.com/esa-plans-second-attempt-at-planetary-defense-mission/

[19]The 2009 FP7 work programme of 2009 included a list of external events including "space debris, hostile laser or Anti SATellite systems (ASAT), jamming, viruses, natural or man-made electromagnetic disturbances". The work programme of 2010 supported 15 R&D projects. While the 2011 work programme included NEOs, space debris and weather were respectively covered in 2013. Horizon 2020 included a section on 'Protection of European assets in and from space' (H2020-PROTEC) in the 2014–2015 work programme, and a section called 'Secure and safe space environment' in the work programme 2018–2020. Within FP7, additional financial contributions were allocated to the PEOPLE Programme distributing grants for training and career development, under which the SpaceDebEMC (Space Debris Evolution, Collision risk and Mitigation), and the STARDUST (the Asteroid and Space Debris Network), benefited.

Table 5.1 Estimated level of coverage by size of object and orbit of the initial architecture (2017) and expected architecture (2021)

MS architecture (orbit and object size)	2017 (initial architecture) Total observed (%)[a]	2017 (initial architecture) Total well-observed (% of the total)[b]	2021 (expected architecture) Total observed (%)[a]	2021 (expected architecture) Total well-observed (% of the total)[b]
LEO (>7 cm)	19%	14%	35%	19%
LEO (>50 cm)	79%	72%	95%	80%
LEO (>1 m)	96%	95%	98%	97%
MEO (>40 cm)	18%	7%	62%	7%
GEO (>50 cm)	40%	30%	66%	42%

[a]Observed objects are objects that were observed at least once during the 14-day period of the simulation

[b]Well-observed objects are those observed objects that were observed every day in LEO and every 3 days in MEO/GEO

large projects. It is important to note that since 2014, the EU has also funded the SST support framework (roughly at around €28 million per year) through H2020-SPACE grants and also through funds allocated to the Copernicus and Galileo programmes. Contributions to the SST support framework are not included in this figure. The decrease observed since 2013 actually indicates that EU support has switched from R&D projects to operational capacity building.

In addition to initiatives funded through programmatic investments, the EU has launched the EU Space Surveillance and Tracking Support Framework (http://www.eusst.eu/) aimed at supporting the long-term sustainability of space activities of European stakeholders.[20] The overarching purpose of the Consortium is to leverage the networking of already existing national capabilities to deliver European SST services. To achieve that goal, the Consortium is expected to produce services driven by civilian user requirements, complementary to R&D activities conducted within Horizon 2020, to engage with international partners, particularly with the USA and to promote the update of capabilities after the networking phase.

In 2018, the European Commission and the European Parliament published a report on the implementation of the EU SST Framework Programme for the 2014–2017 period comparing the capabilities of the sensor function as of today and in 2021 (Table 5.1)[21]:

On the basis of a documentation review, stakeholders' consultations (including member states, users and other public stakeholders) and with the support of

[20]Official Journal of the European Union (2014). Decision No 541/2014/Eu Of The European Parliament And Of The Council. Retrieved from EUR-Lex: https://eur-lex.europa.eu/eli/dec/2014/541(1)/oj

[21]European Commission (2014). Decision No 541/2014/Eu Of The European Parliament And Of The Council. Retrieved form EUR-Lex: https://eur-lex.europa.eu/resource.html?uri=cellar:fbafc703-4eb8-11e8-be1d-01aa75ed71a1.0021.02/DOC_1&format=PDF

independent experts, the European Commission identified the following key achievements of the support framework[22]:

- *Availability of the EU SST services* (i.e. collision avoidance, in-orbit fragmentation and re-entry services) since 1 July 2016, through the EU SST portal to all European institutional users and spacecraft owners and operators free of charge and on a 24/7 basis
- *Outreach to users* including identification of potential users, documentation of their needs and awareness raising of space risks and the need to protect space infrastructure
- *Cooperation and collection of shared know-how* with the establishment of regular communication between NOCs and increased cooperation between national experts through working groups
- *Mapping and pooling of European assets* with 33 sensors contributing to the initial EU SST operations, a complete mapping of national and European sensors and beginning of national sensors upgrades
- *Outreach to other member states* to collaborate with or to join the SST Consortium

Stakeholders interviewed by ESPI unanimously acknowledged the multiple achievements of the consortium so far, particularly for the complex systems networking and the initial steps toward a full-fledged European Union SST service, while complying with both civil and military frameworks. An important achievement of the consortium, difficult to measure but often praised by consortium partners, has been the reinforcement of European coordination and the confidence-building among partners. Achieving consensus among France, Germany, Italy, Spain and the UK—which speak today with a clear and unified voice—has been a challenge that the consortium successfully took on. Today, EU member states established national databases and are currently building a common European database of unclassified data. The consortium also continues its efforts to promote the exchange of classified data through bilateral agreements and the construction of a secure data sharing exchange network. From this standpoint the consortium is certainly achieving its purpose.

Nevertheless, stakeholders also underlined that various issues remain and that serious challenges lie ahead of the consortium to take a more prominent role, and achieve more ambitious objectives in line with the rising space security stakes in Europe. Key stakes ahead of the consortium identified by the European Commission assessment include[23]:

- *Effectiveness and European added-value optimisation* to avoid duplication of efforts and support an efficient development of EU SST capabilities

[22]Ibid.

[23]Ibid.

- *Achievement of an acceptable level of European autonomy* based on further networking and development effort in line with a level of ambition to be decided by the European Union and with, as a first step, the delivery of a common EU database of orbital objects building on national data
- *Development of EU SST services in compliance with users' needs* requiring additional outreach effort to raise awareness and collect feedback but also the development of common operational procedures and standards
- *Synergies with other components of security in outer space* to cover the range of space hazards over the entire space mission life cycle
- *Governance optimisation* to accommodate a broader member state participation, an enhanced role of the European Commission for guidance and monitoring at the strategic, policy and organisational levels, and to explore the role of EU SatCen as EU SST services front-desk

With regard to financial management, other issues will have to be addressed by future arrangements revision including:

- Simplification of funding sources to avoid unnecessary burdens affecting the results achieved by the consortium
- Budget allocation, to adapt current rules (i.e. sharing on an equal base between the members of the consortium, independently from their current capabilities) to the enlargement of the consortium

5.3.2 Space Environment Protection and Preservation

Space Environment Protection and Preservation (SEPP) is another subdomain of security in outer space. This encompasses all undertakings aimed at keeping the space environment safe to operate in and proactively promoting its preservation.

5.3.2.1 European Countries

Several member states are adopting national legislation aimed at regulating space activities. In seeking to establish a legal framework for their activities, member states have developed requirements, conditions and restrictions for licences facilitating entities that are eager to engage in space activity.[24] Several jurisdictions also include enforcement mechanisms, actively contributing to Space Environment Protection and Preservation. Already, Austria, Belgium, France, Germany, Luxembourg, Norway, Sweden, and the UK have passed such laws.

In international fora, a handful of European states have been vocal about the need for greater coordination on Space Security. Good examples are the Inter-Agency

[24]Ibid.

Space Debris Coordination Committee (IADC), the United Nations Committee on the Peaceful Uses of Outer Space (UNCOPUOS), the Conference on Disarmament (CD) and the Committee on Space Research (COSPAR). Among other major initiatives, European states actively contributed to the report of the Group of Governmental Experts (GGE) on Transparency and Confidence-Building Measures (TCBMs). On the occasion, several experts appointed by European states elaborated a strategy toward promoting mutual understanding and reducing risks of mishaps in outer space.[25] Additionally, the Working Group on the Long-term Sustainability of Space Activities (WG-LTS) was set up by the UN COPUOS in 2010 to investigate possible measures to ensure the safe and sustainable use of outer space for peaceful purposes and for the benefit of all countries. The final product of that work was to compile best-practice guidelines and to have them endorsed at the 61st session of the UNCOPUOS in June 2018. Unfortunately, consensus was not reached. Work on WG-LTS guidelines was initiated under the chairmanship of France.

The role of states in security in outer space is pivotal to addressing sustainability issues, and the need for cooperation among countries is frequently restated in policy documents. In this respect, the French Ministry of Defence declared that "a European approach to this topic of mutual interest will be promoted, taking advantage of existing resources and developing new concrete projects".[26]

European states also conduct national feasibility studies, research projects and experiments on possible technologies that can contribute to space sustainability. Among others, the UK is particularly active on these matters. The University of Surrey coordinates the RemoveDEBRIS project, which is an international consortium aimed at developing deorbiting and capturing solutions for space debris.[27] Consortium members include Airbus Defence and Space (France, Germany and UK), ArianeGroup (France), SSTL (UK), ISIS (Netherlands), CSEM (Switzerland), Inria (France) and Stellenbosch University (South Africa). The Defence Science and Technology Laboratory (DSTL) launched the Daedalus project that will be tested in 2019 and has similar objectives.[28] The British company Clyde Space has its own deorbiting project called AEOLDOS (Aerodynamic End-of-Life DeOrbit System). The Italian company D-Orbit is developing debris removal solutions.[29] Other European legacy companies are also looking into these services at conceptual level.

---

[25]Note: Representatives of China, France, Russia, the UK and the USA as permanent members of the UN Security Council and representatives of Brazil, Chile, Italy, Kazakhstan, Nigeria, Romania, South Africa, South Korea, Sri Lanka and Ukraine.

[26]Ministère de la Défense. (2013). Defence And National Security. French White Paper. Retrieved from https://www.defense.gouv.fr/content/download/215253/2394121/White%20paper%20on%20defense%20%202013.pdf

[27]https://www.surrey.ac.uk/surrey-space-centre/missions/removedebris

[28]The IET. (2017). https://eandt.theiet.org/content/articles/2017/07/satellites-to-be-fitted-with-sails-to-take-them-out-of-orbit-and-prevent-space-junk/

[29]http://www.deorbitaldevices.com/

However, as in any field, pooling national resources into budgets allocated at European level remains the top line of action for greater coordination. As such, European institutions are also active in that domain.

5.3.2.2 European Institutions

The European Space Agency is significantly contributing through the development of more sustainable standards and technologies.[30] ESA's CleanSpace initiative is "a technology project aiming at developing the necessary technologies to support the compliance of future satellites with Space Debris Mitigation (SDM) requirements". The overarching goal is to develop an eco-friendly and sustainable approach to space activities through the development of industrial materials, processes and technologies that limit the environmental impact of space activity by addressing weaknesses and shortages in the entire life cycle of space systems from conceptual phase to end of life and disposal. ESA's CleanSpace revolves around three main core activities:

- *EcoDesign*: deals with environmental impact assessments, compliance with legal frameworks and the development of green technologies.
- *CleanSat*[31]: deals with the development of technical solutions and standards to mitigate space debris proliferation. It is the leading programme of ESA in the field of Space Environment Protection and Preservation.
- *E.Deorbit*: deals with active debris removal and associated capabilities such as target characterisation, capture mechanisms and disposal methods.

International cooperation is continuing, mainly under the UN Committee on the Peaceful Uses of Outer Space (UN-COPUOS) and through the European Space Weather Week conference organised annually. ESA has also been chairing the Space Missions Planning Advisory Group (UN-SMPAG) and co-organised the Planetary Defence Conference. Finally, the Space Debris Conference 2017 also organised by ESA provided an excellent opportunity to address the very closely related SST aspects.

In Brussels, the European Union's political discussions on how to sustain a safe to operate in space environment have been growing steadily in recent years. When the European Union became an active actor in space, developing a strategy at global level for Space Security appeared to be the next necessary step. In this respect, the most sophisticated diplomatic initiative was the ill-fated International Code of Conduct for Outer Space Activities (IcoC), a non-legally binding and voluntary-based document aimed at regulating outer space activities. More specifically, the Code sought "to enhance the safety, security, and sustainability of all outer space

[30]Jessica. (2017) The clean space blog. Retrieved from ESA: http://blogs.esa.int/cleanspace/2017/02/03/cleansat-an-exciting-opportunity-for-the-european-space-industry/

[31]ESA. Clean Space: Cleansat. Retrieved from ESA: http://www.esa.int/Our_Activities/Space_Engineering_Technology/Clean_Space/CleanSat

activities" and to "establish transparency and confidence-building measures, with the aim of enhancing mutual understanding and trust and helping both to prevent confrontation and foster national, regional and global security and stability".[32]

The EU has played an important role in other diplomatic initiatives that include WG-LTS negotiations at UNCOPUOS, the GGE, the UN General Assembly Resolution 71/42 regarding the Prevention of an Arms Race in Outer Space (PAROS), projects related to the "Prevention of the Placement of Weapons in Outer Space, the Threat or Use of Force against Outer Space Objects" (PPWT) and "No First Placement of Weapons in Outer Space" (NFP) and the Principles of Responsible Behaviour for Outer Space (PORBOS) that are also designed to prevent an arms race in outer space[33] among others.[34]

Interestingly, the European approach to SEPP comprises different strategies that interact with each other for greater efficiency. Moreover, these activities are also complemented by a series of measures in the field of Space Infrastructure Security.

5.3.3 *Space Infrastructure Security*

The focal point of Space Infrastructure Security is the protection of the space infrastructure from threats that diminish its proper functioning. More precisely, it covers the exploitation of space systems' vulnerabilities and can be unintentional or intentional. Space Infrastructure Security includes security-by-design, programme security architecture and supply chain security covering the field of engineering, safety standards and regulations.

As European approaches to space security first developed at national level, member states all have their own objectives and ambitions for the protection of their assets. In the field of Space Infrastructure Security, a number of small projects are conducted by national space agencies. A few can mentioned such as the upgrading of the French Syracuse telecom satellites that include an anti-jamming system.[35] More anecdotic projects can also be mentioned such as the Robusta

[32]Code of conduct Working Document 21 (2014). Draft International Code of Conduct For Outer Space Activities. Retrieved from https://eeas.europa.eu/sites/eeas/files/space_code_conduct_draft_vers_31-march-2014_en.pdf

[33]EEAS (2017). Conference on Disarmament – Working Group on the "Way Ahead" – EU Statement on the Prevention of an Arms Race in Outer Space. Retrieved from European Union-External Action: https://eeas.europa.eu/headquarters/headquarters-homepage/28329/conference-disarmament-working-group-way-ahead-eu-statement-prevention-arms-race-outer-space_en

[34]EEAS (2016). Conference on Disarmament – Working Group of the "Way Ahead". Geneva. Retrieved from European Union External Action: https://eeas.europa.eu/headquarters/headquarters-homepage/28329/conference-disarmament-working-group-way-ahead-eu-statement-prevention-arms-race-outer-space_en

[35]Retrieved from CNES (2018): https://syracuse4.cnes.fr/fr/syracuse-4

programme that is a student led nanosatellite project aimed at studying the effects of solar flares on bipolar electronic components.[36]

Member states are also playing a role in Space Infrastructure Security through the involvement of their industries. Indeed, defining security standards and requirements is also to a large extent determined at industry level, although the standardisation of these requirements has not yet been reached. For instance, Leonardo, an Italian aerospace company, was selected for a data security project on the Galileo constellation.[37]

In developing security protocols and technical standards, ESA has played an important role in security in outer space. ESA's involvement in security in outer space includes:

- Definition of a workable regulatory framework encompassing security agreement, security regulations and implementing procedures and facilities to receive, store and produce classified information to set up a regulatory framework (Security Agreement, Security Regulations and Implementing Procedures and Facilities), to build a capability to receive, store and produce classified information and exchange classified information with third parties.
- Development of standards for space project management, assurance and system engineering. For example, a branch related to space sustainability was added to the ECSS set of standards incorporated in ESA projects.
- Establishment of the European Space Security and Education Centre at Redu (ESEC) as a centre of excellence for space cybersecurity services.
- Contributions to the security aspects of EU space programmes such as Galileo and Copernicus.
- ESA's participation in international fora (IADC, IAC, COSPAR, IADC, IAA, UNCOPUOS).

ESA is also supporting R&D in the field of security in outer space through the funding of projects.

Similarly, the European Union funds a great number of research projects in that field but also conducts its own Space Security activities, mostly related to the EU flagship programmes. A good example is the establishment of an independent Security Accreditation Board (SAB) within the European GNSS Agency (GSA) to verify the compliance of the programme with applicable security protocols and regulations.[38] Additionally, the Galileo Open Service will soon feature an Open Service Navigation Message Authentication (OS-NMA) option to address jamming and spoofing attacks and a Galileo High Accuracy Service to enhance the quality and resilience of signals and the Open Service. The European Union is actively supporting Space Infrastructure Security through the funding of R&D projects in

---

[36]Retrieved from CNES: https://robusta.cnes.fr/fr/ROBUSTA/Fr/index.htm

[37]Leonardo. (2017) Press release: http://www.leonardocompany.com/en/-/esa-galileo-cyber

[38]European GNSS Agency (2016). Security Accreditation. Retrieved from European GNSS Agency: https://www.gsa.europa.eu/security/accreditation

that subdomain in member states. Considerations regarding resilience are seriously treated by the European Union, for which space resilience is a key objective. Its role can be expected to evolve with the greater integration of the Galileo system into the European economy. In this respect, the European Commission recently issued a proposal for the setting-up of a new EU Agency for the Space Programme. Accordingly, an increased budget share will be dedicated to defence spending on R&D and procurement, which also means more opportunities to strengthen the EU's involvement in that particular field.

Security in outer space is being addressed in Europe at different levels. The member states are actively conducting R&D activities in the field of SSA, SEPP and SIS to protect their space assets. These efforts are to be considered in a broader context, as a significant share of national budgets is being invested in European endeavours. ESA, acting as an intergovernmental body, is pooling such efforts under the umbrella of its SSA programme but also remains a proactive player in the field of SEPP and SIS. The agency's scope of action is also up for further interpretation, should the institution take on a more prominent role in these matters. Additionally, the European Union, as a supranational actor, is becoming increasingly involved in Space Security, as it operates its own constellations. Other policy developments indicate the EU's willingness to increasingly incorporate security policy considerations in its strategy and will likely take a more instrumental role in conducting security-related projects for that reason.

5.4 Toward an Enhanced Role of Europe in Security in Outer Space

5.4.1 Key Elements for Consideration

Security in outer space is intrinsically a complex issue because of the diversity of potential threats to the various European space infrastructures and to the provision of associated services, as well as of the complexity of mitigation or remediation measures.

As far as Europe is concerned, the multiplicity of stakeholders brings an additional layer of complexity. In one way or another, this issue impacts all European space actors. Therefore, they would undoubtedly benefit from sharing experiences and concerns, as well as from pooling resources to, at the minimum, cope with the security challenges that are transverse to all space systems, such as the proliferation of space debris or space weather hazards. This encompasses:

- Private operators and public bodies
- Civil and military entities
- National agencies, intergovernmental and supranational organisations

What is first at stake is the capacity of Europe to efficiently mobilise the substantial resources required to deal with this complex issue, which implies:

- Reaching a broad political consensus on the objectives to be set for a "joint" European security in outer space policy. Such policy should be based on a joint overarching threat analysis and on internationally recognised standards.
- Devising, accordingly, technological roadmaps for the development and the maturation of required technologies and systems, at affordable conditions, and with the appropriate level of European autonomy.
- Assessing the associated programmatic provisions and schedules in all available or potential sources of funding (MFF, Security Fund, national and ESA programmes, private funding).
- Achieving a sufficient level of coordination of activities among various stakeholders in order to avoid gaps and useless duplication of efforts.

Second comes the implementation of such policy, which implies the definition of agreed governance schemes building on the existing areas of expertise readily available throughout Europe, as well as on the integration of established best practices and respective competencies of key stakeholders. These governance schemes shall, for each system concerned, propose an appropriate organisation and provide for the breakdown of responsibilities among the various entities in the processes, aimed at ensuring:

- The definition of security requirements based on specific threat analysis
- The setting up and implementation of risk management schemes
- The elaboration and implementation of operational procedures
- The monitoring of compliance of operations
- The coordination of the various entities

A central challenge for these governance schemes will be to find the appropriate mechanism to accommodate a more prominent European leadership for which the European Union is the most suitable candidate, further building on the existing centres of expertise and the need for member states to maintain control over national systems and data. Notwithstanding the considerable progress achieved by Europe in both space and security and defence domains over the period 2014–2020, an in-depth reassessment of stakeholders' views, in particular with regard to the respective roles of member states and the EU in space-related matters, would be timely and opportune in order to move further.

With a view to building a reinforced, comprehensive, coherent and cooperative European approach to security in outer space, various sensitive issues are at stake:

- Capacity building for SSA (integrating all related components: SST, SWE and NEO)
- Space Environment Protection and Preservation including, in particular, relevant technology developments, common standards and guiding principles for diplomacy and international cooperation

- Space programme security architecture and risk management procedures for European space programmes

The success of the European approach to security in outer space will rely on key enabling factors:

- Setting up consensual European leadership in a domain that, so far, has been driven primarily by member states. For that, two scenarios can be envisaged:
  - Bottom-up approach building on existing intergovernmental frameworks (in particular ESA and the EU SST consortium) and seeking a progressive enlargement of partners and scope toward, first, a comprehensive SSA framework and, ultimately, the construction of a European Space Traffic Management structure
  - Top-down approach with the introduction of a full-fledged programme establishing the European Union as the main owner and operator of European civil SSA capacities with the necessary reinforcement of security protocols and risk management
- Development and distribution of key industrial capacities and adequate involvement of industry as a core contributor to capacity building in the following strategic areas of activity (i.e. such involvement would require the setting up of stringent security protocols and should be conditioned on the expected benefits in terms of cost savings, improved competitiveness and business development, including on international markets):
  - Data production, relying on privately operated sensors and processing capabilities and reinforcing/substituting governmental capacities in carefully selected fields of a lesser sensitivity such as the GEO orbit, space weather or NEO.
  - Value-added services, elaborating on publicly available data, possibly complemented by private data, for governmental and commercial users. Ultimately, such services could be embedded in an EU programme to augment SSA services, including for security and defence governmental users.
  - Standards, for in-orbit operations and spacecraft design supporting a European approach to the preservation of a safe and secure space environment.
- Sharing of clearly identified ambitions to achieve European autonomy and freedom of action in space and, subsequently, mobilisation of the required resources. As the crux of every political issue, the level of funding allocated by the different stakeholders including the European Union, member states and ESA will be a determining factor of Europe's capacity to meet the rising challenges in the field of security in outer space. Here, two aspects shall be considered:
  - Use of complementary public funding including, in particular, the new European Defence Fund which is a particularly appropriate mechanism to support EU and national efforts for development and acquisition of systems.

  - Leverage private investment, provided that more prominent involvement of private actors is sought.
- International diplomacy and cooperation, in synergy with European efforts in other areas, for both:
  - Operational concerns about the way forward to an international space traffic management framework (e.g. data sharing, in-orbit operations, SWE and NEO warnings)
  - Fostering of a global approach to space activities in line with Europe's vision on the preservation of a safe and secure space environment and for the peaceful use of outer space

5.4.2 Preparation of a New Framework for 2021–2027

An important step toward the preparation of a renewed European framework for security in outer space lies in the current preparation of the future Multiannual Financial Framework 2021–2027 of the European Union, which will set a number of important parameters for midterm developments in this domain including ambitions, means and organisation. To provide a baseline for negotiations with member states, the European Commission issued in June 2018 a draft regulation for the Space Programme of the European Union.[39] When finalised and adopted,[40] the new regulation will repeal previous decisions for EGNOS/Galileo, Copernicus and the SST support framework covering the period 2014–2020 and provide a legal basis for activities on the period 2021–2027.

Stakes are high.

From a general standpoint, the draft regulation proposes the integration of the different space programmes of the EU into a single EU programme with several components (governed by a single regulation), the horizontal extension of GSA mandate to cover executive and security activities for all components of the integrated programme (GSA to become the EU Agency for the Space Programme) and an increase of the overall programme budget to EUR 16 Billion (segmented as follows: EGNOS/Galileo: EUR 9.7 billion, Copernicus: EUR 5.8 billion, SST and GOVSATCOM: EUR 0.5 billion). The regulation establishes that "the Commission shall implement the Union space policy and shall assume responsibility for the Union space programme, *including in the field of security*. Even though tasks related

---

[39]European Commission. (2018). Proposal for a Regulation of the European Parliament and of the Council establishing the space policy programme of the European Union, relating to the European Union Agency for Space and repealing Regulations (EU) No 1285/2013, No 377/2014 and No 912/2010 and Decision 541/2014/EU. Brussels.

[40]Note: at the time of this report publication, the draft regulation is still under negotiation with member states.

to security (e.g. accreditation, SSA, rules and standards implementation, etc.) can be entrusted to other entities such as ESA, member states or agencies of the EU, the European Commission is ultimately accountable for the security of the EU space programme. This responsibility encompasses:

- The protection of infrastructure, both ground and space, and of the provision of services, particularly against physical or cyberattacks
- The control and management of technology transfers
- The development and preservation within the Union of the competence and know-how acquired
- The protection of sensitive non-classified and classified information

5.4.2.1 Space Infrastructure Security

The draft regulation provides a number of provisions related to the security architecture of the EU space programme (Chap. 5) including security activities and governance (Chap. 1), security accreditation (Chap. 2) and management of classified information (Chap. 3). Overall, the draft regulation establishes that the EC shall determine policy measures to ensure a high degree of security for each component of the EU space programme and stresses the importance of building on member states experience and expertise in this field. Here, the most substantial proposed evolution with regard to the 2014–2020 framework is related to the reinforcement of security accreditation procedures, in particular with regard to independence from operational functions of the programme and to the enlargement of GSA role across all components of the programme (so far limited to Galileo and EGNOS programmes).

Noticeably, measures to organise a European response to military threats or deliberate attacks are not addressed by the regulation. At this point in time, such move would be quite premature since it implies some prerequisites, including a European space defence doctrine or a formal governance scheme to deal with such issues. This shall be addressed in the framework of the further development of the EU's Foreign and Security Policy and European Defence Action Plan.

5.4.2.2 Space Environment Protection and Preservation

Space Environment Protection and Preservation is not directly addressed in the draft regulation with concrete measures and activities (e.g. standards, technology developments, etc.); however, a number of mentions delineate the vision of the European Commission in this field:

- Consistency with existing space hazards mitigation measures and with an initiative leading to a non-legally binding instrument to be negotiated within the framework of the United Nations[41]
- Monitoring of international initiatives and developments in the area of the space traffic management to be taken into consideration in the context of the midterm review of the MFF
- Synergies of SST with initiatives of active removal and passivation measures of space debris with a view to reducing risks of collision

With these mentions, the regulation highlights the importance of these topics for Europe but does not elaborate on the ways to ensure that EU will be in a position to weigh in space-related international negotiations in order to effectively contribute to future potential initiatives in the field of space diplomacy (United Nations instrument), Space Traffic Management (monitoring of international developments) and active/passive debris removal.

5.4.2.3 Space Situational Awareness

Space Situational Awareness is proposed as a component of the EU space programme supporting "a global approach toward the main space hazards" and includes activities in the field of SST, space weather and NEOs. This SSA component aims:

- "to enhance SST capabilities at the EU level, reduce the risks of spacecraft collisions during all its operational phases from the launch to the decommissioning, survey uncontrolled re-entries of spacecraft or space debris in order to provide early warnings to mitigate damage to EU citizens and terrestrial infrastructure and seek to prevent proliferation of space debris;"
- "to monitor the relevant observational parameters related to space weather events"
- "to establish and inventory of NEO European capacities, to network them with an aim of supporting the exchange of NEO data and information".

The introduction of an SSA component within the EU space programme with a dedicated budget (corresponding to a share of the 500 M€ allocated to GOVSATCOM/SSA) represents a substantial step toward a more ambitious and integrated capacity-building effort. This new SSA component will build on past activities in the field of SST (essentially on the EU SST support framework) and in the fields of Space Weather and NEOs (essentially on Space Weather prediction

[41] As a way to foster increased international cooperation, establishing standards of responsible behaviour across the full range of space activity, strengthening commitments to non-interference in the peaceful exploration and use of outer space, facilitating equitable access to outer space and increasing transparency of outer space activities.

services for GNSS, H2020 grants for R&D in these fields and on a number of complementary activities).

With regard to SST, a few changes are proposed. The most noticeable change lies in the enlargement of the scope to encompass a financial support to the development of new sensors (i.e. today, the SST support framework should only "encourage" the development of new sensors and financial support should be addressed either nationally or through a European research and development programme). The SST component shall, now, "support the establishment, development and operation of a network of ground and/or space-based sensors of the Member States, including sensors developed through ESA and EU sensors nationally operated". The notion of "EU sensors nationally operated" is further developed in the introduction of the draft regulation: "in case new assets are financed by the programme, the Union should be co-owner of these assets with due consideration of various ownership and governance models". The possibilities of co-owned sensors and of EU sensors operated by member states are the main steps forward in the direction of a more prominent role of the EU in this field. European leadership could also be supported by additional changes in the governance structure that could arise from future arrangements revision.

5.4.3 Proposed Further Developments in Light of ESPI Conclusions

In its current state, the proposed regulation does not introduce radical changes in the domain of security in outer space but still makes some noticeable steps toward the consolidation of a European programmatic approach. Although much is left to implementing arrangements to be established or revised accordingly by the different stakeholders, the regulation certainly provides for a more cohesive framework in a number of important areas.

Overall, the increasing importance awarded to security features of the EU space programme is striking. In this regard, the draft regulation fits well within the strategy framework set forth by the European Union in both space (Space Strategy for Europe) and security and defence sectors (Global Strategy for the European Union's Foreign and Security Policy, European Defence Action Plan) which call for further developments in the domain of security in outer space. However, if the current strategy clearly sets a number of key principles, it does not provide for concrete objectives. In this respect, setting up a comprehensive European policy in the domain of space security seems a necessary step to ensure smooth and controlled implementation. Such policy should clarify the approach to a number of pressing issues.

5.4.3.1 European Ambitions and Means

First, the policy should establish clear, measurable, objectives to be achieved in the short-medium- and long-term. In particular, it should:

- Clarify the notion of "European autonomy and freedom of action".
- Set the required level of autonomy for Europe in these matters.
- Set relevant standards applicable to civil and military systems, respectively.
- Prioritise the associated developments of critical capabilities.

Accordingly, the overall effort to achieve these objectives should be translated into a broadly agreed roadmap allowing for budget planification and phased scheduling of developments. Actually, roadmapping of activities is necessary to avoid a "Go As You Pay" approach, which is not appropriate for a controlled deployment of capacities.

Considerations for the Next MFF
• The introduction of a multiannual plan defining clear objectives for the SST component is an important feature of the proposal. Eventually, the European Commission shall endorse this plan as part of its responsibility over the security of the EU space programme. • The level of budget included in the EC proposal for the period 2021–2027 will increase available resources. However, additional funding from Horizon Europe (i.e. FP9) and other instruments, such as the European Defence Fund, will be essential to reach the level of funding necessary to progressively bridge the capability gap through the deployment of additional EU sensors. • Avoiding unnecessary duplication of efforts and fostering complementarity should be set as core principles in order to optimise budget efficiency.

5.4.3.2 Governance

The governance scheme that has been set through the Consortium has allowed to accommodate the requirements of different communities by recreating the conditions of an intergovernmental model within the supranational framework of the EU. In such governance scheme, the role of the European Commission is rather limited, even though it is the sole funding source and holds the responsibility over the security of the EU space programme. In essence, the intergovernmental model is meant to mitigate two conflicting objectives:

- On the one hand, leverage cooperation among the most motivated member states to foster a regional approach for the sake of the protection of the European Union space assets.
- On the other hand, preserve national interests by safeguarding national sovereignty over capabilities development and control over SST systems and data.

The argument for an intergovernmental model also lies in the inherent dual nature of SST and in the discrepancy between member states' capabilities and strategic interests across Europe. Indeed, the model promotes cooperation between capable states and use/upgrade of existing national capabilities (systems, know-how, facilities) resulting from prior investments of a handful of European countries to protect, first and foremost, their national/military space assets. This situation, which cannot be overlooked, is at the core of the governance debate which eventually boils down to the question of the weight of national concerns and interests against European added-value and to the sharing of responsibilities between member states and the EU.

The draft regulation proposes to maintain this model, seeking enlargement of contributing member states and progressive reinforcement of the EU role through co-ownership of new sensors, along with revision of the governance scheme and the introduction of a multiannual plan monitoring key performance indicators. Although these are necessary steps forward on the way to a more prominent European leadership, such intergovernmental scheme might reach its limits with specific risks of:

- *Divergence of interests among stakeholders*, hindering the capacity to implement a coordinated policy in the field of security in outer space
- *Duplication of efforts and reduced cost-effectiveness*, if motives to develop specific national capabilities surpass the willingness (and readiness) to focus on distribution and complementarity across Europe

To tackle these issues, and with the announced objective to involve more contributing states (including some with limited SST capacities or that may not operate satellites, but have a national space agency, or a security protocol of data processing and sharing), there is a pressing need to build further on EU added-value and reinforce European leadership, management and coordination. This might imply the introduction of some supranational provisions giving mandate to the EU in specific areas, in particular to negotiate international arrangements. Interestingly, SST is not the only domain where EU member states seek improved cooperation within the Union framework and new instruments such as the Permanent Structured Cooperation on Defence (PESCO) introduced in 2017 may provide new ways to better accommodate national interests and EU added-value.

Anyhow, getting close to the 2020 turning point, a reassessment of member states concerns and of the role that the EU intends to play in the field of space security would be timely. Such discussion should take into account the substantial progress:

- Of EU space programmes
- Of confidence-building among consortium partners
- SST systems networking and service delivery
- Of the most appropriate institutional framework (including GSA, ESA, national agencies, etc.)
- And of the developing mandate and new instruments of the EU in security and defence

Considerations for the Next MFF
• Working on a clear identification of each party's concerns and requirements would certainly support the identification of an appropriate solution accommodating national/European and military/civil perspectives. • An enhanced role of the EU, with specific mandates, could be considered to reinforce European leadership, in particular on the international scene. • A Permanent Structured Cooperation for security in outer space, supported by the European Defence Fund and allowing future participation of third states (e.g. UK, USA), could be considered.

5.4.3.3 Exploitation

Smooth and efficient exploitation of such sophisticated and critical systems requires a number of conditions:

- Adequate and stable budget properly assessed according to clearly identified operational objectives.
- Pooling of expertise available either at national or European institutional level. This raises in particular the question of the involvement of ESA, which has invested for its own needs in these matters and has gained substantial internal expertise. It remains to be seen how such capabilities could be exploited in a closely military-related environment and how it might interact with the proposed enlargement of the GSA.

5.4.3.4 Role of the Private Sector

Another important consideration is the involvement of the private sector as a user and provider of SSA capabilities but also as an industry whose business and competitiveness is impacted by developments in the field of security (e.g. standards, regulations). With the intensification of global space activity and an expected growth of the space economy, the provision of SSA data and services will likely become a new sizeable commercial market. This trend is marked by a growing number of SSA private providers. For example, ExoAnalytic Solutions, Rincon, Lockheed Martin, LeoLabs or ArianeGroup sell data produced by private sensors and Analytical Graphics, Boeing, Schafer Corp. and Applied Defence Solutions use commercial SSA data to develop and sell value-added services to governmental and commercial operators.[42]

So far, and despite noticeable national developments (e.g. integration of ArianeGroup's GEOTracker service by France), the involvement of industry in space security matters does not seem to be perceived as a priority in Europe. In the

[42]Weeden, B. (2017). Space Situational Awareness *Fact Sheet.* Retrieved form Space World Foundation: https://swfound.org/media/205874/swf_ssa_fact_sheet.pdf

meantime, the USA adopted a policy encouraging the involvement of private companies. The rationale behind this move is to:

- Give way to a potentially promising commercial market.
- Foster competitiveness from supposedly more cost-effective economic agents.
- Avoid diverting valuable public assets from their strategic missions.

Such approach supposes a strict delineation of the limits of the respective roles of public (military) actors focusing on strategic security-related activities and of the private actors providing services on a commercial basis in a number of less-sensitive areas. From this perspective, on the European side, the elaboration of a comprehensive space-security policy is probably a prerequisite to a clear identification of the potential role of the European industry in the security in outer space, taking into account impacts of future developments (e.g. SSA services, security standards, STM, etc.) on business.

However, advanced reflections should be initiated given the short-term risk of emergence of a de facto US industry monopoly on these services, or the creation of parallel SST data channels, making EU efforts redundant. From this standpoint, involving the private sector is at the crossroads of space security and industry competitiveness objectives.

Considerations for the Next MFF

• In the short term, a first step to favour a greater involvement of private actors could consist in the establishment of an enhanced dialogue with commercial stakeholders, for example, in the frame of the preparation of the SSA multiannual plan with the objective to:

– Map European private capabilities.
– Evaluate business opportunities and challenges in this domain.
– In general, assess the needs and views of private industry on potential future developments in this domain.

Chapter 6
Security in Outer Space: Perspectives on Transatlantic Relations

6.1 Introduction

6.1.1 Rationale

In the 60 years since the launch of the first satellite in 1957, the global space economy has grown tremendously.[1] Space-based services now bring substantial socio-economic benefits to end-users, across numerous strategic domains and economic sectors.[2] Simultaneously, the increasingly pervasive use of space-based services by governments, businesses and consumers has resulted in a growing dependence of society and economy on space infrastructure.

In Europe, the significant progress made by Galileo, EGNOS and Copernicus programmes and the introduction of new initiatives in the field of GOVSATCOM have further emphasised the importance of ensuring that all conditions for the delivery of operational space-based services conforming to user needs are met. This is a prerequisite to fostering the uptake of space services and, consequently, maximising the benefits of the space infrastructure. This is also a key area of space policy development in light of the growing dependence of the European economy and society on space assets, including critical infrastructure such as telecommunications, energy and transport. The conditions to ensure an appropriate quality of service, include a proven or certified level of performance, long-term availability of services and a secure service that can justifiably be trusted. From an infrastructure standpoint, these conditions translate into (1) operational capacities that meet user performance requirements, (2) continuity of programmes to ensure infrastructure

[1]Space Foundation (2018). *The Space Report 2018: The Authoritative Guide to Global Space Activity.*

[2]European Commission (2017). Dependence of the European Economy on Space Infrastructures: Potential Impacts of Space Assets Loss, (March 2017).

E. Burger, G. Bordacchini, *Yearbook on Space Policy 2017*, Yearbook on Space Policy, https://doi.org/10.1007/978-3-030-05417-5_6

maintenance and upgrade, and (3) appropriate measures to protect the infrastructure against threats. The last encompasses multiple and transverse aspects covering:

- Access to space to ensure an autonomous capacity to deploy space systems safely
- The integrity of the space segment, including all systems in orbit
- The ground segment that includes stations and equipment for the operation and delivery of services
- Downlinks and uplinks used to operate the space system and receive its data

Each of these components is vulnerable to a range of security concerns (passive man-made hazards, active man-made threats and natural hazards). Experts have routinely cautioned governments and space businesses that space is increasingly vulnerable, posing an intensifying challenge to guaranteeing a safe, secure and sustainable environment to deploy, operate and exploit space assets.

The deteriorating security situation has been acknowledged by the European Union, which made "the protection and resilience of critical European space infrastructure" a flagship objective of the Space Strategy for Europe adopted in October 2016.[3] The strategy underlines the objectives of the EU to enhance the security of its Space Programme, ensure the protection and resilience of critical European space infrastructure (in particular through the development of Space Situational Awareness (SSA) capabilities) and reinforce synergies between civil and security space activities. For these objectives, the EU stresses the importance of cooperation, first at European level with ESA, EUMETSAT and member states but also at international level, in particular with the USA. The importance given to space security and international cooperation in this field is shared by the European Union External Actions Services (EEAS) and is addressed in the Global Strategy for the European Union's Foreign and Security Policy and in the European Defence Action Plan. These elements were again addressed in the draft regulation for the Space Programme of the European Union published in June 2018[4] by the European Commission in preparation for the next Multiannual Financial Framework (MFF) 2021–2027. This document proposes several provisions in line with the existing strategic framework and recalls that "[Space Situational Awareness] should have regard to cooperation with international partners, in particular the United States of America".[5]

Across the Atlantic, in the USA, space has always had a prominent defence and national security dimension, and securing space assets has long been a strategic

[3]European Commission (2016). Space Strategy For Europe. COM (2016) 705 final. Brussels: European Commission. Retrieved from https://ec.europa.eu/transparency/regdoc/rep/1/2016/EN/COM-2016-705-F1-EN-MAIN.PDF

[4]European Commission. (2018). Proposal for a Regulation of the European Parliament and of the Council establishing the space policy programme of the European Union, relating to the European Union Agency for Space and repealing Regulations (EU) No 1285/2013, No 377/2014 and No 912/2010 and Decision 541/2014/EU. Brussels.

[5]European Commission (2018). 'Establishing the space programme of the Union and the European Union Agency for the Space Programme and repealing Regulations (EU) No 912/2010, (EU) No 1285/2013, (EU) No 377/2014 and Decision 541/2014/EU', COM(2018) 447 final (June 6, 2018).

priority, as witnessed by the development of an ambitious Space Situational Awareness (SSA) programme by the US Department of Defense (DoD). This programme relies on an extensive and advanced network of surveillance sensors[6] including space- and ground-based radars and optical telescopes operated by military and civilian entities, together making up the US Space Surveillance Network (SSN). The outstanding tracking capabilities of this network are expected to be further improved by the end of 2018[7] as part of the US Space Fence programme.[8] Noticeably, the USA has adopted a model of global SSA cooperation with close to 100 unclassified SSA sharing agreements with commercial and international partners,[9] which also supports US capabilities by improving the quality and quantity of SSA information. More recently, a major development in the USA approach to security in outer space was the adoption of Space Policy Directive-3 (SPD-3), which establishes the first "approach to Space Traffic Management (STM) that addresses current and future operational risks."[10] The policy also announced that the US Department of Commerce (DoC) would be responsible for relations with commercial and international partners for STM and SSA, shifting some of the responsibilities now handled by US DoD.

Recent developments in space security in Europe and the USA have reshuffled the cards at strategic, policy and operational levels. In this context, an examination of the new state of affairs on each side of the Atlantic and of mutual transatlantic relations seems timely. More specifically, this research aimed to provide elements of an answer to the following questions:

- What is the state of affairs in space security in Europe, the USA and transatlantic relations?
- How might recent developments impact transatlantic relations in the short- to long-term?
- How can partners revisit or reinforce cooperation frameworks for mutual benefit?

[6]Space Traffic Management: Towards a Roadmap for implementation, IAA Cosmic Study (2017) 80.

[7]For completeness, it should be noted that the current Space Fence project is to replace the previous Air Force Space Surveillance System, one of the SSN components active from 2008 to 2013.

[8]The Space Fence will utilise S-band ground-based radars. http://www.lockheedmartin.com/us/products/space-fence.html

See also: http://www.airforcemag.com/MagazineArchive/Pages/2017/August%202017/A-Closer-Watch-on-Space.aspx

[9]U.S. Strategic Command and Norway sign agreement to share space services, USSTRATCOM Public Affairs, April 2017 http://www.stratcom.mil/Media/News/News-Article-View/Article/1142970/us-strategic-command-norway-sign-agreement-to-share-space-services-data/

[10]Space Policy Directive-3, National Space Traffic Management Policy, (June 18, 2018).

6.1.2 Objectives of the Research

The overarching objective of the research was to raise awareness of the key stakes in transatlantic relations in the field of space security by providing useful perspectives on recent and potential future developments in this domain. Specifically, the research aimed to achieve the following:

- Review the US and European approaches to address challenges to space infrastructure security, including:
  - Strategic and policy rationales, key priorities and objectives
  - Activities and capabilities in the field of security in outer space
- Investigate past and current transatlantic relations in the field of space security and analyse the conditions to reinforce cooperation in this domain.
- Examine the potential implications of ongoing policy developments on future transatlantic relations in space security.

With the objective of providing a common understanding of space security challenges and analysing a potential way forward in transatlantic relations in security in outer space, the research aimed to provide an assessment of common space security challenges, followed by a common understanding of each partners' priorities, insights on the respective approaches to the issue and ultimately sound evaluation of drivers and obstacles to cooperation.

6.1.3 Research Scope and Key Concepts

"Space Security" is understood here primarily as "Security in Outer Space"—the protection of human life and the Earth environment against natural threats and risks coming from space. Given the strong interrelation between the three dimensions of space security, "Outer Space for Security" (the use of space systems for security and defence purposes) and "Security from Outer Space" (the protection of human life and the Earth environment against natural threats and risks coming from space) were addressed sporadically when relevant.

"Security in Outer Space" itself encompasses three areas of action that were investigated and that correspond to three core objectives: (1) monitor the space environment, (2) mitigate threats to space infrastructure and (3) reduce vulnerability of space infrastructure.[11]

[11]Measures directly targeting the source of the threats, such as actions against cybercrime or signal jamming, are not included this report. Measures that are external to the space sector, such as actions against cybercrime, disarmament policies, and radio spectrum management are not addressed here. By the same token, broader security and defence matters such as strategic stability in outer space,

- *Space Situational Awareness (SSA)*: Current and predictive knowledge and understanding of the outer space environment including space weather and location of natural and manmade objects in orbit around the Earth
- *Space Environment Protection and Preservation (SEPP)*: Preventive and curative mitigation of the negative effects of human activity in outer space on the safety and sustainability of the outer space environment
- *Space Infrastructure Security (SIS)*: Assurance of infrastructure ability to deliver a service that can justifiably be trusted despite a hazardous environment

Space infrastructure can be described as a network of space-based and ground-based systems interconnected by communication channels and enabled by access to space capabilities. It includes a space segment, a ground segment, a user segment and downlinks and uplinks to interface between these segments. As the present research focuses on "Security in Outer Space", the analysis addresses predominantly security challenges to the "space segment" and, whenever relevant, to the downlink and uplink data. However, the research does not address security challenges specific to ground and user segments, such as Earth natural hazards, physical attacks on facilities or eavesdropping. This research focuses primarily on security challenges to space infrastructure in operation, including system disposal. Security challenges at earlier stages of Space Programmes (e.g. security-by-design, supply chain security) are mentioned whenever relevant but have not been investigated specifically. By extension, the security architecture of Space Programmes (e.g. security accreditation, threat response architecture), which requires to be adapted to the specificities of each infrastructure, is not addressed in the present work, which focuses on common and external dimensions of security in outer space.

6.1.4 Methodology

The research has followed a methodology based on a comparative analysis of the European and American approaches to security in outer space, a review of current transatlantic relations in this domain and the identification of drivers and barriers to a reinforced framework. The research has been supported by an extensive literature review, face-to-face interviews of key stakeholders in Europe and in the USA and a review of the progress by a board of recognised international experts. An overview of the general methodology is provided below (Fig. 6.1):

The review of European and US approaches aimed to disentangle the strategy and policy frameworks, in order to identify elements that influence each partner's attitude to cooperation/coordination. The analysis is based on the collection and review of relevant strategy (i.e. general long-term direction) and policy

deterrence and space superiority, although exceedingly central in the posture of many space powers, will not be directly covered by the report.

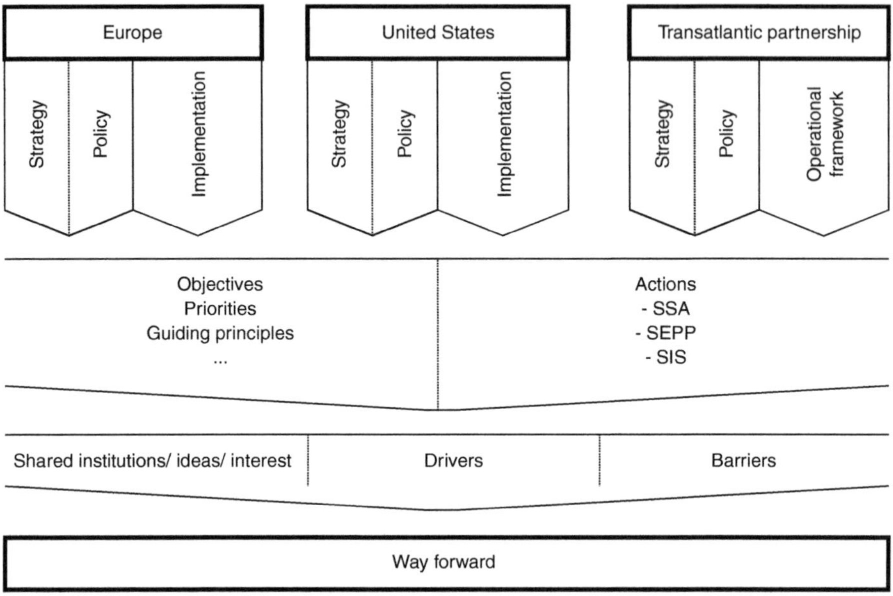

Fig. 6.1 General approach to the problematic

(i.e. statement of position directing responses or actions) documents, upon which activities and decisions are based.

The analysis of strategy and policy frameworks was complemented with a comprehensive overview of implementing activities across the three areas of security in outer space (SSA, SEPP, SIS) taking into account three fields of action:

- Capacity-building programmes
- Legal and regulatory regimes
- Diplomatic and cooperation frameworks

The analysis of European and US approaches was completed with a review of current transatlantic relations and cooperation frameworks. The analysis then focused on the assessment of space security transatlantic collaboration with the overarching aim of supporting the identification of key areas of reinforced cooperation/coordination. This assessment was performed through:

- An analysis of the integration of transatlantic cooperation in respective approaches to space security
- A review of the achievements of, and limits to, current transatlantic cooperation frameworks
- The identification and analysis of drivers of and barriers to reinforced transatlantic cooperation at the strategic, political and operational levels

For this last point, an assessment of the way forward was conducted through a significant number of interviews both in Europe and in the USA. The information collected from the interviewees listed below helped shape the findings and later recommendations by identifying potential areas for reinforced cooperation. The results of the previous analytical work and of the discussions with stakeholders across the Atlantic paved the way for the preparation of main takeaways in the context of space security from a transatlantic perspective and the identification of key areas and potential options for transatlantic cooperation development in the field.

6.2 Transatlantic Relations in Security in Outer Space

6.2.1 A Privileged Partnership

Historically, Europe and the USA share strong roots of value-based partnership presented through ties in various sectors. In the field of economy and trade, they are the two largest economic blocks in world economy, considered as close trading partners in both export and import businesses. When it comes to security and defence, the NATO framework was joined by a number of European countries. Additionally, a great deal of bilateral cooperation agreements was signed between European countries and the USA. As far science and R&D are concerned, a fruitful and favourable academic environment and the help of funding schemes allowed the cross-fertilisation of ideas between both. Very recently, however, a general deterioration of these relations was witnessed. This deterioration can be contextualised in a number of ups and downs in transatlantic relations history. In space, Europe and the USA also have a long history of cooperation, which has taken place at various levels, in different frameworks and—thematically—in a plethora of fields. So far, relations in the space domain remained largely unaffected by fluctuations on the more general geopolitical scene. Nevertheless, tensions witnessed today affect a number of areas including economy and trade and security and defence and implications in the space security domain are still undetermined at this stage.

The issue of security in space is already one area of transatlantic space cooperation frameworks but remains limited by comparison to the rising stakes in this domain. Some areas of transatlantic space cooperation are not primarily concerned with security in outer space but inherently include activities that have explicit space security considerations—SSA data sharing contributing to the SSA capabilities across the Atlantic, GNSS interoperability and access to service improving the resilience of the satellite navigation capabilities owned by either party.

An overview of the strategic positions of Europe and the USA concerning security in outer space highlights differences in vital interests in pursuit of cooperative frameworks relevant to security in outer space. Whereas the USA aims to maintain strategic superiority (dominance), supported by soft leadership and redundancy of capability through collaboration, the European approach largely reflects pursuit of autonomy and freedom of action, followed by expanding interests in maintaining its position as one of the global space powers, yet being unable to operate self-sufficient capabilities in space situational awareness supporting the security dimension of European space interests.

6.2.2 Transatlantic Cooperation in Security in Outer Space

- The SSA domain involves SSA data sharing agreements signed between the USSTRATCOM and European countries (9) and intergovernmental organisations (EUMETSAT, ESA). These agreements are approached with different objectives but nevertheless constitute the main instrument of transatlantic cooperation in security in outer space. They are augmented by recurrent practical exercises and favourable academic environments for cooperation in SSA-related research.
- Within SEPP, strategic considerations display more complementarity—mainly through interest in space debris mitigation efforts. Clearest example of transatlantic cooperation in this regard is the Inter-Agency Space Debris Coordination Committee (IADC), which emerged from initially bilateral ESA-NASA coordination meetings on space debris mitigation.
- SIS dimension of transatlantic space cooperation includes patterns in improving space system resilience (GNSS interoperability with access to signals, EO data sharing), ISS programme security architecture cooperation and participation of European and US experts on development of ISO space-related standards.

Bilateral SSA data sharing agreements between the USA and its European (also non-European) counterparts constitute the primary instrument of transatlantic cooperation in space situational awareness. Additionally, several recurrent practical exercises at the operational level between transatlantic partners have taken place in recent years. By sharing interests in the need for protection of space infrastructure from threats and hazards of the near-Earth space, the incentive for information and data sharing is a natural logical step to improving capabilities to protect one's own space infrastructure.

The US stance on SSA cooperation has evolved over time. The Commercial and Foreign Entities (CFE) Pilot Program formalised and streamlined the US cooperation approach to SSA sharing in 2004 and was upgraded into a full-fledged SSA Sharing Program managed by the USSSTRATCOM in 2009.[12] The main US objective of entering into these agreements was indeed to promote international security and safety cooperation among allies, encourage the transparency of outer space activities and enhance the quality of the US service delivery system. This policy envisioned the combined development of space doctrine with principles, goals and objectives that opened the possibility of collaborative sharing of space

[12]The Secure World Foundation argues that the key motivation for this formalization in 2009, was the Iridium-Cosmos collision, which occurred in February 2009, see https://swfound.org/media/3584/ssa_sharing_program_issue_brief_nov2011.pdf

Table 6.1 National SSA DATA sharing agreements of the USA

European countries and organisations (11)	Other countries (8)
Belgium, Denmark, France, Germany, Italy, the Netherlands, Norway, Spain, the UK, ESA, EUMETSAT	Australia, Brazil, Canada, Israel, Japan, South Korea, Thailand, the United Arab Emirates

capabilities in case of crisis or conflict. It sought to expand mutually beneficial agreements with key partners to utilise existing and planned capabilities that could improve the security of US national space capabilities.

As of October 2018, the US Strategic Command manages the data sharing programme established through 19 (inter)governmental, 2 classified and more than 70 commercial (e.g. satellite operators or launch providers) SSA data sharing agreements.[13] The majority of agreements pursued at the government level were signed with European countries or international organisations, as seen in Table 6.1. Among the commercial actors participating in SSA sharing with USSTRATCOM, private European space companies (such as Eutelsat or SES) can be easily identified.

These SSA sharing agreements are essential for a number of European institutions and operators for the management of in-orbit operations. Given the limited SSA capabilities of European countries, the European approach to SSA has evolved into an international European effort. The dedicated SSA programme is run by the European Space Agency (which is more focussed on Space Weather and NEOs as member states have been rather hesitant in dealing with the national security considerations of SST through ESA),[14] under the EU umbrella. The SST consortium with so far five participating countries (France, Germany, Italy, Spain, UK) was established after the European Union launched the EU SST Support Framework in 2014. The latest European development, a proposal for a new Regulation of the European Parliament and of the Council establishing the Space Programme of the Union and the European Union Agency for the Space Programme,[15] presented by the Commission in June 2018, tackles also SSA, which has been now developed into a component of the Space Programme. Its multilateral nature—among European countries, but also through international cooperation with other countries, mainly the USA—has been clearly articulated in the text.

[13]USSTRATCOM, Thailand sign agreement to share space services, data, USSTRATCOM, October 2018. Retrieved from http://www.stratcom.mil/Media/News/News-Article-View/Article/1659776/usstratcom-thailand-sign-agreement-to-share-space-services-data/

[14]McCormick, C. (2015). Space Situational Awareness in Europe: The Fractures and the Federative Aspects of European Space Efforts, Astropolitics: The International Journal of Space Politics & Policy, Volume 13, 2015, Issue 1, ISSN 1477–7622.

[15]Proposal for a REGULATION OF THE EUROPEAN PARLIAMENT AND OF THE COUNCIL establishing the space programme of the Union and the European Union Agency for the Space Programme and repealing Regulations (EU) No 912/2010, (EU) No 1285/2013, (EU) No 377/2014 and Decision 541/2014/EU, European Commission, 2018. Retrieved from https://eur-lex.europa.eu/resource.html?uri=cellar:33f7d93e-6af6-11e8-9483-01aa75ed71a1.0003.03/DOC_1&format=PDF

In recent years there has been an increased frequency of new SSA data sharing agreements between the USA and its European partners, which suggests deepening of SSA cooperation across the Atlantic. **In general, lopsided capabilities between the US and European space surveillance network give way to a rather unbalanced exchange of information, and the approach to these agreements of the different partners follows different objectives**. This exchange of information enables European countries to operate with more credible data and services to secure their space systems while giving the USA some form of redundancy and duplicity of capability as a safety measure in case of disruption of this capability or where domestic coverage is considered inadequate. So, whereas European stakeholders approach SSA sharing agreements with a strategic interest in mind, fearing that their SSA capabilities would be seriously weakened should SSA sharing become unavailable, US SSA capabilities do not rely that critically on data acquired through SSA sharing instruments. A list of transatlantic SSA sharing agreements has been made publicly available by USSTRATCOM and is provided in Table 6.2.

Based on this list and related sources, some observations about the nature of SSA sharing agreements in the transatlantic context can be made:

- With the exception of Belgium, all other national SSA sharing agreements (excluding ESA and EUMETSAT) showcase either the defence ministry or armed forces office as a responsible counterpart for SSA sharing at the "European side" of each SSA sharing agreement. This underscores that although SSA-related R&D is often conducted at agency level (DLR, CNES, etc.), operation of SST assets continues to be left mostly to military actors.
- Media releases by USSTRATCOM following signing of SSA sharing agreements traditionally highlight the importance of SSA sharing with the following quote: "The information is crucial for launch support, satellite manoeuvre planning, support for on-orbit anomalies, electromagnetic interference reporting and investigation, satellite decommissioning activities and on-orbit conjunction assessments."[16]
- Statements by USSTRATCOM also highlight that formalised agreements streamline the process of requesting information gathered by JSpOC.[17] This confirms the aforementioned assessment that the amount of data shared is somewhat imbalanced in favour of the provision of US data/service to its international partners.
- SSA sharing agreements explicitly establish channels for direct information exchange. Additionally, they also open doors for future activities under the agreement framework, such as practical capacity-building-oriented experiments

[16]E.g. http://www.stratcom.mil/Media/News/News-Article-View/Article/1497343/usstratcom-denmark-sign-agreement-to-share-space-services-data/ or http://www.stratcom.mil/Media/News/News-Article-View/Article/1607213/usstratcom-brazil-sign-agreement-to-share-space-services-data/

[17]DoD Agrees to Share Space Data with South Korea, U.S. Department of Defense, (September 2014), http://archive.defense.gov/news/newsarticle.aspx?id=123097

Table 6.2 SSA Data sharing agreements between the USA and European countries

European partners	Direct interlocutor	Identified benefits for USSTRATCOM	Nature of cooperation
Netherlands (2018)[a]	Royal Netherlands Air Force	Deepening of partnership with Dutch military, step *toward establishing international norms in the space domain*	Data sharing
Norway (2018)[b]	Norwegian Ministry of Defence	*Reinforcement and Protection of EO capabilities in the Artic*	Data sharing, Technical cooperation procedures
Denmark (2018)[c]	Defence Ministry of Denmark/ Defence Command	Supporting Danish responsibilities in mapping in the Artic	Data sharing, cooperation in sensors and platform development
Germany—MoU (2017)[d]	German Air Force	Liaison Officer (LNO) participating in general planning and mutual training opportunities, Member of the EU SST Consortium	Data sharing, promotion of mutual understandings
Spain—MoU (2016)[e]	Spanish Air Force	*Member of EU SST Consortium, host country of SatCen*	Data sharing
Belgium (2014)[f]	Belgium Federal Office for Science Policy	Supporting Belgium's space regulations	Data sharing
UK(2014)[g]	UK Royal Air Force	Exchange officer based at JSpOC,[h] Five Eye partner, Member of the EU SST Consortium	Data sharing
France (2014)[i]	French Ministry of Defence	*Interoperability with Graves systems, CAESAR*	Data sharing
ESA (2014)[j]		*Better insight over ESA's space operations*	Data sharing
EUMETSAT (2014)[k]—in partnership with NOAA		*Protection of EUMETSAT assets working in collaboration with NOAA on the Initial Joint Polar System*	Data sharing, scientific cooperation
Italy	Italian Ministry of Defence	*Member of EU SST Consortium*	Data sharing

[a]US Strategic Command and Royal Netherlands Air Force sign agreement to share space services data, USSTRATCOM Public Affairs, (September 2018). Retrieved from http://www.stratcom.mil/Media/News/News-Article-View/Article/1647946/us-strategic-command-and-royal-netherlands-air-force-sign-agreement-to-share-sp/

[b]US Strategic Command and Norway sign agreement to share space services, USSTRATCOM Public Affairs, (April 2017)

[c]USSTRATCOM, Denmark sign agreement to share space services, data, USSTRATCOM Public Affairs, (April 2018)

[d]USSTRATCOM, German air force sign liaison officer agreement, USSTRATCOM Public Affairs, (June 2017)

[e]USSTRATCOM, Spain sign memorandum to share space services, data, USSTRATCOM Public Affairs, (February 2017)

(continued)

[f]US Strategic Command, Belgium sign agreement to share space services, data, USSTRATCOM Public Affairs, (February 2017)
[g]DOD Signs Space Data Sharing Agreement with UK, USSTRATCOM Public Affairs, (September 2014)
[h]US Strategic Command Fact Sheet Joint Space Operations Center/614th Air Operations Center, USSTRACTOM, (April 2018)
[i]US Strategic Command, France Enhance Space Data Sharing Agreement, USSTRATCOM Public Affairs, (April 2015)
[j]USSTRATCOM signs Space-Data Sharing Agreement with ESA, USSTRATCOM Public Affairs, (October 2014)
[k]USSTRATCOM enters into Space-Data Sharing Agreement with EUMETSAT, USSTRATCOM Public Affairs, (August 2014)

or cooperation in the development of future (joint or individual but complimentary) SSA sensors or facilities.

In addition to bilateral forms of cooperation on SSA data sharing, USSTRATCOM has initiated the recurrent organisation of SSA experiments and exercises with its international partners. Four such experiments have taken place so far, with the latest taking place in 2017 at the Lockheed Martin Facility in Virginia. European partners have participated regularly at these events. During the fourth SSA experiment in 2017, each participating nation maintained a space operations centre (SpOC) to command and control (C2) their respective SSA assets and experimented with fully integrated, notional Federation SpOC (FedSpOC) to demonstrate the value of a combined and integrated C2 capability.[18]

6.3 Way Forward: Parallel Routes Toward Common Objectives

6.3.1 Comparative Analysis of European and US Approaches

When comparing the US and European approaches to security in outer space, a series of parameters should be considered. The following table presents a brief synthesis of the overall assessment resulting from the respective space and security policy and operational frameworks on each side of the Atlantic. This comparative analysis is necessarily based on a simplified assessment of different concepts to capture key areas of convergence and divergence (Table 6.3).

Based on the comparative analysis above, the overview of strategic considerations of security in outer space in the USA and Europe can be summarised as follows (Table 6.4).

[18]USSTRATCOM Hosts Fourth SSA Experiment with International Partners, USSTRATCOM Public Affairs, (October 2017). Retrieved from http://www.stratcom.mil/Media/News/News-Article-View/Article/1340856/usstratcom-hosts-fourth-ssa-experiment-with-international-partners/

Table 6.3 Summary table of US and European approaches to space security

Parameter	USA	Europe
Strategy and policy/institutional organisation		
Strategy and ambitions	National security; military superiority; global leadership; promotion of commercial market	Protection of economy and society; meeting security requirements for services; achieve autonomy
Organisational culture	Responsibility sharing among institutional actors (intricate decision-making process as a consequence of a higher number of national actors); top-down approach to military/civil domains	Loosely coordinated multiple levels of actors and decision-makers (national, intergovernmental, supranational)
Ongoing policy-making milestones	New national space security strategy; initiation of a national STM policy, initiation of establishment of the Space Force/command within the DoD; review of private space regulations	New regulation proposal by the EC (SSA component of the Space Programme; EU Agency for the Space Programme, etc.); new Defence Space Strategies (UK & France); greater coordination in security and defence policies under consideration
Operational implementation/activities and capabilities		
Capabilities	Self-sufficient (unmatched SSA capabilities, coverage to be complemented)	Reliance on non-European SSA data sources; improvement of SSA capabilities expected as a result of a greater effort
Major undertakings	Improvement of SSA capabilities (various programmes, Space Fence); expansion of SSA sharing agreements; further integration of commercial capabilities	EU SST Consortium; proposed SSA component of the Union Space Programme; EU Agency for the Space Programme (security accreditation across the Union Space Programme); national assets development; EU co-owned SST sensors
Key actors	Sharing of responsibilities between DoD and DoC (SSA/STM); other national institutions (NASA, NOAA, FCC, FAA)	European countries (mainly MoDs with support from national agencies); EU and its agencies (evolving role under consideration: EC, EEAS, GSA, SatCen); ESA (capability-building)
Commercial policy/role of private industry		
Private capabilities	Developing commercial activity in SSA data and related services	Mostly contractors (R&D projects, development and manufacturing)
Involvement of private actors	Congressional support for commercial endeavours; mature technologies catering for operational defence needs; USSTRATCOM relationships with industry; expansion of Office of Space Commerce within the DoC	Repeated calls for more industry-led initiatives; weak market dynamics for space security

(continued)

Table 6.3 (continued)

Parameter	USA	Europe
Legal and regulatory framework/international cooperation		
International initiatives	Active in international fora; reluctance to enter legally binding international regulations; agreement on pursuit of TCBMs	Active in international fora; preference for pursuit of TCBMs; ill-fated ICoC (possible upcoming renewed initiative); supportive of binding international regulations
Domestic regulatory aspects	Existing regulatory framework in process of updating and reorganising; security protocols developed at operational level	National space legislations; security accreditation in EU programmes; ECSS standards applicable at operational level
Views of transatlantic cooperation	Predominantly SSA data sharing: potential to complement and enhance domestic capabilities	Predominantly SSA data sharing: increasing political will to secure own infrastructure with domestic critical capabilities

Table 6.4 Overview of European and US approach to security in outer space

European approach: synthesis	US approach: synthesis
The European approach to security in outer space builds upon a multilayered institutional structure of European space activities, which creates a European governance model that can come across as difficult to fathom by external actors. Security considerations with relation to space were first initiated at national level, integrated within Space Programmes of European countries. From the inception of their governmental space systems, member states have set up national policies. Security in space gained significant attention at the international institutional level through the work of ESA and EUMETSAT, as well as at the top political level of the EU, following the recognition of strategic European interests in maintaining and operating space systems. Strategic concerns and the importance of space security are well acknowledged in key political documents adopted at EU level. Even though the nature of these space security-related concerns still remains mostly civilian, the pivotal role of space as an enabler for other key policies, such as agriculture, digital or defence, is well understood. Further European integration calls for harmonisation and setting-up of a new governance model, which is complex since it implies sharing sensitive information and some transfer of sovereignty	The US approach to security in outer space is relatively more straightforward since supranational governance issues are not an area of concern. There is a stronger military dimension in space security issues, which is highlighted by the critical importance of space systems in foreign and national security policies. Security aspects of space activities have been leveraged in the USA for a longer period of time (in comparison with Europe) and have historically reflected the relevant position that space has had across various administrations. US capabilities related to security in outer space largely outweigh those of other spacefaring nations including Europe as a single actor in space security. One recent trend suggests that the US Administration is more inclined to pursue a more commercially focused space policy and notably in national security areas. Greater focus is also visible in rather assertive statements on the importance of space control, dominance and leadership in space, with variations in the wording depending on the incumbent administration. On the other hand, the importance of partnerships has also been stressed and appears to be a focal point in current US Space Policy. A more prominent role of civil space actors in SSA data gathering and sharing is also expected

It is clear that the mechanisms and drivers at work on both sides of the Atlantic for the definition and the implementation of respective space security policies greatly differ. Nevertheless, both the USA and Europe are concerned by the emergence of new threats that might compromise their interests in space and on the ground. As a consequence, outer space security is gaining priority in respective civil and defence space agendas. In this respect, despite the discrepancies highlighted above in the respective approaches to outer space security, many factors contribute to the expectation of a higher degree of convergence in these matters in the near and medium terms.

Shared Perception of a Growing Vulnerability

The USA and Europe share a largely common assessment of space security challenges ahead. However, given cultural differences ranging from strategic vision to operational considerations, they prioritise these challenges differently in their respective space policy frameworks, in particular in the civil and military segments of space activities. As a consequence, this initial joint assessment leads to quite different concrete results in terms of a vision of the way forward. The need to recognise that potential misunderstanding will occur has to be understood and resolved.

Strong Reinforcement of the US Space Security Policy

The US National Space Policy prioritises space security as a prerequisite for the protection of its national security. In this context, the US approach to outer space security is driven, first and foremost, by the perception of a growing vulnerability of key national space assets. This assessment drives the need to protect space infrastructures against a number of risks, in particular military threats (e.g. ASAT, cyberattacks and intentional interferences), and to prepare a tactical response. The proposal of President Trump to create a Space Force—a military organisation separate from the Air Force for the conduct of in-space military operations—is a symbol of this vision. However, although this proposal seems quite in line with the overall US approach to space and security issues, it is raising a number of concerns, and there is a vivid public debate at the moment around such option, which will ultimately be arbitrated by the Congress.

In parallel, the announcement of the national Space Traffic Management initiative under the authority of the Department of Commerce is the civilian complement.

In fact, if at first glance the US governance model displays a seemingly fully integrated structure, several interviewees stressed that US bureaucratic processes can be intricate as a consequence of the multiplicity of engaged stakeholders. At this point in time, it is clear that there will be a strong military side in the USA concerning the ongoing development of the various outer space security challenges. The balance that will be found with civilian options and the potential diplomatic implications are still to be determined.

Rising Sensitivity to Space Security Issues in Europe

In comparison, in Europe, public debates on outer space security issues are just starting and no consensus has emerged among policy-makers to tackle them as short-term, pressing operational priorities.

European stakeholders are fully aware of the multiple challenges ahead, but the European space security ecosystem remains complex and multilayered, structured along a general principle of national leadership and sovereignty. The prominent place of security concerns in the EU strategy for space, the inclusion of the SST/SSA programme under the umbrella of the European Union, the achievements of the EU SST Consortium to pool resources and the multiple initiatives of ESA are clear indications that these issues are now an integral part of the European space policy. However, priorities continue to differ between the various stakeholders, resulting in delays in actually translating the perception of vulnerability into concrete operational cooperative decisions addressing the full range of potential threats.

In this context, the announcement of the first UK and French Space Defence Strategies and the latest statements of the German government are strong signals. These moves highlight a change of attitude toward space security.

- The German Defence Minister Ursula von der Leyen has supported the forging of "a common strategic culture for Europe",[19] and the German government has stated that the development of an "independent space surveillance capacity of the EU" is key,[20] giving momentum to a bolder approach to European autonomy in this matter.
- The French Minister for Armed Forces, Mrs. Florence Parly, made the case for a space defence policy building on European cooperation in September 2018 when denouncing the behaviour of the Russian spy satellite Luch-Olymp orbiting around the French-Italian military telecom satellite Athena-Fidus.[21]
- In a similar vein, French President Emmanuel Macron has stated that "European defence cannot be solely structured around relationships with the U.S." and that "equilibriums, automatisms on which alliances were built [since the Cold War] must be revisited"—notably in the field of cybersecurity, chemical weapons,

[19]German Government. (2018). *Speech by Federal Minister of Defence Dr Ursula von der Leyen on the occasion of the opening of the 54th Munich Security Conference in Munich.* Retrieved from: https://www.bmvg.de/resource/blob/22180/a4b7d92394e5ff6b7689c79cc71fa9d9/20180216-download-eroeffnungsrede-englisch-data.pdf. (16 February 2018).

[20]Deutscher Bundestag. (2018). *Deutschland gut in der Raumfahrtforschung.* Retrieved from: https://www.bundestag.de/presse/hib/-/566972

[21]Le Monde. (2018). *La France accuse la Russie de tentative d'espionnage par satellite.* Retrieved from: https://www.lemonde.fr/international/article/2018/09/07/paris-revele-une-tentative-d-espionnage-russe-sur-un-satellite-franco-italien-en-2017_5351908_3210.html. (7 Septembre 2018).

classical warfare, territorial conflicts, space security and artic areas.[22] He has announced the establishment of a Defence Space Strategy to be ready by 2019.[23]
- UK Foreign Minister Gavin Williamson has announced the launch of a UK Defence Space Strategy.[24]

The rising political significance of space security could very well trigger a reassessment of European ambitions and accelerate the consolidation of the European approach to the issue across the different stakeholders (governance, objectives, architecture and more).

6.3.2 A Fertile Ground for a Reinforced Partnership

State of Play: A Complex Mix of Arrangements

Current transatlantic relations in space security encompass a complex mix of dialogues and cooperation frameworks involving different US and European stakeholders. They are organised through multiple channels, including:

- *Bilateral government-to-government channels*: these revolve predominantly around 11 data sharing agreements of USSTRATCOM with European countries and international governmental organisations (ESA, EUMETSAT). This model of cooperation is augmented by regular practical exercises for partner nations and liaison officers posted to the USA to work in close cooperation with their counterparts. Bilateral cooperation on other space projects (e.g. in space exploration or space science) beyond security in outer space can further foster transatlantic relations in outer space activities.
- *Europe-wide to US channel*, which includes:
  - Permanent EU-US Space Dialogue at diplomatic level (most recently held in January 2018 with the participation of various US and European officials. (Space security was put forward as a major agenda item).
  - Case-by-case cooperation between US and European organisations (e.g. NOAA/EUMETSAT, NASA/ESA) on specific programmatic topics: space exploration, remote sensing, meteorology and the ISS programme.

[22]Le Monde. (2018). *Les orientations diplomatiques d'Emmanuel Macron: «sécurité» en Europe et «crise humanitaire» en Syrie*. Retrieved from: https://www.lemonde.fr/international/article/2018/08/27/europe-syrie-libye-macron-devoile-sa-feuille-de-route-diplomatique_5346644_3210.html. (27 July 2018).

[23]French Government. (2018). *Discours du Président de la République Emmanuel Macron à l'Hôtel de Brienne*. Retrieved from: http://www.elysee.fr/declarations/article/discours-du-president-de-la-republique-emmanuel-macron-a-l-hotel-de-brienne/ (13 July 2018).

[24]The Guardian. (2018). *Defence secretary unveils strategy to protect UK satellites*. Retrieved from: https://www.theguardian.com/politics/2018/may/21/defence-secretary-unveils-strategy-to-protect-uk-satellites (21 May 2018).

- GNSS issues, concerning mainly EU-US relations on the compatibility and complementarity of the GPS and Galileo systems. After initial US concerns following the EU decision to proceed with Galileo development, both parties reached an agreement on GPS/Galileo interoperability in 2004 and continue to cooperate in satellite positioning, navigation and timing at bilateral as well as multilateral level. At the same time however, talks on US access to Galileo's PRS have not yet been finalised given the sensitive nature of this issue.

- *Multilateral channels*: US and European stakeholders have been displaying converging views and positions on space security at space-related international fora, such as UN COPUOS, Conference on Disarmament, IADC, ITU, Cospas-Sarsat, ICG and others. Mutual interest in pursuit of international space TCBMs was visible, e.g. in the recent "long-term sustainability of outer space activities" agenda within the UNCOPUOS. The example of ESA—NASA cooperation on debris mitigation, which evolved from bilateral coordination meetings into a full-fledged multilateral platform in the form of the IADC, is particularly relevant here. Recently, NATO has expressed interest in further elaborating a proper space policy. At the moment, it fully relies on its member states when it comes to space capabilities and neither owns nor operates its own space assets.
- *Government-to-industry and industry-to-industry cooperation*: this channel is becoming increasingly relevant in the wake of the expansion of private capabilities, in particular in the field of SSA. A number of USSTRATCOM SSA data sharing agreements with commercial operators (including European) illustrates this trend, as well as the case of the Space Data Association, a commercial operators' non-profit cooperation relying on governmental and private services to foster exchange of information and best practices concerning responsible space behaviour. Involvement of private industry in space security is bound to increase.

Toward Reinforced Cooperation Seeking Mutual Benefit

Despite these various successful collaborative frameworks, or maybe because of their multiplicity, no formal and inclusive framework has yet been established at political level between the USA and Europe. As a consequence, so far, transatlantic cooperation has been confined to a case-by-case-basis and, as far as outer space security is concerned, to bilateral, government-to-government agreements.

Several factors can explain the somewhat fragmented transatlantic cooperation framework. Among them, the dual and multilayered organisational structure of both US and European governance may very likely be a major obstacle to a more integrated approach. Indeed, while Europe still faces difficulties in translating stakeholders' positions into Europe-wide policies and programmes, interviewed US officials also stressed that, although the integration of the US line of command suggest a capacity to federate and coordinate on the US side, decision-making still faces hurdles to align positions and interests of the different organisations and departments involved.

Interviewees uniformly underlined that establishing a more comprehensive and inclusive relationship, based on reinforced cooperation seeking mutual benefits, will require several preliminary steps:

- A joint assessment of the situation
- Mutual understanding of each partner's priorities
- Shared interest in cooperation based on mutual benefits
- The capacity to converge on common objectives
- A balanced contribution from the actors

Notwithstanding these prerequisites, recent and planned developments in the field of space security policy on both sides seem to provide a fertile ground for the USA and Europe to broaden the scope of their current space-related relations and to consider what could be the drivers for a reinforced cooperation in space security:

- In the USA, the Space Strategy 2018 and Space Policy Directive-3 will profoundly alter the way space security is envisioned and organised across military and civil/commercial branches. Upcoming developments will have consequences for international partners. Beyond changes in arrangements, these developments also open up opportunities for new areas of cooperation such as Space Traffic Management.
- In Europe, the declared ambition to strengthen European autonomy in space security will result in an increase in resources for capacity-building and in the consolidation of the European approach. At the same time, expected further space policy and programmatic developments might encourage a fresh look at cooperation with third countries, in particular with the USA.

The future of transatlantic relations in space security will also be influenced by the development of general political and diplomatic relations between the USA and Europe. From this perspective, and although cooperation in space is usually unaffected by political ups and downs, transatlantic relations are currently experiencing a serious deterioration in a number of fields including security and defence. The implications of this deterioration in the domain of space security remain unclear at this stage. Overall, interviewed US and European stakeholders and experts highlighted the following opportunities and challenges for reinforced transatlantic cooperation in security in outer space (Table 6.5).

The USA and Europe conduct activities in similar domains (SSA, SEPP and SIS), but the current state of affairs including policy, resources and capabilities is strongly lopsided. As a consequence, the resulting imbalanced relationship impacts Europe's bargaining power in future potential negotiations, as well as its attractiveness for a reinforced partnership. While US capabilities relevant to security in outer space (in particular SSA capabilities) are technically mature and designed for self-sufficiency (although coverage can always be complemented), European capabilities are still under development, both at national and European levels. Therefore, Europe still relies on US SSA data and services for several of applications.

Table 6.5 Interviewees' views on opportunities and threats for reinforcement of transatlantic cooperation operational implementation/activities and capabilities

Opportunities	Challenges
• Intensifying threats requiring joint efforts (interdependence of actors) • Enhanced capabilities on each side creating an opportunity to foster reciprocity and burden sharing • Convergence on a number of new policy issues • transatlantic framework as a test bed for enlarged global collaborative schemes • Commercial policies fostering open competitive markets and industry-to-industry cooperation	• Overall deterioration of transatlantic relations at the political level • Different sensitivities to threats (military/civil ratio) • Transatlantic cooperation dependent on better coordination of national/regional actors within the respective governance frameworks • Fairness of commercial competition • Duplication of efforts

On the European side, SSA-related activities are mostly handled through dual-use publicly owned and operated facilities, although no responsibility-sharing scheme similar to that currently taking shape in the USA between DoD and DoC has been announced so far. At the same time, recent developments suggest an increasing role of the European Union in ownership and management of SST assets along the lines of the draft resolution submitted in October 2018 by the European Commission to its member states for the next Multiannual Financial Framework. Another illustration of this trend is the positioning of SatCen as the front desk for the distribution of information and services exploiting the SST data currently delivered by the Consortium.

In a nutshell, there are obvious discrepancies between the US and European approaches to outer space security, as well as a strong imbalance in the level of resources allocated to such activities. However, recent trends from both shores of the Atlantic suggest a convergence on the assessment that outer space security requires additional efforts and resources, as well as a redefined institutional framework (Space Fence, Space Force, SSA component of the EU Space Programme). The conjunction of these factors might open a window of opportunity to consider the potential mutual benefits of closer transatlantic coordination/cooperation/collaboration in these matters.

Commercial Policy/Role of Private Industry

Over the last few years, the US private sector has been particularly active in developing the basis for a commercial offer for SSA services. Given the nature of activities to be performed, such as visualisation, advanced computing, big data analytics and artificial intelligence, the expertise of industry is envisioned to help leverage technical know-how in senior technologies to more effectively address quickly emerging space threats.

In addition, US policy-makers frequently stress the merits of making commercial options available as a means of globally improving the security of operations in orbit. Indeed, the present analysis confirms that US industry has gained a high degree of maturity in the field of SSA data gathering and advanced exploitation through a growing number of SMEs and other legacy companies. In Europe, the number of private initiatives targeting this line of business is much more limited, partially owing to weaker domestic market dynamics and/or unclear perspectives regarding critical data policy.

In fact, SSA services are at the convergence of two key priorities of the long-established US space strategy: space commerce and national security. As a consequence, the US Congress has displayed great political support for these private initiatives, stressing the potential benefits from the expected higher efficiency of private management, as well as the will to support the competitiveness of the domestic industry. This materialised in particular in the creation of the Office of Space Commerce at DOC, which has no equivalent in Europe.

Legal and Regulatory Framework/International Cooperation
The USA and Europe share the vision that outer space security issues will be better addressed through pooling of resources, in particular with each other. A prerequisite for this is to be able to converge on some legal and regulatory provisions.

For the time being, the difference in the appreciation of the stakes related to space security, as well as the need to comply with the current overarching diplomatic framework, complicates the process of finding common grounds for balanced cooperation. Additionally, the international legal implications of such endeavour must be assessed against the provisions of the various national space legislations and strategies in Europe.

6.4 What Scope for a Reinforced Transatlantic Partnership?

6.4.1 Shifting US Posture Toward National Leadership

The growing importance placed on security in outer space by the USA and Europe in their respective strategy and policy frameworks is based on the shared assessment that space infrastructures are exposed to increasingly serious security challenges. In an ever more congested and contested space environment, space security challenges are:

- *Multiple and diverse* in nature and origin and as a consequence require a coherent set of diverse preventive, operative and curative measures (i.e. holistic approach).
- *Interrelated and interdependent*, creating a situation of mutual dependence and responsibility between the different actors involved in space activities (countries, civil/military, public/commercial, agencies/industry, etc.).
- *Ubiquitous and inclusive*, although the degree of exposure and vulnerability of space systems to specific threats (e.g. collision, cyberattacks, spoofing, jamming) may vary.

- *Intensifying*, driven by endogenous and exogenous trends including:
  - Increasing space activity in terms of the number of launches and objects in orbit but also the number of governmental and commercial actors operating space systems
  - New concepts, technologies and capabilities (e.g. mega-constellations, miniaturised systems, in-orbit servicing)
  - An ever more connected space infrastructure to ground networks and systems
  - The increasing importance of space infrastructure, which makes it a key target for a variety of actors pursuing different objectives
  - The rehabilitation of a "space warfare" doctrine encompassing activities to develop "space control" capabilities

This situation, which is expected to further deteriorate in the future, ultimately creates risks for the economy, society and security at large.

With the recent adoption of Space Policy Directive-3 (i.e. National Space Traffic Management Policy), the USA made an important step forward in recognising the severity of issues at stake and the urgency of setting up a framework to prevent and mitigate space security threats. More specifically, the policy recognises that "the future space operating environment will be shaped by a significant increase in the volume and diversity of commercial activity in space" and that "as the number of space objects increases, [the current] limited traffic management activity and architecture will become inadequate."[25] In this context, the US policy aims to "develop a new approach to space traffic management that addresses current and future operational risks."[26]

The new policy directive marks a shift in the US posture and underlines a clear political willingness to accelerate activities through national-led engagements.

The policy does not necessarily challenge the relevance of multilateral efforts in space security. In fact, the policy recalls that "it is a shared interest and responsibility of all spacefaring nations to create the conditions for a safe, stable, and operationally sustainable space environment",[27] recalling that a fully effective approach can only be envisioned as the outcome of a coherent and inclusive global effort. Notwithstanding, such a national-led approach can be understood as a reaction to the limited progress achieved at the international level on space security and sustainability topics. Indeed, the multilateral efforts in which both the USA and Europe actively participate (e.g. IADC space debris mitigation guidelines, Long-Term Sustainability guidelines, International Code of Conduct, Prevention of an Arms Race in Outer Space—PAROS, etc.) are significantly slowed down by the recurring difficulty of making actors with diverging views and interests converge on necessarily constraining international measures.

---

[25]U.S. Government. (2018). Space Policy Directive-3, National Space Traffic Management Policy, (June 18, 2018).

[26]Ibid.

[27]Ibid.

No transatlantic collaboration at pan-European level can be envisioned in the field of the military activities that are currently being considered by the US Administration in the field of Space Security. None of the partners would probably consider this as an option at this point in time.

On the civilian side, the establishment of a national Space Traffic Management framework is the next step implied in US space security policy. If this example is followed by other countries, the development of multiple, possibly divergent, national and regional STM frameworks in parallel to international negotiations may occur. The implementation of this US policy, which will have consequences across the entire spectrum of preventive, operative and curative space security measures, has major implications for Europe and transatlantic relations and is clearly the main driver for revisiting transatlantic cooperation in line with developments on the European scene.

6.4.2 Space Traffic Management: Stakes and Implications for Europe

The US Space Traffic Management policy aims to tackle a number of security challenges related to the expected boom in space traffic, especially related to the deployment of mega-constellations and of non-manoeuvrable spacecraft (e.g. CubeSats). With the aim of providing a structure for the control and management of in-orbit operations, STM is characterised by the development of a normative approach through "best practices, technical guidelines, safety standards, behavioural norms, pre-launch risk assessments, and on-orbit collision avoidance services".[28]

The development and implementation of a full-fledged Space Traffic Management architecture is a difficult task. Beyond intricate policy and governance aspects, STM involves the supervision of a complex chain of information/data to support operational decision-making.

More specifically, the development of a STM architecture raises the following major stakes:

- *SSA data enhancement* to reach the appropriate accuracy required to safely plan, coordinate and synchronise in-orbit activities and mitigate collision risks
- *SSA data policy* to set up appropriate information management structures (collection, fusion, distribution) safeguarding data integrity, reliance and confidentiality
- *Specification of STM best practices and norms* to enhance the safety, stability and sustainability of operations in the space environment across different stakeholders (military, civil, commercial)

[28]U.S. Government. (2018). Space Policy Directive-3, National Space Traffic Management Policy, (June 18, 2018).

The results find that transatlantic relations should be revisited and potentially reinforced along these lines taking into account the complementarity between political and operational cooperation frameworks.

SSA Data Enhancement and Data Policy

"Timely, accurate, and actionable data are essential for effective SSA and STM."[29] For this reason, the USA seeks to enhance SSA capabilities through the development of new sensors, the improvement of SSA data sharing (i.e. interoperability and greater data sharing) and the purchase of SSA data and services. Enhancing SSA data precision and accuracy necessarily implies relying on multiple data sources including:

- National SSA data (based on domestic systems)
- SSA data from third parties (based on foreign/commercial SSA systems)
- Satellite operators' data (incl. satellite orbital parameters and manoeuvres)

Improving currently available SSA data through multiple data sources, or "crowdsourcing", will also bring new challenges related to data availability, reliability and integrity. Effective management of these issues will involve (1) a revisit of data sharing agreements with international and private partners and (2) a clear policy for the integration of commercial SSA data and services.

The distribution of information to diverse stakeholders will also require establishing a robust data policy to ensure confidentiality across military and civil branches and between partners.

An essential aspect of this issue will be to delineate military and civil domains in terms of data access. Developments observed in the USA suggest that a top-down approach is currently the preferred option with the military branch in charge of domestic SSA capabilities and responsible for the establishment of a public catalogue corresponding to a subset of information and data available to them. In Europe, such a top-down approach enabling a delineation of military/civil domains has not been implemented, as so far, SST cooperation efforts have been mainly dedicated to networking national capabilities that are based on integrated dual frameworks.

Implications for Europe and transatlantic cooperation
Data sharing agreements will remain the backbone of transatlantic cooperation. A revisit of these agreements toward more balanced cooperation for mutual benefit should be based on: • An enhancement of European SSA capabilities to close the capability gap and increase Europe's bargaining power • A suitable balance between the desire for European autonomy and transatlantic complementarity in the prioritisation of European SSA capabilities development • The progressive emergence of European leadership through the attribution of specific mandates to the European Union to lead SSA international cooperation Enhanced data sharing will imply revisiting arrangements regarding data policy. This implies: • A compatible data policy

(continued)

[29]Ibid.

Implications for Europe and transatlantic cooperation
• Provisions to ensure the integrity of the data and address cyberthreats • Provisions to ensure the confidentiality of the data along a shared delineation of military/civil domains
Europe needs a clear policy to frame the role of the private sector as a key stakeholder in an integrated approach to SSA/STM. Beyond the involvement of private operators as providers and users of SSA data and STM services, such a policy should also address: • The integration of commercial SSA data in European capabilities and the necessary provisions to ensure effective data fusion • The conditions to be met to enable the emergence of an open transatlantic market for SSA/STM-related data and services

Specification of STM Best Practices and Norms

Switching from an informative to a normative approach to STM implies the specification of norms of behaviour through nonbinding best practices, as a preparatory first step, and eventually through standards to be integrated in the regulatory regime governing space activities. From this standpoint, standards and best practices will have to be considered in a consistent manner encompassing preventive, operative and curative measures across the full life cycle of space systems:

- Spacecraft and launcher design and manufacturing
- Launch operations and separation
- In-orbit operations
- End of life and de-orbiting

In this respect, the US Administration intends to "support the development of operational standards and best practices to promote safe and responsible behaviour in space. A critical first step in carrying out that goal is to develop U.S.-led minimum safety standards and best practices to coordinate space traffic [. . .] and to use them to inform and help shape international consensus, practices, and standards".[30]

Even though the USA will work on the development of a national approach, in coordination with its own industry, such an approach ultimately needs to be coordinated at international level to be fully effective given the strong interdependence between global actors.

In particular, it remains to be seen whether the ultimate goal of Space Traffic Management is the setting up of a centralised authority defining the details of avoidance manoeuvres to be implemented and directing the operators to execute them, or if it will be left up to the operators to define and implement the most appropriate actions.

[30]U.S. Government. (2018). Space Policy Directive-3, National Space Traffic Management Policy, (June 18, 2018).

Implications for Europe and transatlantic cooperation
• Europe must play a role in the development of best practices and standards that will shape a future international approach to STM to remain a key actor in the global space scene and to mitigate the competitive bias that such norms might create for industry
• To participate in the elaboration of STM norms, Europe must, at least, converge on its own approach to STM. This requires, as a first essential step, to set up a dedicated forum gathering all relevant European key stakeholders: EEAS, EU, ESA, member states, manufacturers, operators as well as military organisations
• The development of a set of common (or at least compatible) norms of behaviour in space could form a second backbone for transatlantic cooperation

6.4.3 Conclusion

A global effort is necessary to achieve security in outer space. To be positioned as a key player in the international space scene, Europe must contribute its share to this endeavour. Much needs to be done on the European side to ensure balanced cooperation with other key players including, in particular, the USA.

An important near-term step for Europe is to define the desired degree of autonomy to be achieved and subsequently, to assess the means necessary to meet these requirements. This will also highlight the need to establish a European space security policy and to optimise the current intricate framework to foster coherence, cost-effectiveness and leadership. This is a matter of driving the emerging political will across European countries toward common views, joint objectives and rational responsibility sharing.

The rise of space security as a priority of USA and European space policies, respectively, and the changing political and operational environments on each side of the Atlantic create new opportunities to revisit the currently loosely organised transatlantic relations. On the European side, fostering reinforced transatlantic cooperation in space security, as intended by the Space Strategy for Europe, would require:

- Taking full advantage of the new opportunities offered by a more ambitious European effort in SSA. Also a desirable reinforcement of regional cooperation in this field is essential to improve Europe's bargaining power though:
  - Consideration of the added-value of European SSA capabilities for the USA (complementarity, resilience, interoperability) in European SSA capability-building roadmaps and data policies
  - Fostering the emergence of European commercial actors able to compete/cooperate in an open transatlantic SSA market
- Preparing a European approach to Space Traffic Management. This implies, first of all, setting up a dedicated forum to coordinate the views, needs and possible contributions of different stakeholders such as the EU, ESA, member states, manufacturers and operators.

Part III
2017 Facts and Figures

Chapter 7
Chronology of Space Activities in 2017

7.1 Launch Log

Launches Are Listed in Chronological Order and in Groups According to the Country of the Launch Site

All launch dates are given using Greenwich Mean Time (GMT); hence the date at the launch site may differ from the date listed here by 1 day.
All content, information and consequent details derive from the European Space Policy Institute's proprietary database.

Legend
Launch Outcome:

S: Success (A launch attempt is considered a success when the launcher performed nominally and as planned.)
PF: Partial failure (A launch attempt is considered a partial failure if the launcher does not perform nominally and, as a result, one or more payloads are damaged or injected into the wrong orbit.)
F: Failure (A launch attempt is considered a failure when one or more payloads are lost.)

Markets:

C: Commercial (A mission is considered commercial when the spacecraft's capabilities are intended to be sold on a market.)
E: Education (A mission is considered an education mission when the spacecraft's development and capabilities are intended to be used for a pedagogical purpose.)
GC: Governmental civil (A mission is considered as governmental or civil when the spacecraft's capabilities are intended to provide public services or to serve a function of the government.)

E. Burger, G. Bordacchini, *Yearbook on Space Policy 2017*, Yearbook on Space Policy, https://doi.org/10.1007/978-3-030-05417-5_7

M: Military (A mission is considered military when the spacecraft's capabilities are intended to be used for a military purpose.)

D: Dual use (A mission is considered to be dual use when the spacecraft's capabilities support both military and civilian purposes without the capacity to identify one of these as a primary function.)

Mission:

AIS: Automatic Identification System
Astr: Astronomy
Earth Obs: Earth Observation
Navig: Navigation
Signal Int: Signal Intelligence
Space Station Infr: Space Station Infrastructure
SSA: Space Situation Awareness
Techno Demo: Technology Demonstration
Telecom: Telecommunication
Biology
Cargo Transfer
Crew Transfer
Meteorology
Early Warning
Earth Science
Satellite Data Relay

Launch date	Spaceport	Launcher	Launch outcome	Spacecraft name	Customer country/ region	Manufacturer country/region	Mass at launch (kg)	Orbit	Mission	Market
5-Jan	**Xichang**	**CZ-3B/G2**	**S**	Tongxin Jishu Shiyan 02	China	China	3000,00	GEO	Techno/ Demo	GC
9-Jan	**Jiuquan**	**KZ-1A**	**S**	Jilin 1 03	China	China	95,00	LEO	Earth Obs.	C
				Kaidun 01	China	China	3,00	LEO	Techno/ Demo	E
				Xingyun Shiyan 01	China	China	3,00	LEO	Techno/ Demo	E
2-Mar	**Jiuquan**	**Kaituozhe-2**	**S**	Tiankun 01	China	China	100,00	LEO	Techno/ Demo	GC
12-Apr	**Xichang**	**CZ-3B/G2**	**S**	ShiJian 13	China	China	4600,00	GEO	Telecom.	C
20-Apr	**Wenchang**	**CZ-7**	**S**	SilkRoad-1 01 (Silu 1)	China	China	4,50	LEO	Earth Obs.	GC
				Tianzhou 01	China	China	12,910,00	LEO	Cargo Transfer	GC
15-Jun	**Jiuquan**	**CZ-4B**	**S**	Huiyan	China	China	2500,00	LEO	Astr.	GC
				NuSat 03	Others	Others	37,00	LEO	Earth Obs.	C
				OVS (2 satellites)	China	China	50,00	LEO	Earth Obs.	C
18-Jun	**Xichang**	**CZ-3B/G2**	**PF**	ZhongXing 09A/-ChinaSat 9A	China	China	5100,00	GEO	Telecom.	C
2-Jul	**Wenchang**	**CZ-5**	**F**	ShiJian 18	China	China	7600,00	GEO	Techno/ Demo	GC
29-Sep	**Xichang**	**CZ-2C(3)**	**S**	Yaogan 30-01 (3 satellites)	China	China	300,00	LEO	Signal Int.	M
9-Oct	**Jiuquan**	**CZ-2D(2)**	**S**	VRSS 02	Others	China	1000,00	LEO	Earth Obs.	GC
5-Nov	**Xichang**	**CZ-3B/YZ1**	**S**	Beidou 3 M1	China	China	1014,00	MEO	Navig.	D
				Beidou 3 M2	China	China	1014,00	MEO	Navig.	D

(continued)

Launch date	Spaceport	Launcher	Launch outcome	Spacecraft name	Customer country/ region	Manufacturer country/region	Mass at launch (kg)	Orbit	Mission	Market
14-Nov	**Taiyuan**	**CZ-4C**	**S**	Fengyun 3D	China	China	2200,00	LEO	Meteorology	GC
				HEAD 1	China	China	45,00	LEO	AIS	C
21-Nov	**Taiyuan**	**CZ-6**	**S**	Jilin 1 (3 satellites)	China	China	95,00	LEO	Earth Obs.	C
24-Nov	**Xichang**	**CZ-2C(3)**	**S**	Yaogan 30-02 (3 satellites)	China	China	300,00	LEO	Signal Int.	M
3-Dec	**Jiuquan**	**CZ-2D(2)**	**S**	Ludikancha Weixing 01	China	China	3000,00	LEO	Earth Obs.	GC
10-Dec	**Xichang**	**CZ-3B/G2**	**S**	Alcomsat 1	Others	China	5225,00	LEO	Telecom.	GC
23-Dec	**Jiuquan**	**CZ-2D(2)**	**S**	LKW 2	China	China	2000,00	LEO	Earth Obs.	GC
25-Dec	**Xichang**	**CZ-2C(3)**	**S**	Yaogan 30-03 (3 satellites)	China	China	300,00	LEO	Signal Int.	M
28-Jan	**CSG**	**Soyuz-ST-B Fregat-MT**	**S**	Hispasat 36W-1	Europe	Europe	3221,00	GEO	Telecom.	C
14-Feb	**CSG**	**Ariane-5ECA**	**S**	Intelsat 32e	USA	Europe	6000,00	GEO	Telecom.	C
				Telkom 03S	Others	Europe	3550,00	GEO	Telecom.	C
7-Mar	**CSG**	**Vega**	**S**	Sentinel 2B	Europe	Europe	1130,00	LEO	Earth Obs.	GC
4-May	**CSG**	**Ariane-5ECA**	**S**	Koreasat 07	Others	Europe	3680,00	GEO	Telecom.	C
				SGDC 1	Others	Europe	5735,00	GEO	Telecom.	C
18-May	**CSG**	**Soyuz-ST-A Fregat-M**	**S**	SES 15	Europe	USA	2302,00	GEO	Telecom.	C
1-Jun	**CSG**	**Ariane-5ECA**	**S**	Eutelsat 172B	Europe	Europe	3551,00	GEO	Telecom.	C
				ViaSat 02	USA	USA	6418,00	GEO	Telecom.	C
28-Jun	**CSG**	**Ariane-5ECA**	**S**	GSat 17	India	India	3477,00	LEO	Telecom.	GC

				Hellas-Sat 3/Inmarsat-S-EAS	Europe	Europe	5780,00	GEO	Telecom.	C
2-Aug	**CSG**	**Vega**	**S**	OPTSAT-3000	Europe	Others	368,00	LEO	Earth Obs.	M
				VENμS	Others	Others	264,00	LEO	Earth Obs.	GC
29-Sep	**CSG**	**Ariane-5ECA**	**S**	BSat 4A	Japan	USA	3520,00	GEO	Telecom.	C
				Intelsat 37e	USA	USA	6438,00	GEO	Telecom.	C
8-Nov	**CSG**	**Vega**	**S**	Mohammed VI A	Others	Europe	1110,00	LEO	Earth Obs.	GC
12-Dec	**CSG**	**Ariane-5ES**	**S**	Galileo (4 satellites)	Europe	Europe	733,00	MEO	Navig.	GC
15-Feb	**Satish Dhawan**	**PSLV-XL**	**S**	Al-Farabi 01	Others	Others	2,00	LEO	Techno/ Demo	E
				BGUSat	Others	Others	4,00	LEO	Techno/ Demo	E
				Cartosat 2D	India	India	714,00	LEO	Earth Obs.	GC
				DIDO 02	Europe	Europe	3,00	LEO	Biology	C
				Flock-3p (88 satellites)	USA	USA	5,00	LEO	Earth Obs.	C
				INS 1A	India	India	8,40	LEO	Techno/ Demo	GC
				INS 1B	India	India	9,40	LEO	Techno/ Demo	GC
				Lemur 2 (8 satellites)	USA	USA	4,00	LEO	AIS	C
				Nayif 01	Others	Others	1,00	LEO	Techno/ Demo	E
				PEASSS	Europe	Europe	4,00	LEO	Techno/ Demo	C
5-May	**Satish Dhawan**	**GSLV Mk.2**	**S**	GSat 09	India	India	2195,00	GEO	Telecom.	GC

(continued)

Launch date	Spaceport	Launcher	Launch outcome	Spacecraft name	Customer country/region	Manufacturer country/region	Mass at launch (kg)	Orbit	Mission	Market
5-Jun	**Satish Dhawan**	**GSLV Mk.3**	**S**	GSat 19	India	India	3136,00	GEO	Telecom.	GC
23-Jun	**Satish Dhawan**	**PSLV-XL**	**S**	Aalto 01	Europe	Europe	4,00	LEO	Earth Obs.	E
				Blue Diamond	Europe	Europe	6,00	LEO	Telecom.	C
				Cartosat 2E	India	India	714,00	LEO	Earth Obs.	GC
				CE-SAT 01	Japan	Japan	50,00	LEO	Earth Obs.	C
				CICERO 6	USA	USA	10,00	LEO	Earth Obs.	C
				Compass 02	Europe	Europe	4,00	LEO	Techno/ Demo	E
				D-Sat	Europe	Europe	4,00	LEO	Techno/ Demo	C
				Green Diamond	Europe	Europe	6,00	LEO	Telecom.	C
				InflateSail	Europe	Europe	4,00	LEO	Techno/ Demo	GC
				Lemur 2 (8 satellites)	USA	USA	4,00	LEO	AIS	C
				LituanicaSAT 02	Europe	Europe	4,00	LEO	Techno/ Demo	E
				Max Valier Sat	Europe	Europe	15,00	LEO	Astr.	GC
				NIUSAT	India	India	15,00	LEO	Earth Obs.	GC
				NUDTSat	China	China	2,00	LEO	Earth Science	E
				PACSCISAT	USA	USA	4,00	LEO	Techno/ Demo	C
				Pegasus	Europe	Europe	2,00	LEO	Earth Science	E

				Red Diamond	Europe	Europe	6,00	LEO	Telecom.	C
				ROBUSTA 01B	Europe	Europe	1,00	LEO	Techno/ Demo	E
				skCUBE	Europe	Europe	1,00	LEO	Techno/ Demo	E
				SUCHAI	Others	Others	1,00	LEO	Techno/ Demo	E
				UCLSat	Europe	Europe	3,00	LEO	Earth Science	E
				URSA MAIOR	Europe	Europe	3,00	LEO	Techno/ Demo	E
				Venta 01	Europe	Europe	5,00	LEO	AIS	GC
				VZLUsat 01	Europe	Europe	2,00	LEO	Earth Science	E
31-Aug	**Satish Dhawan**	**PSLV-XL**	**S**	IRNSS 1H	India	India	1425,00	GEO	Navig.	D
14-Jan	**Kagoshima**	**SS-520 (3 stages)**	**F**	TRICOM 1	Japan	Japan	3,00	LEO	Techno/ Demo	GC
24-Jan	**Tanegashima**	**H-2A-204**	**S**	Kirameki 02	Japan	Japan	4000,00	GEO	Telecom.	M
17-Mar	**Tanegashima**	**H-2A-202**	**S**	IGS-Radar 05	Japan	Japan	1600,00	LEO	Earth Obs.	D
1-Jun	**Tanegashima**	**H-2A-202**	**S**	QZS 02	Japan	Japan	4000,00	GEO	Navig.	GC
19-Aug	**Tanegashima**	**H-2A-204**	**S**	QZS 03	Japan	Japan	4700,00	GEO	Navig.	GC
9-Oct	**Tanegashima**	**H-2A-202**	**S**	QZS 04	Japan	Japan	4000,00	GEO	Navig.	GC
23-Dec	**Tanegashima**	**H-2A-202**	**S**	GCOM-C	Japan	Japan	1950,00	LEO	Earth Obs.	GC
				SLATS	Japan	Japan	400,00	LEO	Techno/ Demo	GC
25-May	**Onenui**	**Electron**	**F**	Humanity star	Others	Others	10,00	LEO	Other	C
22-Feb	**Baikonur**	**Soyuz-U**	**S**	Progress-MS 05	Russia	Russia	7285,00	LEO	Cargo Transfer	GC

(continued)

Launch date	Spaceport	Launcher	Launch outcome	Spacecraft name	Customer country/ region	Manufacturer country/region	Mass at launch (kg)	Orbit	Mission	Market
20-Apr	**Baikonur**	**Soyuz-FG**	**S**	Soyuz-MS 04	Russia	Russia	7220,00	LEO	Crew Transfer	GC
8-Jun	**Baikonur**	**Proton-M Briz-M (Ph.3)**	**S**	EchoStar 21	USA	USA	6871,00	GEO	Telecom.	C
14-Jun	**Baikonur**	**Soyuz-2-1a**	**S**	Progress-MS 06	Russia	Russia	7285,00	LEO	Cargo Transfer	GC
				Radioskaf RS (2 satellites)	USA	USA	4,80	LEO	Techno/ Demo	E
				Sfera 53 02	Russia	Russia	13,00	LEO	Earth Science	GC
				TNS 0 02	Russia	Russia	4,80	LEO	Techno/ Demo	GC
23-Jun	**Plesetsk**	**Soyuz-2-1v Volga**	**S**	Kosmos 2519	Russia	Russia	300,00	LEO	Unknown	M
				Kosmos 2521/ Sputnik Inspektor	Russia	Russia	10,00	LEO	Unknown	M
				Kosmos 2523	Russia	Russia	100,00	LEO	Unknown	M
14-Jul	**Baikonur**	**Soyuz-2-1a Fregat-M**	**S**	CICERO (3 satellites)	USA	USA	10,00	LEO	Earth Obs.	C
				Corvus-BC (2 satellites)	USA	USA	11,00	LEO	Earth Obs.	C
				Flock-2k (48 satellites)	USA	USA	5,00	LEO	Earth Obs.	C
				Flying Laptop	Europe	Europe	120,00	LEO	Techno/ Demo	GC
				Iskra-MAI-85	Russia	Russia	4,00	LEO	Techno/ Demo	E

				Kanopus-V-IK 01	Russia	Russia	473,00	LEO	Earth Obs.	GC
				Lemur 2 (8 satellites)	USA	USA	4,00	LEO	AIS	C
				Mayak	Russia	Russia	4,00	LEO	Techno/ Demo	GC
				MKA-N (2 satellites)	Russia	Russia	11,70	LEO	Earth Obs.	GC
				NanoACE	USA	USA	4,00	LEO	Techno/ Demo	C
				NORSAT 1	Europe	Others	30,00	LEO	AIS	GC
				NORSAT 2	Europe	Others	15,00	LEO	AIS	GC
				TechnoSat	Europe	Europe	18,00	LEO	Techno/ Demo	GC
				UTE-UESOR	Others	Others	1,00	LEO	Techno/ Demo	E
				WNISAT 01R	Japan	Japan	10,00	LEO	Earth Obs.	C
28-Jul	**Baikonur**	**Soyuz-FG**	**S**	Soyuz-MS 05	Russia	Russia	7220,00	LEO	Crew Transfer	GC
16-Aug	**Baikonur**	**Proton-M Briz-M (Ph.3)**	**S**	Blagovest 11L	Russia	Russia	5000,00	GEO	Telecom.	GC
11-Sep	**Baikonur**	**Proton-M Briz-M (Ph.3)**	**S**	Amazonas 05	Europe	USA	5900,00	GEO	Telecom.	C
13-Sep	**Baikonur**	**Soyuz-FG**	**S**	Soyuz-MS 06	Russia	Russia	7220,00	LEO	Crew Transfer	GC
22-Sep	**Plesetsk**	**Soyuz-2-1b Fregat-M**	**S**	Glonass-M 46	Russia	Russia	1415,00	MEO	Navig.	M
28-Sep	**Baikonur**	**Proton-M Briz-M (Ph.4)**	**S**	AsiaSat 9	China	USA	6141,00	GEO	Telecom.	C

(continued)

Launch date	Spaceport	Launcher	Launch outcome	Spacecraft name	Customer country/ region	Manufacturer country/region	Mass at launch (kg)	Orbit	Mission	Market
13-Oct	**Plesetsk**	**Rokot-KM**	**S**	Sentinel 5P	Europe	Europe	900,00	LEO	Earth Obs.	GC
14-Oct	**Baikonur**	**Soyuz-2-1a**	**S**	Progress-MS 07	Russia	Russia	7285,00	LEO	Cargo Transfer	GC
28-Nov	**Vostochniy**	**Soyuz-2-1b Fregat-M**	**F**	AISSat 03	Europe	Others	6,50	LEO	AIS	GC
				Baumanets 02	Russia	Russia	100,00	LEO	Techno/ Demo	GC
				Corvus-BC (2 satellites)	USA	USA	11,00	LEO	Earth Obs.	C
				D-Star One	Europe	Europe	4,00	LEO	Telecom.	C
				IDEA-OSG 01	Others	Others	22,00	LEO	SSA	C
				Lemur 2 (10 satellites)	USA	USA	4,00	LEO	AIS	C
				LEO Vantage 02	Others	USA	70,00	LEO	Telecom.	C
				Meteor-M2 01	Russia	Russia	2750,00	LEO	Meteorology	GC
				SEAM	Europe	Europe	4,00	LEO	Earth Science	E
2-Dec	**Plesetsk**	**Soyuz-2-1b**	**S**	Lotos-S 02	Russia	Russia	6500,00	LEO	Signal Int.	M
17-Dec	**Baikonur**	**Soyuz-FG**	**S**	Soyuz-MS 07	Russia	Russia	7220,00	LEO	Crew Transfer	GC
26-Dec	**Baikonur**	**Zenit-3F**	**S**	Angosat-1	Others	Russia	1550,00	LEO	Telecom.	GC
14-Jan	**Vandenberg**	**Falcon-9 v1.2**	**S**	Iridium-NEXT (10 satellites)	USA	Europe	860,00	LEO	Telecom.	C
21-Jan	**Cape Canaveral**	**Atlas-5(401)**	**S**	SBIRS-GEO 03	USA	USA	4500,00	GEO	Early Warning	M
19-Feb	**KSC**	**Falcon-9 v1.2**	**S**	Dragon CRS-10	USA	USA	7747,00	LEO	Cargo Transfer	GC

				SAGE 03	USA	USA	527,00	LEO	Space Station Infr.	GC
				STP-H5	USA	USA	400,00	LEO	Space Station Infr.	M
1-Mar	**Vandenberg**	**Atlas-5(401)**	**S**	NOSS-3/Intruder (2 satellites)	USA	USA	3250,00	LEO	Signal Int.	M
16-Mar	**KSC**	**Falcon-9 v1.2(ex)**	**S**	EchoStar 23	USA	USA	5600,00	GEO	Telecom.	C
19-Mar	**Cape Canaveral**	**Delta-4M+ (5,4) (upg.)**	**S**	WGS 09	USA	USA	5987,00	GEO	Telecom.	M
30-Mar	**KSC**	**Falcon-9 v1.2**	**S**	SES 10	Europe	Europe	5300,00	GEO	Telecom.	C
18-Apr	**Cape Canaveral**	**Atlas-5(401)**	**S**	Aalto 02	Europe	Europe	2,00	LEO	Earth Science	E
				Altair Pathfinder	USA	USA	6,00	LEO	Techno/ Demo	C
				Aoxiang 01	China	China	2,00	LEO	Earth Science	E
				BeEagleSat	Others	Others	2,00	LEO	Earth Science	E
				CSUNSat 01	USA	USA	2,00	LEO	Techno/ Demo	E
				CXBN 02	USA	USA	3,00	LEO	Astr.	E
				Cygnus OA-7	USA	USA	7492,00	LEO	Cargo Transfer	GC
				DUTHSat	Europe	Europe	2,00	LEO	Earth Science	E
				Ex-Alta 01	Others	Others	4,00	LEO	Earth Science	E
				HAVELSAT	Others	Others	2,00	LEO	Earth Science	E

(continued)

Launch date	Spaceport	Launcher	Launch outcome	Spacecraft name	Customer country/ region	Manufacturer country/region	Mass at launch (kg)	Orbit	Mission	Market
				Hoopoe	Others	Others	2,00	LEO	Earth Science	E
				IceCube	USA	USA	4,00	LEO	Earth Science	GC
				i-INSPIRE 02	Others	Others	2,00	LEO	Earth Science	E
				KySat 03	USA	USA	1,00	LEO	Techno/ Demo	E
				Lemur 2 (4 satellites)	USA	USA	4,00	LEO	AIS	C
				LilacSat 01	China	China	2,00	LEO	Techno/ Demo	E
				LINK	Others	Others	2,00	LEO	Earth Science	E
				NJUST 01	China	China	2,00	LEO	Earth Science	E
				nSIGHT 01	Others	Others	2,00	LEO	Earth Science	E
				Phoenix	Others	Others	2,00	LEO	Earth Science	E
				PolyITAN 02-SAU	Others	Others	2,00	LEO	Earth Science	E
				qbee50-LTU-OC	Europe	Europe	2,00	LEO	Earth Science	E
				QBITO	Europe	Europe	2,00	LEO	Earth Science	E

				QBUS (3 satellites)	USA	USA	2,00	LEO	Earth Science	E
				SHARC	USA	USA	2,00	LEO	Other	M
				SNUSAT (2 satellites)	Others	Others	2,00	LEO	Earth Science	E
				SOMP 02	Europe	Europe	2,00	LEO	Earth Science	E
				SpaceCube	Europe	Europe	2,00	LEO	Earth Science	E
				SUSat	Others	Others	2,00	LEO	Earth Science	E
				UNSW-EC0	Others	Others	2,00	LEO	Earth Science	E
				UPSat	Europe	Europe	2,00	LEO	Earth Science	E
				X-CubeSat	Europe	Europe	2,00	LEO	Earth Science	E
				ZA-AeroSat	Others	Others	2,00	LEO	Earth Science	E
1-May	**KSC**	**Falcon-9 v1.2**	**S**	NROL 76/USA 276	USA	USA	5000,00	LEO	Earth Obs.	M
15-May	**KSC**	**Falcon-9 v1.2(ex)**	**S**	Inmarsat 5 F4	Europe	USA	6070,00	GEO	Telecom.	C
3-Jun	**KSC**	**Falcon-9 v1.2**	**S**	ANUSAT 01	Japan	Japan	1,00	LEO	Techno/ Demo	E
				BRAC Onnesha	Japan	Japan	1,00	LEO	Techno/ Demo	E
				Dragon CRS-11	USA	USA	7747,00	LEO	Cargo Transfer	GC

(continued)

Launch date	Spaceport	Launcher	Launch outcome	Spacecraft name	Customer country/ region	Manufacturer country/region	Mass at launch (kg)	Orbit	Mission	Market
				Mazaalai	Japan	Japan	1,00	LEO	Techno/ Demo	E
				MUSES	USA	USA	200,00	LEO	Space Station Infr.	C
				NICER	USA	USA	372,00	LEO	Space Station Infr.	GC
				Nigeria EduSat 01	Japan	Japan	1,00	LEO	Techno/ Demo	E
				ROSA	USA	USA	400,00	LEO	Space Station Infr.	M
				Toki	Japan	Japan	1,00	LEO	Techno/ Demo	E
23-Jun	**KSC**	**Falcon-9 v1.2**	**S**	BulgariaSat 1	Europe	USA	3669,00	GEO	Telecom.	C
25-Jun	**Vandenberg**	**Falcon-9 v1.2**	**S**	Iridium-NEXT (10 satellites)	USA	Europe	860,00	LEO	Telecom.	C
5-Jul	**KSC**	**Falcon-9 v1.2(ex)**	**S**	Intelsat 35e	USA	USA	6761,00	GEO	Telecom.	C
14-Aug	**KSC**	**Falcon-9 v1.2**	**S**	ASTERIA	USA	USA	12,00	LEO	Techno/ Demo	GC
				CREAM	USA	USA	1300,00	LEO	Space Station Infr.	GC
				Dellingr	USA	USA	8,00	LEO	Space Science	GC
				Dragon CRS-12	USA	USA	7747,00	LEO	Cargo Transfer	GC
				Kestrel Eye 2M	USA	USA	50,00	LEO	Earth Obs.	M

				OSIRIS-3U	USA	USA	4,00	LEO	Earth Science	E
18-Aug	**Cape Canaveral**	**Atlas-5(401)**	**S**	TDRS 13	USA	USA	3454,00	GEO	Satellite Data Relay	GC
24-Aug	**Vandenberg**	**Falcon-9 v1.2**	**S**	FORMOSAT 5	Others	Others	475,00	LEO	Earth Obs.	GC
26-Aug	**Cape Canaveral**	**Minotaur-4 Orion-38**	**S**	DHFR	USA	USA	4,00	LEO	Techno/ Demo	M
				ORS 05	USA	USA	140,00	LEO	SSA	M
				Prometheus 2 (2 satellites)	USA	USA	2,00	LEO	Techno/ Demo	M
7-Sep	**KSC**	**Falcon-9 v1.2**	**S**	X-37B OTV 05/X-37B 01F3	USA	USA	5400,00	LEO	Techno/ Demo	M
24-Sep	**Vandenberg**	**Atlas-5(541)**	**S**	SBIRS HEO 04/ Trumpet-F/O 04	USA	USA	5200,00	HEO	Signal Int.	M
9-Oct	**Vandenberg**	**Falcon-9 v1.2**	**S**	Iridium-NEXT (10 satellites)	USA	Europe	860,00	LEO	Telecom.	C
11-Oct	**KSC**	**Falcon-9 v1.2**	**S**	EchoStar 105/SES 11	Europe	Europe	5200,00	GEO	Telecom.	C
15-Oct	**Cape Canaveral**	**Atlas-5(421)**	**S**	SDS 4 02/Quasar 21	USA	USA	5900,00	GEO	Satellite Data Relay	M
30-Oct	**KSC**	**Falcon-9 v1.2**	**S**	Koreasat 05A	Others	Europe	3500,00	GEO	Telecom.	C
31-Oct	**Vandenberg**	**Minotaur-C-XL-3210**	**S**	Flock-3m (4 satellites)	USA	USA	5,00	LEO	Earth Obs.	C
				SkySat (6 satellites)	USA	USA	120,00	LEO	Earth Obs.	C
12-Nov	**Wallops**	**Antares-230**	**S**	AeroCube 07B/OCSD B (2 satellites)	USA	USA	3,00	LEO	Techno/ Demo	GC

(continued)

Launch date	Spaceport	Launcher	Launch outcome	Spacecraft name	Customer country/ region	Manufacturer country/region	Mass at launch (kg)	Orbit	Mission	Market
				Asgardia 01	Europe	Europe	3,00	LEO	Techno/ Demo	C
				CHEFsat	USA	USA	4,00	LEO	Techno/ Demo	M
				Cygnus OA-8	USA	USA	7492,00	LEO	Cargo Transfer	GC
				EcAMSat	USA	USA	11,00	LEO	Biology	GC
				ISARA	USA	USA	5,00	LEO	Techno/ Demo	GC
				Lemur 2 (8 satellites)	USA	USA	4,00	LEO	AIS	C
				PropCube 02	USA	USA	1,00	LEO	Earth Science	M
				TechEdSat 06	USA	USA	4,00	LEO	Techno/ Demo	E
18-Nov	**Vandenberg**	**Delta-7920-10C**	**S**	Buccaneer RMM	Others	Others	4,00	LEO	Techno/ Demo	GC
				EagleSat	USA	USA	1,00	LEO	Techno/ Demo	E
				MakerSat 0	USA	USA	1,00	LEO	Techno/ Demo	E
				MiRaTA	USA	USA	4,00	LEO	Techno/ Demo	E
				NOAA 20	USA	USA	2540,00	LEO	Meteorology	GC
				RadFxSat	USA	USA	1,00	LEO	Techno/ Demo	E

15-Dec	**Cape Canaveral**	**Falcon-9 v1.2**	**S**	Dragon CRS-13	USA	USA	7747,00	LEO	Cargo Transfer	GC
				SDS	USA	USA	200,00	LEO	Space Station Infr.	GC
				TSIS	USA	USA	200,00	LEO	Space Station Infr.	GC
23-Dec	**Vandenberg**	**Falcon-9 v1.2**	**S**	Iridium-NEXT (10 satellites)	USA	Europe	860,00	LEO	Telecom.	C

7.2 Chronology of Major Space Policy and Related Developments in 2017

Europe	Globally
18th January: the UK Space Agency and the French space agency CNES signed a joint statement to enhance space activities cooperation, in particular around projects on climate action and Mars exploration, as well as space applications and the European space programme	*30th January to 10th February*: the 54th session of UNCOPUOS Scientific and Technical Subcommittee
23rd January: ASI President Roberto Battiston, Italian Ambassador Mauro Massone, and Kenya's Defence Secretary Kirimi Kaberia prepared in the context of the first Italy-Kenya Joint Steering Committee the operational and technical aspects of this renewed agreement. The agreement covers the development of a regional Earth observation centre, accessibility of Earth observation data, telemedicine as well as educational and training activities, and support generally to the Kenyan National Space Agency	*7th February*: USSTRATCOM and Belgium (BELSPO) signed an agreement to share SSA services and information
8th February: UK Space Agency launch a campaign grouped under the LaunchUK title, aiming to promote the national commercial launch industry. A programme worth £50 million was also announced, and it will support satellite launch activities from British spaceports, along with a £10 million grant programme to support the national commercial launch sector. The British government has furthermore published a series of guidance resources on regulatory aspects of operating spaceports in the UK, launching from the UK, and on spaceplane and suborbital flight activities	*27th March to 7th April*: the 56th session of its Legal Subcommittee
8th February 2018: ESA and Airbus Defence and Space have agreed with NASA to build a second module for a second mission with astronauts for launch as early as 2021; the European Service Module will power NASA's Orion spacecraft beyond the Moon and back	*30th March*: Successful demonstration of previously used and refurbished rocket stage by SpaceX for SES-10 launch—first reusable orbital launch since the retirement of the US Space Shuttle
20th February: ESA adopts Open Access Policy	*5th April*: USSTRATCOM and Norway signed an agreement to share SSA services and information
20th February: Polish Space Strategy adopted by the government	*14th April*: ISRO record-breaking launch of 104 satellites on a single rocket

(continued)

Europe	Globally
22nd February: Italian President Sergio Mattarella's state visit to China, an accord was signed by ASI President Battiston and Wang Zhaoyao, Director-General of the China Manned Space Agency (CMSA) concerning scientific experimentation on long-duration astronaut missions in the context of the Chinese space station programme. It additionally will lead to the creation of a Joint Cooperation Committee by ASI and CMSA which will meet yearly and address various aspects of the collaboration	*18th April*: US President Donald Trump signed the Weather Research and Forecasting Innovation Act, which provides new direction from the US Congress for NOAA with regards to weather research and forecasting as well as observation programmes, and it is seen as the first comprehensive policy on weather since previous legislation from 1992
6th March: The European GNSS Agency (GSA) has signed a contract with Eutelsat Communications for the development, integration and operation of the next-generation EGNOS payload signing 18-year contract. It covers the preparation and service provision phases of the EGNOS Geostationary (GEO) space-based augmentation system (SBAS) payload service (GEO-3)	*18th April*: Canada's Innovation, Science and Economic Development (ISED) Minister Navdeep Bains announced the renewal of the Government of Canada's Space Advisory Board
16th March: Jean-Yves Le Gall, President of CNES has been elected chair of the ESA Council, actively beginning the 1st of July. Otylia Trzaskalska-Stroinska and Alice Bunn respectvely from the polish and british delegations have been elected as vice-chairs	*18th April*: Mongolia launched its first satellite—MongolSat-1
23rd March: Adoption of "the elaboration of a Military Satellite Navigation Policy to scope the potential European secure use of positioning, navigation & timing information," the first of its kind adopted by the EDA Steering Board	*15th May*: Japan Aerospace Exploration Agency (JAXA) and the European Space Agency (ESA) held the Inter-Agency Meeting to strengthen and deepen the bilateral cooperation
23rd March: Chair of the DLR Executive Board Pascale Ehrenfreund, Programme Director of DLR Energy Research Bernhard Milow, and President of the Japanese National Institute of Advanced Industrial Science and Technology (AIST) Ryoji Chubachi signed a Memorandum of Understanding concerning scientific collaboration around energy research	*23rd May*: Joint Stock Company Glavkosmos and Limited Liability Company International Space Company Kosmotras established a Joint Stock Company "GK LAUNCH SERVICES." It will become an operator of commercial launch services for satellites with the use of Soyuz family launch vehicles and the launchers developed on the basis of RS-20 rockets from the Russian launch sites
24th March: a GOVSATCOM Common Staff Requirements and a connected Business Case were approved by the European Defence Agency (EDA) Steering Board	*23rd May*: the US White House, under the final budget request, proposed to cancel the NASA Asteroid Redirect Mission
5th April: DLR signed two agreements with JAXA and one with Japan's Electronic Navigation Research Institute on aeronautics research, specifically dealing with air traffic management, electric aviation, and research on reduction of the sonic boom	*25th May*: New Zealand first rocket launched to space

(continued)

Europe	Globally
9th May: ASI and the Argentinian Comisión Nacional de Actividades Espaciales (CONAE) signed an agreement around commercial exploitation of satellite data from the SIASGE radar technology programme, which includes the Argentinian SAOCOM and Italian COSMO-SkyMed satellite constellations	*3rd June*: Successful launch of the first Ghanaian satellite—GhanaSat-1
10th April: JAXA and CNES made an implementing arrangement on MMX, Martian Moons Exploration	*7–16th June*: the 60th plenary session of COPUOS
13th April: EUMETSAT Jason-3 teamwork was awarded from NASA as an "exemplary model for multiple partners working together"	*30th June*: US President Trump's signed an executive order which re-established the National Space Council, previously disbanded in 1993
15th May: Copernicus service in Support to EU External Action (SEA) started to deliver services in operational mode. As Copernicus Entrusted Entity, the European Union Satellite Centre (SatCen) coordinates this service	*4th July*: India and Israel signed three space cooperation agreements that cover specific areas of engineering cooperation and technology transfers in electrical propulsion for small satellites, developing optical links between satellites in geostationary Earth orbit and low-Earth orbit, and the development of atomic clocks used in positioning, navigation, and timing (PNT) satellites
7th June: European Commission launched the European Defence Fund. The Fund will coordinate, supplement and amplify national investments in defence research, in the development of prototypes and in the acquisition of defence equipment and technology	*10th July*: SpaceX Dragon delivered to the ISS a Chinese DNA experiment among the 25 NanoRacks experiments
15th June: the Outline Description for the Governmental Satellite Communications Pooling and Sharing demonstration project was accepted, led by Spain	*31st July*: The United Arab Emirates officially launched Mars campaign aiming to perform Mars Mission (Al Amal probe) by 2020. Later in the year, the UAE also launched nation's Astronaut Programme
15th June: ASI furthered its collaboration with NASA through the signing of an implementing arrangement providing ASI associate membership in the Solar System Exploration Research Virtual Institute managed by NASA Ames. Additionally, NASA's then acting administrator Robert Lightfoot and ASI President Roberto Battiston signed an agreement around ASI's cooperation in the IXPE mission which will take measurements on the polarisation of cosmic X-rays	*29th August*: Taiwan launched the first home-built satellite—Formosat-5—with SpaceX
19–25th June: Le Bourget Air and Space Show, Paris.	*4th September*: The United Nations Office for Outer Space Affairs (UNOOSA) and the United Nations Institute for Training and Research (UNITAR) have agreed to collaborate to further their common goals, particularly

(continued)

Europe	Globally
	the use of space-based technology for the achievement of the 2030 Agenda for Sustainable Development and its 17 Sustainable Development Goals
20th June: Luxembourg and ESA announced enhanced cooperation on space resources. ESA will undertake an analysis of the feasibility assessment and technical maturity of asteroids exploration and utilisation	*11th September*: UNOOSA and JAXA selected a team from the Universidad del Valle de Guatemala for the second round of the UNOOSA-JAXA KiboCUBE programme. The team plans to use its CubeSat to test equipment for monitoring the concentration of harmful cyanobacteria (algae blooms) over inland bodies of water, Guatemala's first ever satellite in orbit
23rd–25th June: ESA and 16 leaders of the European satellite industry signed a joint statement regarding cooperation on Satellite for 5G through Horizon 2020 and the 5G Infrastructure Public Private Partnership	*11th September*: Declaration of the Iranian's Aerospace Research Institute of the intention to send human in space by 2025
27th June: At the EUMETSAT's 87th meeting in Darmstadt, Germany, the Council reached the formal operations extension of the Jason-2 programme from the 1 June 2018 until the end of 2019 (ocean surface topography mission supporting oceanography originally launched in 2008). The programme marks a transition from research activities to operations, involving collaboration between Europe's EUMETSAT and CNES and the US NASA and NOAA	*20th September*: JAXA and DLR held the Inter-Agency Meeting to strengthen and deepen the bilateral cooperation and the establishment of the joint working groups to discuss new collaborative areas
20th July: Luxembourg is the first European country to adopt a Space Law for the regulation of the ownership of space resources	*25th–29th September*: International Astronautical Congress, Adelaide, Australia
24th July: DLR introduced, along with Germany's Federal Ministry for Economic Affairs and Energy, the DLR's new Strategy 2030. The strategy introduces ten cross-sectoral projects on cyber security, broadband rollout, and future fuels, among others, and it will see the DLR establish seven new research institutes, and the German federal government will provide 42 million € annually in funding	*27th September*: At the 68th International Astronautical Congress in Australia, Roscosmos and NASA signed a joint statement "on Researching, Exploring Deep Space" which focuses in particular on research which may lead to the development of NASA's concept Deep Space Gateway
24th August: The UKSA Corporate Plan for 2017–2018 was released, and it details the Agency's four priorities for the period, including the publication of a Space Strategy, enabling UK commercial space access and fourthly reform of the British space regulatory regime	*5th October*: First meeting of US National Space Council where US Vice President Mike Pence directed NASA to create plans for human lunar missions which would lead on to future Mars missions

(continued)

Europe	Globally
13th September: Unveiling of the renewed EU Industrial Policy Strategy that brings together all existing and new horizontal and sector-specific initiatives into a comprehensive industrial strategy	*5th October*: MDA completed acquisition of Digital Globe and rebranded the combined company as Maxar Technologies
21st September: DLR, JAXA, and AIST signed two cooperation agreements again around energy research, focusing in particular on high-performance batteries for space missions	*10th October*: 50th anniversary of the Outer Space Treaty enter into force
13th October: The latest Copernicus Earth observation satellite (Sentinel 5P) was launched into space	*2nd November*: The Disarmament and International Security Committee (First Committee) forwarded 58 draft resolutions and decisions to the UNGA for adoption over nuclear weapons and other weapons of mass destruction
17–18th October: ESA Council took place in Paris, France. Unanimously, an Agreement between NASA and ESA on ExoMars 2020 was approved, including NASA's Deep Space Network in the mission's ground segment and extending Mars proximity relay communications using NASA's MRO and MAVEN orbiters. The ESA Director of Navigation, Paul Verhoef, presented Galileo status and plans for the Second Generation	*2nd November*: Russia-China agreement on cooperation in Lunar and Deep Space
8th November: In the margins of the EU-ESA informal space ministerial meeting organised in Tallinn, Estonia, by the Estonian Presidency of the European Union and the Spanish Chairmanship of the ESA Council, Alain Ratier, the EUMETSAT Director-General and Johann-Dietrich Wörner, the ESA Director-General, signed a statement on the joint contributions of ESA and EUMETSAT to the space strategy for Europe	*8th November*: The United Nations Office for Outer Space Affairs (UNOOSA) and the United Arab Emirates (UAE) Space Agency signed a MoU to cooperate on capacity-building initiatives on both the technical and legal aspects of the peaceful uses of outer space
26th November: The UK published the new Industrial Strategy	*22nd November*: The United Nations Office for Outer Space Affairs (UNOOSA) has signed a MoU with both the Government of Italy and the Italian Space Agency to increase cooperation in the peaceful uses of outer space
29th November: Luxembourg–Japan memorandum of cooperation on exploration and commercial utilisation of space resources LU–UAE agreement before that—in October 2017	*29th November*: North Korea ballistic missile test flew ten times higher than the International Space Station
December: An Implementing Arrangement was signed between the EDA and ESA toward supporting the European Commission in the GOVSATCOM initiative	*29th November*: First Angolan satellite—AngoSat-1—launched from Baikonur

(continued)

Europe	Globally
1st December: the Competitiveness Council of the Council of the European Union convened, presenting pressed for closer ties between the space sector and the digital economy. The Council of the European Union also adopted the conclusions of the midterm reviews of Galileo, EGNOS, and Copernicus	*30th November*: China and USA hold third Civil Space Dialogue, exchanging plans for human and robotic space exploration and discussing engagement through multilateral mechanisms
7th December: At the 353rd session of CNES, Board of Directors were approved the French contribution to the ESA JUICE mission and a resolution on the renewal of draft industrial contracts for the Guiana Space Centre, regarding the launch operations support services from 2018 to 2024	*4th December*: UN General Assembly adopted 53 draft resolutions and 4 draft decisions of these First Committee texts
11th December: Four additional Galileo satellites were successfully launched into orbit with a European launcher Ariane 5 from Europe's spaceport in Kourou. The constellation counts 22 satellites	*4th December*: The United Nations General Assembly (UNGA) passed seven Resolutions relating to the use and exploration of outer space during in its 72nd session: the Resolution 72/26—*Prevention of an arms race in outer space*, the Resolution 72/27—*No first placement of weapons in outer space*, and the Resolution 72/56,*Transparency and confidence-building measures in outer space activities*
12th December: DLR, Japan's National Institute for Environmental Studies (NIES), and JAXA signed a collaboration agreement on validating satellite systems used for greenhouse gases measurements	*6th December*: The United Nations Office for Outer Space Affairs (UNOOSA) and the United Nations Development Programme (UNDP) have committed to work more closely together in the use of space technology for the development around the world, signing a Memorandum of Understanding (MOU) for this increased cooperation
23rd December: The Italian Commission of Productive Activities of the Chamber of Deputies approved the draft law "Measures for the coordination of space and aerospace policy and provision concerning the organisation and functioning of the Italian Space Agency," the so called DDL Space. It leads to a reorganisation of the Space Agency, and it attributes the coordination of the space policies to the Presidency of the Council of the Ministers	*7th December*: UN General Assembly adopted the Resolution 72/77—*International cooperation in the peaceful uses of outer space*; the Resolution 72/78, *Adoption of the Declaration on the 50th anniversary of the Treaty on Principles Governing the Activities of States in the Exploration and Use of Outer Space, including the Moon and Other Celestial Bodies*; and the Resolution 72/79—*Consideration of the 50th anniversary of the United Nations Conference on the Exploration and Peaceful Uses of Outer Space*
	11th December: President Trump signed Space Policy Directive 1, which builds on the first National Space Council meeting in instructing NASA to again conduct human missions to the Moon

(continued)

Europe	Globally
	12th December: The US National Defense Authorization Act for Fiscal Year 2018 (NDAA-18) was signed into law by US President Donald Trump and marks a period of evolution in the US military's space organisation
	12th December: The Cabinet Office of Japan released an Implementation Plan of the Basic Plan on Space Policy, revised for fiscal year 2017; its first objective is to "Ensure space security"
	24th December: UN General Assembly adopted the Resolution 72/250—*Further practical measures for the prevention of an arms race in outer space*

7.3 Country Profiles 2017

This section provides core data on ESA member states and their respective national-level space activities.

Austria

FFG Promoting Innovation.

Population	8.797 million
GDP	369.68 billion €
Authority/authorities	The Austrian Space Programme is financed by the Federal Ministry for Transport, Innovation and Technology (BMVIT) and managed by the Aeronautics and Space Agency (ALR), which is integrated into the Austrian Research Promotion Agency (FFG), the central organisation for fostering science and technology activities in Austria, and serves as a docking station to the international aerospace world for Austrian business and science
Major activities	ALR of FFG implements national aerospace policy and represents Austria in numerous European and international aerospace institutions—from ESA and EUMETSAT through EU committees and other international and national organisations responsible for aerospace policy. Austrian representation in the European Space Agency is of special strategic importance in this context. In addition to ESA programmes, the Austrian Space Applications Programme ASAP and the Space Programmes under the 7th EU Framework Programme are of special relevance for space research projects

(continued)

Budget	**Estimated total space expenditure: 66 million €, including** Contribution to ESA budget: 47.4 million € Contribution to EUMETSAT budget: 11.3 million €
Staff	15 (ALR)
Direct Employment in Space Manufacturing Industry[a]	460

[a]All "Direct Employment in Space Manufacturing Industry" figures are taken from ASD-Eurospace. "Facts and Figures—The European Space Industry in 2016"

Belgium

Belgian Science Policy Office

belspo

Population	11.348 million
GDP	437.20 billion €
Authority/authorities	Two-level competence system: – Federal: space in the frame of international agreements – Regions and communities: basic and applied research The Belgian Federal Science Policy Office (BELSPO) with its Department of Space Research and Applications manages the scientific aspects of the space policy. The Belgian High Representation for Space Policy (BHRS), acting under the auspices of the federal minister in charge of Science Policy, defends Belgian interests in the international institutions dealing with space (ESA and EU levels). Other federal ministries involved with space are the Ministries of Defence and Foreign Affairs. On the regional level, the Ministries for Economic Development, Transport and Environment have stakes in space applications and industrial development
Major activities	Strategy and operational goals: • To create and/or strengthen expertise and industrial capabilities in high-growth areas of the space sector • To increase the return on investment of public money • In space at the ESA level, in the frame of the EU as well as in bilateral cooperation projects • To support industry to be able to tackle new markets, aiming at a positive impact on economy • To give to the public authority the space means necessary to define and to implement its policies • To represent Belgium's interests in the definition and implementation of the European space policy • Cooperation with ESA, EUMETSAT, EC and other aerospace-related agencies Within ESA, Belgium traditionally has major interests in Launchers, Telecommunications, Integrated Applications and Technology Support programmes, Earth Observation

(continued)

Budget	**Estimated total space expenditure: 193.4 million €, including** Contribution to ESA budget: 203.4 million € Contribution to EUMETSAT budget: 13.5 million €
Staff	BELSPO Department of Space Research and Applications: 20
Direct Employment in Space Manufacturing Industry	1540

Czech Republic

Ministry of Transport

Population	10.589 million
GDP	191.64 billion €
Authority/authorities	The Ministry of Transport of the Czech Republic is the coordinator of all space activities in the Czech Republic. For the coordination purposes, it established the Coordination Council for Space Activities under its leadership, which includes all relevant ministries and institutions. The Ministry of Transport is also directly responsible for the major space activities in the Czech Republic, including participation in ESA and EU space policy, and international cooperation in space. The Ministry of Transport is responsible for the implementation of the Galileo programme in the Czech Republic, and it bears the responsibility for satellite systems' follow-up applications (e.g. intelligent transport systems)
Major activities	The Czech Republic subscribed ESA Optional Programmes such as MTG, MetOp-SG, EOEP, GSC, ARTES, EGEP, GSTP, MREP, ELIPS, SSA, FLPP and PRODEX
Budget	**Estimated total space expenditure: 73.9 million €, including** Contribution to ESA budget: 32.7 million € Contribution to EUMETSAT budget: 5.1 million €
Direct Employment in Space Manufacturing Industry	190

Denmark

Ministry of Higher Education and Science

Population	5.767 million
GDP	288.98 billion €
Authority/authorities	In June 2015, the Ministry of Higher Education and Science was formally assigned responsibility for matters relating to the regulation of Danish space activities and for coordinating space activities across ministries and public institutions. In 2016 a new national space strategy was published, and an

(continued)

	Inter-Ministerial Space Committee, consisting of representatives from the ministries which have responsibilities relating to the space sector, was given the important task of coordinating and following up on the strategy and the initiatives it includes
Major activities	Denmark supports space technology research and development through ESA's ELIPS, GSTP, PRODEX, ARTES, FLPP and GNSS Evolution programmes
Budget	**Estimated total space expenditure: 40.8 million €, including** Contribution to ESA budget: 30.5 million € Contribution to EUMETSAT budget: 7.6 million €
Staff	30 (DTU)
Direct Employment in Space Manufacturing Industry	260

Estonia

EAS Enterprise Estonia

Population	1.315 million
GDP	23 billion €
Authority/authorities	Space activities in Estonia are coordinated by the Space Affairs Council (SAC), established in 2010 by the Ministry of Economic Affairs and Communications. The main task of SAC is initiation and governance space-related programmes and activities on national and international level and coordination of the utilisation of resources. SAC is supported by secretariat, containing representatives of Enterprise Estonia and Estonian Research Council. Since 2006 Estonian public business development and support agency Enterprise Estonia (www.eas.ee) is assigned as implementing body (ESA delegation) for the ESA-Estonian agreements. Enterprise Estonia also carries out the tasks of Estonian Space Office, being intermediary between Estonian companies/institutions and international space community. Main tasks concluded mapping and auditing space-related competences and capabilities, organising trainings for entrepreneurs and public sector officials, taking companies to expositions and study trips, encouraging the technology transfer, etc. Enterprise Estonia is a member of EURISY since 2008 and member of IAF since 2009. Space science activities in Estonia are carried out by Tartu Observatory (www.to.ee), having long tradition in astronomy, cosmology and Earth observation. Nowadays Tartu Observatory serves as Estonian space technology development and testing centre

(continued)

Major activities	Estonian space focus of space activities is on Earth observation; navigation; technology, especially related to ICT; and science. Estonian space activities are mainly related to Horizon 2020 projects and cooperation with ESA and PECS programme, where Estonia has 27 projects. Estonia became the 21st ESA member state in September 2015. Estonia is actively preparing for Galileo and Copernicus programme implementation in public and private sector
Budget	**Estimated total space expenditure: 5.1 million €, including** Contribution to ESA budget: 2.5 million € Contribution to EUMETSAT budget: 0.4 million €
Staff	10 (related to space in Enterprise Estonia) 80 (at Tartu Observatory)
Direct Employment in Space Manufacturing Industry	40

Finland

Tekes

Population	5.45 million
GDP	223.84 billion €
Authority/authorities	Finland's public sector space activities are funded by the Ministry of Employment and Economy (ESA membership), Tekes (ESA and national programmes), Academy of Finland (EISCAT, ESO, NOT) under the Ministry of Education and Culture, and Finnish Meteorological Institute (EUMETSAT) under the Ministry of Transport and Communications. National coordination is provided by the Finnish Space Committee under the Ministry of Economy
Major activities	In addition to cooperation with ESA and EC (e.g. Space Programme under the Horizon 2020 EU Framework Programme), Tekes funds space technologies and applications nationally. The Finnish focus of space activities is on space science, Earth observation, satellite telecommunications, satellite navigation and space technologies. Tekes also funds scientific instruments for spacecraft
Budget	**Estimated total space expenditure: 55 million €, including** Contribution to ESA budget: 19 million € Contribution to EUMETSAT budget: 7.2 million €
Staff	10 (Tekes)
Direct Employment in Space Manufacturing Industry	165

France

Population	67.106 million
GDP	2291.70 billion €
Authority/authorities	Space activities in France are under parent Ministries of Higher Education and Research and of Defence. Among different actors, France relies on its national space agency CNES (Centre National d'Etudes Spatiales) for space policy proposal and application
Major activities	France inscribes its space activities in the frame of European space activities and intent to play there a leading role. Activities are conducted in strong interface with EC, ESA, EUMETSAT and other Members States or at international level. National orientations include five acting domains, namely, Ariane (autonomous European access to space), Observation, Sciences, Telecommunications and Defence
Budget	**Estimated total space expenditure: 2753 million €, including** Contribution to ESA budget: 855.9 million € Contribution to EUMETSAT budget: 78.6 million €
Staff	2400 (CNES)
Direct Employment in Space Manufacturing Industry	15,700

Germany

Population	82.659 million
GDP	3263.350 billion €
Authority/authorities	Acting on behalf of the Federal Government, the Space Administration designs and implements Germany's Space Programme, which integrates all German space activities on the national and European plane. These activities include Germany's national Space Program, DLR's 'Space' research and development programme and Germany's contributions to the European Space Agency (ESA) as well as the European Organisation for the Exploitation of Meteorological Satellites (EUMETSAT). In addition, the Administration designs and supervises space and security research projects under the sixth and seventh EU research framework program The Space Administration's principal client is the Federal Ministry for Economics and Technology. However, the Space Administration also works for other ministries, mainly in application related fields such as Earth observation, navigation and satellite communication. The Federal Government's Space Programme provides both business and science

(continued)

	with a reliable political framework for independent planning and action, thus ensuring that public funds are used efficiently. The national programme is implemented by the Space Administration. Its major function is to promote and support Germany's strategic goals in the European programmes of ESA and the EU by purposefully equipping Germany's industry and scientific institutions for competition within the EU as well as for their tasks within the ESA framework
Major activities	Germany has a national civil programme, which includes bi- or trilateral cooperations, a participation in the ESA programmes and a defence programme (e.g. SARLupe, MUSIS). The major areas of interest for Germany's domestic programmes are Earth observation, navigation, telecommunications, space transportation systems, space infrastructures, microgravity research, space science, space technologies and space exploration
Budget	**Estimated total space expenditure: 1973 million €, including** Contribution to ESA budget: 858.4 million € Contribution to EUMETSAT budget: 102.7 million €
Staff	8000 (DLR)
Direct Employment in Space Manufacturing Industry	7900

Greece

Population	10.722 million
GDP	177.73 billion €
Authority/authorities	The General Secretariat for Research and Technology (GSRT), under the responsibility of the Ministry of Education, Research and Religious Affairs, is responsible for Greek space activities
Major activities	The technological activities of interest to Greece are mostly pursued through ESA EOEP, ARTES (focus on integrated applications) and GSTP
Budget	**Estimated total space expenditure: 30 million €** Contribution to ESA budget: 14.6 million € Contribution to EUMETSAT budget: 7.4 million €
Direct Employment in Space Manufacturing Industry	34 (GSRT)

Hungary

Population	9.784 million
GDP	123.49 billion €
Authority/authorities	The Hungarian Space Office (HSO), integrated into the body of the Ministry of National Development, manages, coordinates and represents Hungarian space activities. Hungary's space activity is supervised by the Minister, who makes his decisions based on the advices of the Scientific Council on Space Research in scientific issues and the Hungarian Space Board in strategic questions
Major activities	Participation in microgravity, Earth observation, life and material sciences and GTSP programmes of ESA
Budget	**Estimated total space expenditure: 10.2 million €, including** Contribution to ESA budget: 7.3 million € Contribution to EUMETSA budget: 3.3 million €
Direct Employment in Space Manufacturing Industry	100

Ireland

Population	4.802 million
GDP	294.11 billion €
Authority/authorities	Enterprise Ireland (EI) is the Irish Government's business development agency. EI manages industrial and technology development programmes relating to space activity on behalf of the Department of Jobs, Enterprise and Innovation (DJEI). EI promotes ESA and EU space programmes in Ireland, primarily to the industrial sector, as well as to the academic community and end-user organisations
Major activities	ESA programmes [Telecommunications (ARTES), Earth Observation, Launchers, Navigation, PRODEX, ELIPS and Technology (GSTP)]; Horizon 2020 Space; EUMETSAT
Budget	**Estimated total space expenditure: 21.1 million €, including** Contribution to ESA budget: 17.8 million € Contribution to EUMETSAT budget: 5.2 million €
Staff	740 (Enterprise Ireland)
Direct Employment in Space Manufacturing Industry	70

Italy

Population	60.542 million
GDP	1716.93 billion €
Authority/authorities	The Italian Space Agency, Agenzia Spaziale Italiana (ASI), defines, coordinates and manages national space programmes and the Italian participation to European and international space projects on behalf of the Ministry of Education, University and Research (MIUR) and in coordination with the Ministry for the International Affairs. Other Ministries, such as the Ministry of Economic Development, the Ministry of Defence and the Ministry of Transport and Infrastructures on navigation, support the space sector in agreement with ASI
Major activities	ASI proposes its national strategy within its Strategic Vision Document and coordinates the realisation of competitive programmes and infrastructures in order to sustain intellectual and industrial growth through three main programmatic lines: science in space, technology for space and services from space. Its main investments, at national and international level, are made in Earth Observation, Launchers, Science/Exploration of the Universe and Telecommunications fields. Another important aspect for Italy in space is Education and Outreach. For this reason ASI is involved in communication activities and spread of knowledge both in schools and universities
Budget	**Estimated total space expenditure: 916 million €, including** Contribution to ESA budget: 550 million € Contribution to EUMETSAT budget: 59.3 million €
Staff	240 (ASI)
Direct Employment in Space Manufacturing Industry	5140

Luxembourg

Population	596 thousand
GDP	55.34 billion Euro
Authority/authorities	The Ministry of the Economy, Directorate of ICT and Space Affairs, is in charge of space affairs in Luxembourg. Luxinnovation, the National Agency for Innovation and Research, manages the interface between industry, the public research sector and ESA. And the meteorological service,

(continued)

	department of the Ministry for Sustainable Development and Infrastructure, represents Luxembourg in the EUMETSAT Council
Major activities	Technology R&D activities in Luxembourg are mainly focused on future telecommunication applications. Until now, these activities have been funded through the ARTES program. The adhesion of Luxembourg to ESA will offer new opportunities for supporting technology R&D activities. Luxembourg is active in telecommunications, Earth observation, navigation, security and space technologies
Budget	**Estimated total space expenditure: 38.5 million €, including** Contribution to ESA budget: 22.3 million € Contribution to EUMETSAT budget: 0.6 million €
Staff	5 (3 Ministry of the Economy + 2 Luxinnovation)
Direct Employment in Space Manufacturing Industry	35

The Netherlands

Population	17.127 million
GDP	737.05 billion Euro
Authority/authorities	The Netherlands Space Office (NSO) acts as the Dutch agency for space affairs. The NSO was established by the Dutch government in order to develop the Netherlands' space programme and to bring that programme to action. The NSO is the front office of the Dutch space community for international space organisations like ESA, NASA and JAXA as well as the central point of contact for the space community within the Netherlands. The NSO also works to innovatively and openly bring the story of spaceflight science, usage and exploration to teachers, students and the general public. The director of the NSO reports to the steering committee of the Ministry of Economic Affairs; Ministry of Education, Culture and Science; Ministry of Transport, Public Works and Water Management; and the Netherlands Organisation for Scientific Research (NWO)
Major activities	In the ESA context, the Netherlands participate in all major programmes with a special interest in Earth observation and environmental measurements and in robotics. The Netherlands Institute for Space Research (SRON) develops and uses innovative technology for ground-breaking research in space, focusing on astrophysical research, Earth science and planetary research. In addition to this, SRON has a line of research into new and more sensitive sensors for X-rays and infrared radiation. Space policy in the Netherlands is primarily focused on international cooperation in European

(continued)

	contexts within ESA, EUMETSAT and the EU. The Netherlands also has a national programme guided by the ministries of Economic Affairs; of Education, Culture and Science; and of Traffic, Public Works and Water Management
Budget	**Estimated total space expenditure: 25.4 million €, including** Contribution to ESA budget: 72 million € Contribution to EUMETSAT budget: 23 million €
Staff	30 (NSO) Approx. 200 (SRON)
Direct Employment in Space Manufacturing Industry	1155

Norway

Norsk Romsenter
NORWEGIAN SPACE CENTRE

Population	5.227 million
GDP	353.70 billion €
Authority/authorities	The Norwegian Space Centre (NSC), under the Ministry of Trade, Industry, and Fisheries, is responsible for organising Norwegian space activities, particularly with respect to ESA and the EU, and for coordinating national space activities
Major activities	In addition to ESA programmes (in particular Earth observation, telecommunications and launchers), Norway has national support programmes and commercial activities (Telenor). Moreover, Norway operates the Andøya rocket range and the Svalbard and Antarctica ground stations. Norway has also a bilateral agreement with Canada on the use of RADARSAT-2 data
Budget	**Estimated total space expenditure: 91.7 million €, including** Contribution to ESA budget: 63.5 million € Contribution to EUMETSAT budget: 13.8 million €
Staff	NSC: 40
Direct Employment in Space Manufacturing Industry	410

Poland

Population	38.422 million
GDP	465.60 billion €
Authority/authorities	Polish space activities are under the leading responsibility of the Ministry of Economic Development, in cooperation with the Ministry of Science and Higher Education, the Ministry of National Defence, the Ministry of Foreign Affairs, the Ministry of Environment and other appropriate ministries. The Ministry of Development represents Poland in ESA and in the EU institutions dealing with issues related to space sector
Major activities	The recently established Polish Space Agency (POLSA) is the implementing entity for Polish space activities under the responsibility of the Ministry of Economy, Department of Industry. Prior to its establishment, research and development on space technology in Poland has been carried out mainly by the Space Research Centre (SRC) of the Polish Academy of Sciences (PAS). The five main SRC research areas are physics of the Sun, study of planets and small solar system bodies, interplanetary space physics and astrophysics, plasma physics and planetary geodesy and geodynamics. The SRC is the body cooperating with ESA
Budget	**Estimated total space expenditure: 76.5 million € including** Contribution to ESA budget: 34.6 million € Contribution to EUMETSAT budget: 12.8 million €
Direct Employment in Space Manufacturing Industry	215

Portugal

Population	10.300 million
GDP	193.07 billion €
Authority/authorities	The management and coordination of space activities in Portugal is carried out by the Foundation for Science and Technology (FCT), mainly through its Space Office. The FCT is the national funding agency for scientific and technological projects acting under the responsibility of the Portuguese Ministry of Science, Technology and Higher Education. The Ministry of Science is coordinating the Portuguese membership to ESA and is also providing the funding to ESA's mandatory activities and several optional programmes

(continued)

Major activities	Technology research and development activities are conducted in the frame of ESA TRP, Exploration, ARTES, EOEP and GSTP programmes
Budget	**Estimated total space expenditure: 20.1 million €, including** Contribution to ESA budget: 17 million € Contribution to EUMETSAT budget: 6.3 million €
Direct Employment in Space Manufacturing Industry	160

Romania

Population	19.644 million
GDP	187.94 billion €
Authority/authorities	The Romanian Space Agency (ROSA) is the coordinator of Romania's national and international space activities. ROSA is a public institution entirely self-funded, operating under the decisions of the Ministry of Education and Research—National Authority for Scientific Research and Innovation (ANCSI). As a coordinator of national space research and applications programme, ROSA designs and coordinates the implementation of the National Space Programme. Following its objectives, the Agency is authorised to establish research and development centres
Major activities	Main areas of interest are space exploration, space applications, technology and security. On behalf of the Government, ROSA is the national representative in the cooperative agreements with international organisations, such as ESA and Committee on Space Research (COSPAR), as well as bilateral governmental agreements. Together with the Ministry of Foreign Affairs, ROSA is representing Romania in the sessions of the United Nations Committee on the Peaceful Use of Outer Space (COPUOS) and its Subcommittees
Budget	**Estimated total space expenditure: 44.9 million €, including** Contribution to ESA budget: 30 million € Contribution to EUMETSAT budget: 4.6 million €
Direct Employment in Space Manufacturing Industry	65

Spain

Population	46.440 million
GDP	1113.85 billion €
Authority/authorities	The Centre for the Development of Industrial Technology (CDTI), under the Ministry of Economy and Competitiveness, channels the funding and support applications for national and international R&D&i projects of Spanish companies, including the Spanish space activities in coordination with the Ministry of Industry, Energy and Tourism. The National Institute of Aerospace Technology (INTA) is the other important Spanish actor in the space field. INTA is a public research organisation specialised in aerospace research and technology development reporting to the Ministry of Defence. The INTA has four main action lines related to design, development, integration, verification and testing of small- and medium-sized platforms and payloads, ground segment and satellite tracking and operations
Major activities	In addition to ESA and EUMETSAT programmes, Spain has several national space programmes in the field of Earth observation (SEOSAT/INGENIO), communication satellites (Hispasat, Amazonas), defence space systems (SPAINSAT, XTAR-EUR, HELIOS, SECOMSAT, Pleiades and SEOSAR/PAZ), small satellites, space exploration (MSL-REMS) and ground control stations
Budget	**Estimated total space expenditure: 278 million €, including** Contribution to -ESA budget: 151.2 million € Contribution to EUMETSAT budget: 37.3 million €
Staff	320 (CDTI)
Direct Employment in Space Manufacturing Industry	3425

Sweden

Population	10.073 million
GDP	477.38 billion €
Authority/authorities	The Swedish National Space Board (SNSB), a central governmental agency under the Ministry of Higher Education and Research, is responsible for national and international activities relating to space and remote sensing, primarily research and development. The Swedish space programme is carried out by means of extensive international cooperation, in particular through Sweden's membership of ESA. SNSB's responsibility for

(continued)

	international activities includes the Swedish involvement in ESA as well as bilateral cooperation within space. Most of the activities funded by the Swedish National Space Board are carried out in cooperation with other countries
Major activities	The Swedish space programme is mostly carried out through international cooperation. Sweden has a high level of activities in the technological ESA programmes GSTP and ARTES. And nationally, special focus is put on on-board computers, telecom equipment, MEMS-technology, separation systems and adapters, turbines, AOCS and EP systems, green propellant, formation-flying platforms and the Ariane nozzle
Budget	**Estimated total space expenditure: 113.7 million €, including** Contribution to ESA budget: 72.3 million € Contribution to EUMETSAT budget: 15.1 million €
Staff	20
Direct Employment in Space Manufacturing Industry	1035

Switzerland

Schweizerische Eidgenossenschaft
Confédération suisse
Confederazione Svizzera
Confederaziun svizra

Population	8.372 million
GDP	601.32 billion €
Authority/authorities	The Swiss Space Office (SSO) under the authority of the State Secretariat for Education, Research and Innovation (SERI) of the Federal Department of Economic Affairs, Education and Research (EAER) is the administrative body in charge of planning and implementing the Swiss space policy. The SSO chairs the Interdepartmental Committee for Space Affairs (IKAR) and acts as the executive secretariat for the Federal Commission for Space Affairs (CFAS), which advises the Federal Council on matters relating to a coherent and forward-looking space policy by taking into consideration the European and global development and the national interests (among others of political, application-oriented, scientific, technological and industrial nature). It also issues recommendations to the Federal Council regarding the implementation of this policy and advises the Federal Council in terms of the evaluation of proposals for projects with important scientific or technological relevance

(continued)

Major activities	The Swiss participation to ESA's PRODEX programme is an important bridge to experimental space research and has led to a remarkable improvement in the country's space research capabilities. Other ESA programmes include General Budget, Science Programme and CSG, Earth Observation, Launchers, Human Spaceflight and Exploration, Technology and Telecommunication and Navigation
Budget	**Estimated total space expenditure: 154.7 million €, including** Contribution to ESA budget: 145.1 million € Contribution to EUMETSAT budget: 18.5 million €
Direct Employment in Space Manufacturing Industry	950

United Kingdom

UK SPACE AGENCY

Population	66.051 million
GDP	2327.73 billion €
Authority/authorities	The UK Space Agency, an executive agency of the Department for Business, Innovation and Skills (BIS), and reporting to the Minister of State for Universities and Science, is responsible for the strategic decisions on all UK space activities. The UK Space Agency also interfaces with other departments on security and military programmes. Although it does not manage these programmes directly, the UK Space Agency is kept informed and involved in decisions relating to the programmes as necessary
Major activities	The focus of UK space activities is on Innovative Technologies that have the potential to maximise public and commercial mission objectives. The UK is largely supporting Space Science, Earth Observation, ARTES, Galileo MREP, the ELIPS programme, and increasing participation on the ISS. Technologies are developed both by means of national activities funded through UK Industry such as NSTP and by ESA's TRP, GSTP, CTP, ARTES and ETP programmes
Budget	**Estimated total space expenditure: 620.5 million €, including** Contribution to ESA budget: 300 million € Contribution to EUMETSAT budget: 70.4 million €
Staff	70 (UKSA)
Direct Employment in Space Manufacturing Industry	3765

7.4 Space Policy Publications from the Year 2017

Books and Reports

1. Al-Ekabi C. (2017), "Space Policies Issues and Trends", Report 63, European Space Policy Institute.
2. Abul Failat Y., Ferreira-Snyman A. (2017) "*Outer Space law: legal police and practice*", Horsell, Surrey.
3. Cavallaro U., (2017), "*Women Spacefarers. Sixty Different Paths to Space*", Springer Praxis Books, Springer, Cham.
4. Chang Díaz F., Seedhouse E., (2017), "*To Mars and Beyond, Fast!*", Springer, New York.
5. Dawson L. (2017) "*The Politics and Perils of Space Exploration. Who Will Compete, Who Will Dominate?*" Springer Praxis Books, Spüringer, Cham.
6. English N., (2017) "*Space Telescopes*", Springer, New York.
7. Genta, Giancarlo (2017), "*Next Stop Mars: The Why, How, and When of Human Missions*", Springer Praxis Books, Springer, Cham.
8. Godefroy, A. B. (2017), "*The Canadian Space Program*", Springer, New York.
9. Jakhu Ram, Pelton Joseph, Nyampong Yaw Otu Mankata, "*Space Mining and Its Regulation*" (2017) Springer, New York.
10. Janz, B. B. (Ed) (2017), "*Place, Space and Hermeneutics*", Springer, New York.
11. Leepuengtam T (2017) "*The Protection of Intellectual Property Rights in Outer Space Activity*", Elgar.
12. Melvin L. (2017) "*Chasing Space: An Astronaut's Story of Grit, Grace, and Second Chances*", Amistad.
13. Minier, V., Bonnet, R., Bontems, V., de Graauw, T., Griffin, M., Helmich, F., Pilbratt, G., Volonte, S. (2017), "*Inventing a Space Mission*", Springer.
14. Peake T. (2017) "Ask an Astronaut", Little Brown and Co.
15. Pyle R. (2017) "*Amazing Stories of the Space Age*", Prometheus, New York.
16. Rao R. V. (Ed), Gopalakrishnan, V. (Ed), Abhijeet, K., "*Recent Developments in Space Law*" (2017), Springer, New York.
17. Ruyters G., Betzel, C., Grimm, D. (2017), "*Biotechnology in Space*", Springer, New York.
18. Seedhouse E. (2017), "*Tim Peake and Britain's road to space*", Springer Praxis Books, Cham.
19. Shayler D. J. (2017), "*Assembling and Supplying the ISS. The Space Shuttle Fulfills Its Mission*", Springer Praxis Books, Springer, Cham.
20. Shayler D.J, Burgess C., "*Expanding the Space Frontier in the Late Sixties*", Springer Praxis Books, Springer, Cham.
21. Shayler, D.J. (2017), "*Linking the Space Shuttle and Space Stations. Early Docking Technologies from Concept to Implementation*", Springer Praxis Books, Springer, Cham.

22. Shayler D.J., Burgess C., (2017), "*The Last of NASA's Original Pilot Astronauts*, Springer, New York.
23. Sitruk A, S.Plattard (2017), "the Governance of Galileo", European Space policy Institute, Report 62.
24. Sivolella, D. (2017), "*The Space Shuttle Program*", Springer, New York.
25. Stubbe P. (2017) "*Space Accountability for Space Debris*", Studies in Space Law, Brill Nijhof.
26. Vakoch D.A., Aoki S., Milligan A. (2017) "*Global Space Governance: an international study*", Springer, New York.
27. Van den Abeelen L. (2017), "*Spaceplane HERMES: Europe's Dream of Independent Manned Spaceflight*", Springer Praxis Books, Springer, Cham.
28. Van Pelt M. (2017) "*Dream Missions, Space Colonies, Nuclear Spacecraft and Other Possibilities*", Springer Praxis Books, Springer, Cham.
29. Von Ehrenfried M. "Dutch" (Ed.) (2017), "*Exploring the Martian Moons. A Human Mission to Deimos and Phobos*", Springer Praxis Books, Springer, Cham.
30. Weinersmith K. and Z. (2017) "*Soonish*", Penguin Press, United Kingdom.

Articles

31. Alalinarde M., "*The popularization of space – Link between science, policy, and public perception Star Trek as an early mind-opener for space endeavours*", Space Policy, Volume 41, August 2017, Pages 36–41.
32. Astolfi D., Alessandri A., Zaccarin L., "*Stubborn ISS Redesign for Nonlinear High-Gain Observers*", IFAC-PapersOnLine, Volume 50, Issue 1, July 2017, Pages 15422–15427.
33. Blackman C., "*Telecommunications policy: The medium is the message*", Telecommunications Policy, Volume 41, Issue 10, November 2017, Pages 847–852.
34. Brainard S., Szajnfarbe Zoe, "*Understanding the burden of government oversight on engineering work: Adding empirical data to the debate*", Space Policy, Volume 42, Pages 70–80, November 2017.
35. Carpenter A., "*Theorizing European Space Policy*", Space Policy, Volume 41, Pages 77–78, August 2017.
36. Cresto Aleina S., Viola N., Fusaro R., Saccoccia G., "*Approach to technology prioritization in support of moon initiatives in the framework of ESA exploration technology roadmaps*", Acta Astronautica, Volume 139, October 2017, Pages 42–53.
37. Cuca B., Hadjimitsis D.G., "*Space technology meets policy: An overview of Earth Observation sensors for monitoring of cultural landscapes within policy Journal of Archaeological Science*", Volume 14, August 2017, Pages 727–733.
38. De Man P., "*State practice, domestic legislation and the interpretation of fundamental principles of international space law*" Space Policy, Volume 42, November 2017, Pages 92–102.

39. Denis G., Claverie A., Pasco X., Darnis J.P., De Maupeou B., Lafaye M., Morel E., "*Towards disruptions in Earth observation? New Earth Observation systems and markets evolution: Possible scenarios and impacts*", Acta Astronautica, August 2017.
40. Du R., "*China's approach to space sustainability: Legal and policy analysis*", Space Policy, Volume 42, November 2017, Pages 8–16.
41. Durrani H., "*Our Window on the World: Life in the Orbital Heterotopia of the International Space Station*", Sacknoff Price for Space history, Columbia Law School, 2017.
42. Entrena Utrilla C.M., "*Asteroid-COTS: Developing the cislunar economy with private-public partnerships*", Space Policy, Volumes 39–40, May 2017, Pages 14–19.
43. Ercan C., Kale I., "*The role of space in the security and defence policy of Turkey. A change in outlook: Security in space versus security from space*", Space Policy, Volume 4, November 2017, Pages 17–25.
44. Ercan C., Kale I., *Historical space steps of Turkey: "It is high time to establish the Turkish space agency*", Acta Astronautica, Volume 130, January–February 2017, Pages 67–74.
45. European Space Policy Institute, "*Super Heavy Lift Launch Vehicles: Global Status and European Perspectives*", Executive Brief 18, November 2017.
46. European Space Policy Institute, "*Next Steps to the Moon: What Role for Europe*", Executive Brief 17, November 2017.
47. European Space Policy Institute, "*Spacetugs: Towards New Solution for Satcom Operators*", Executive Brief 16, October 2017.
48. European Space Policy Institute, "*About the Enthusiastic Multiplication of Spaceport Worldwide*", Executive Brief 15, July 2017.
49. European Space Policy Institute, "*Satellite Data to Monitor International Agreements*", Executive Brief 14, June 2017.
50. European Space Policy Institute, "*Are we on the verge of Suborbital Flights*?", Executive Brief 13, June 2017.
51. European Space Policy Institute, "*Making Exploration Great Again*?", Executive Brief 12, May 2017.
52. European Space Policy Institute, "*Delimitation of Outer Space*", Executive Brief 11, March 2017.
53. European Space Policy Institute, "*Italy-China Collaboration*", Executive Brief 10, March 2017.
54. European Space Policy Institute, "*Europe Out of the Box*", Executive Brief 9, February 2017.
55. European Space Policy Institute, "*Trumps Administration Plans for Space: Implication for Europe*", Executive Brief 8, January 2017.
56. European Space Policy Institute, "*China's 2016 White Paper on Space: An Analysis*", Executive Brief 7, January 2017.
57. Forganni A., "*The potential of space tourism for space popularisation: An opportunity for the EU Space Policy?*" Space Policy, Volume 41, August 2017, Pages 48–52.

58. Force K.M., "*The 2017 Manfred Lachs Space Law Moot Court Competition*", International institute of Space Law, 2017.
59. Froehlich A., "*Utilization - Consumption - Appropriation: Asteroid Mining is in the Pipeline*", German Journal of Air and Space Law (ZLW), Volume 66, 2017, 268–459, p.70.
60. Fitzgerald P.P., '*Intercontinental Space Flight: Learning from the Concorde*' (2017), 42 Air and Space Law, Issue 6, pp. 583–600, Volume 42, 2017.
61. Glassmeier, "*Reconstruction of the flight and attitude of Rosetta's lander Philae*", Acta Astronautica, November 2017.
62. Geiselhart R., Noroozi N., "*Equivalent types of ISS Lyapunov functions for discontinuous discrete-time systems*" Automatica, Volume 84, October 2017, Pages 227–231.
63. Gringé L., "*Science fiction works for the development of the aerospace sector*", Space Policy, Volume 41, August 2017, Pages 42–47.
64. Hao L., Tronchetti F., "*Should the Red Dragon arise? Assessing China's options vis-à-vis the enactment of a domestic space resources utilization law*", Space Policy, Volumes 39–40, Pages 9–13, May 2017.
65. Hatton S., "*IISL Dos Study on Space Resource Mining*", international Institute of Space Law, March 2017.
66. Hickman J., "*Exoplanet naming, electoral competition and strategic failure*", Space Policy, Volume 42, November 2017, Pages 62–65.
67. Hobe S., Man P., "*National Appropriation of Outer Space and State Jurisdiction to Regulate the Exploitation, Exploration and Utilization of Space Resources*", German Journal of Air and Space Law (ZLW), Volume 66, 2017, p. 460.
68. Hobe S., "*The IISL Assumes Responsibility for Questions of Cyber Law*", German Journal of Air and Space Law (ZLW), Volume 66, 2017, p. 647.
69. Hoerber T., Athanasopoulos K.H., "*Popularising European space policy: Introduction*", Space Policy, Volume 41, August 2017, Pages 1–4.
70. Johnson C.D., '*The 59th Colloquium on the Law of Outer Space at the 67th International Astronautical Congress*' (2017) 42 Air and Space Law, Issue 1, pp. 89–94, Volume 42, 2017.
71. Jakhu R.S., Steer C., Chen K., "*Conflicts in Space and the Rule of Law*", German Journal of Air and Space Law (ZLW), Volume 66, 2017, p. 657.
72. Kaiser S.A., "*Legal Protection against Contamination from Space Resource Mining*", German Journal of Air and Space Law (ZLW), Volume 66, 2017, p. 282.
73. Kerstens N., Giannopapa C., Dolmans S., Reymen I., "*Down to earth: Popularisation of geo-information services in the Netherlands*", Space Policy, Volume 41, August 2017, Pages 12–19.
74. Kienzler F., "*European Space Policy. European Integration and the Final Frontier*", Space Policy, Volume 41, August 2017, Pages 79–80.
75. Kishi N., "*Management analysis for the space industry*", Space Policy, Volumes 39–40, May 2017, Pages 1–6.
76. Kong D., "*Shaping a legal framework for China's BeiDou Navigation Satellite System*", Space Policy, Volume 42, Pages 26–30, November 2017.

77. Lambrigh W.H., "*Relay leadership and long-term policy: The case of Mars*", Space Policy, Pages 1–7, Volume 42, November 2017.
78. Lele A., "*India's policy for outer space*", Space Policy, Volumes 39–40, May 2017, Pages 26–32.
79. Lehnert C., Karlsson E., Giannopapa C., "*Global risk & global challenges – Space as a game changer for socioeconomic sustainable development*", Acta Astronautica, Volume 140, November 2017, Pages 59–65.
80. Lemmens S., Lips T., Kanzler R., "*Upgrade of ESA's Debris Risk Assessment and Mitigation Analysis (DRAMA) tool: Spacecraft Entry Survival Analysis Module*", Acta Astronautica, 5 December 2017.
81. Lieberman S., "*Understanding our changing relationship with space: An international political economy reading of space popularisation*", Space Policy, Volume 41, August 2017, Pages 53–59.
82. Mazzuccato M., Robinson D.K.R., "*Co-creating and directing Innovation Ecosystems? NASA's changing approach to public-private partnerships in low-earth orbit*", Technologic Forecasting and Social Changes, 27 April 2017.
83. Munsami V, Offiong E.O, "*Should Africa follow the European space governance model for an African space programme?*" Space Policy, April 2017.
84. Mizushima A., Fujii K., Ishido K., '*What Is an Appropriate Interaction Between International Law and Domestic Legal Systems to Promote Space Resources Development?*' (2017) 42 Air and Space Law, Issue 6, pages 543–570, Volume 42, 2017.
85. Natov N., "*Private International Law Aspects of the draft International Code of Conduct for Outer Space*", German Journal of Air and Space Law (ZLW), Volume 66, 2017, p. 290.
86. Oikonomou I., "*'All u need is space': Popularizing EU space policy*", Space Policy, Volume 41, August 2017, Pages 5–11.
87. Orme K.C., '*Guardians of the Galaxy: How Space Tourism Regulation Will Shape Telehealth*' (2017) 42 Air and Space Law, Issue 2, pp. 163–184, Volume 42, 2017.
88. Plummer S., Lecomte P., Doherty M., "*The ESA Climate Change Initiative (CCI): A European contribution to the generation of the Global Climate Observing System Remote Sensing of Environment*", Volume 203, 15 December 2017, Pages 2–8.
89. Poursanidis D., Chrysoulakis N., "*Remote Sensing, natural hazards and the contribution of ESA Sentinels missions*", Volume 6, April 2017, Pages 25–38.
90. Reed J.A., '*Cold War Treaties in a New World: The Inevitable End of the Outer Space and Antarctic Treaty Systems*' (2017) 42 *Air and Space Law*, Issue 4/5, pp. 463–486, Volume 42, 2017.
91. Revell T, "*It's time to update the laws that govern space*", New Scientist, Volume 234, Issue 3130, Page 25, 17 June 2017.
92. Roma A., "*Drones and popularisation of space*", Space Policy, Volume 41, August 2017, Pages 65–67.
93. Szocik K, Tomasz Wójtowicz, Leszek Baran, "*War or peace? The possible scenarios of colonising Mars*", Space Policy, Volume 42, Pages 31–36, November 2017.

94. Shelhamer M., "*Participatory technology assessment for Mars mission planning: Public values and rationales*", Space Policy, Volume 42, Pages 41–53, November 2017.
95. Shelhamer M., "*Why send humans into space? Science and non-science motivations for human space flight*", Acta Astronautica, November 2017.
96. Sivanpillai R., Jones B.K., Lamb M.R., "*Accessing satellite imagery for disaster response through the International Charter: Lessons learned from the 2011 US Midwestern Floods*", Space Policy, Volume 42, November 2017, Pages 54–61.
97. Sankaran R., Raju N., "*A Framework to address burgeoning Commercial Complexities in Space Mining*", German Journal of Air and Space Law (ZLW), Volume 66, 2017, p. 70.
98. Sun R., "*Legal Issues of Space Colonization*", German Journal of Air and Space Law (ZLW), Volume 66, 2017, p. 476.
99. Sundahl M.J., '*Regulating Non-Traditional Space Activities in the United States in the Wake of the Commercial Space Launch Competitiveness Act*' (2017) 42 Air and Space Law, Issue 1, pp. 29–42, Volume 42, 2017.
100. Tanaka K., "*Applicability of remote sensing policies to space situational awareness*", Space Policy, Volume 42, November 2017, Pages 83–91.
101. Thomas A., "*The Astronautical discourse in an English primary school during the Principia ESA mission: A critical analysis*" Space Policy, Volume 41, August 2017, Pages 27–35.
102. Terrier D., Loizos Heracleous, Steven Gonzalez, "*Enabling paradigm change and agility at NASA's Johnson Space Center – Interview with Chief Technology Officer, Douglas Terrier*", Space Policy, Volumes 39–40, May 2017, Pages 20–25.
103. Traphagan J.W., "*Extraterrestrial altruism: Evolution and ethics in the cosmos*", Space Policy, Volume 42, Pages 81–82, November 2017.
104. Vasko C. A., M. Adriaensen, A. Bretel, I. Duvaux-Bechon, C. G. Giannopapa, "*Space assets, technology and services in support of energy policy*", Acta Astronautica, Volume 138, September 2017, Pages 295–300.
105. Vettorel A., '*Global Positioning System Evidence in Court Proceedings and Privacy: The Case of Italy*' (2017) 42 Air and Space Law, Issue 3, pages 295–312, Volume 42, 2017.
106. Vecchio V., Verhaeghe K., '*Report on the Small Satellites Tech, Business & Regulatory Workshop (ECSL/ESA), Held at ESTEC, Noordwijk, the Netherlands, on 13 April 2017*' (2017) 42 Air and Space Law, Issue 4/5, pp. 501–508, Volume 42, 2017.
107. Vecchio V., "*Customary International Law in the Outer Space Treaty - Space Law as Laboratory for the Evolution of Public International Law*", German Journal of Air and Space Law (ZLW), Volume 66, 2017, p. 491.
108. Wachowicz M. E., Bury M., "*Space technology transfer problems in the context of protecting the space heritage*", Space Policy, Volume 42, November 2017, Pages 66–69.
109. Zypries B., "*Space, the Public, and Politics*", Space Policy, Volume 41, August 2017, Pages 73–74.

MIX
Papier aus verantwortungsvollen Quellen
Paper from responsible sources
FSC® C105338

If you have any concerns about our products,
you can contact us on
ProductSafety@springernature.com

In case Publisher is established outside the EU,
the EU authorized representative is:
Springer Nature Customer Service Center GmbH
Europaplatz 3, 69115 Heidelberg, Germany

Printed by Libri Plureos GmbH
in Hamburg, Germany